The Palgrave Macmillan Animal Ethics Series

Series Editors
Andrew Linzey
Oxford Centre for Animal Ethics
Oxford, UK

Clair Linzey
Oxford Centre for Animal Ethics
Oxford, UK

In recent years, there has been a growing interest in the ethics of our treatment of animals. Philosophers have led the way, and now a range of other scholars have followed from historians to social scientists. From being a marginal issue, animals have become an emerging issue in ethics and in multidisciplinary inquiry. This series will explore the challenges that Animal Ethics poses, both conceptually and practically, to traditional understandings of human-animal relations. Specifically, the Series will:

- provide a range of key introductory and advanced texts that map out ethical positions on animals
- publish pioneering work written by new, as well as accomplished, scholars;
- produce texts from a variety of disciplines that are multidisciplinary in character or have multidisciplinary relevance.

More information about this series at
http://www.palgrave.com/gp/series/14421

Cristina Hanganu-Bresch
Kristin Kondrlik
Editors

Veg(etari)an Arguments in Culture, History, and Practice

The V Word

palgrave
macmillan

Editors
Cristina Hanganu-Bresch
University of the Sciences
Philadelphia, PA, USA

Kristin Kondrlik
West Chester University of Pennsylvania
West Chester, PA, USA

ISSN 2634-6672 ISSN 2634-6680 (electronic)
The Palgrave Macmillan Animal Ethics Series
ISBN 978-3-030-53279-6 ISBN 978-3-030-53280-2 (eBook)
https://doi.org/10.1007/978-3-030-53280-2

This Palgrave Macmillan imprint is published by the registered company Springer Nature Switzerland AG.
The registered company address is: Gewerbestrasse 11, 6330 Cham, Switzerland

Series Editors' Preface

This is a new book series for a new field of inquiry: Animal Ethics.

In recent years, there has been a growing interest in the ethics of our treatment of animals. Philosophers have led the way, and now a range of other scholars have followed from historians to social scientists. From being a marginal issue, animals have become an emerging issue in ethics and in multidisciplinary inquiry.

In addition, a rethink of the status of animals has been fuelled by a range of scientific investigations which have revealed the complexity of animal sentiency, cognition and awareness. The ethical implications of this new knowledge have yet to be properly evaluated, but it is becoming clear that the old view that animals are mere things, tools, machines or commodities cannot be sustained ethically.

But it is not only philosophy and science that are putting animals on the agenda. Increasingly, in Europe and the United States, animals are becoming a political issue as political parties vie for the "green" and "animal" vote. In turn, political scientists are beginning to look again at the history of political thought in relation to animals, and historians are beginning to revisit the political history of animal protection.

As animals grow as an issue of importance, so there have been more collaborative academic ventures leading to conference volumes, special journal issues, indeed new academic animal journals as well. Moreover, we have witnessed the growth of academic courses, as well as university posts in Animal Ethics, Animal Welfare, Animal Rights, Animal Law, Animals and Philosophy, Human-Animal Studies, Critical Animal Studies, Animals and Society, Animals in Literature, Animals and Religion—tangible signs that a new academic discipline is emerging.

"Animal Ethics" is the new term for the academic exploration of the moral status of the non-human—an exploration that explicitly involves a focus on what we owe animals morally, and which also helps us to understand the influences—social, legal, cultural, religious and political—that legitimate animal abuse. This series explores the challenges that Animal Ethics poses, both conceptually and practically, to traditional understandings of human-animal relations.

The series is needed for three reasons: (i) to provide the texts that will service the new university courses on animals; (ii) to support the increasing number of students studying and academics researching in animal related fields, and (iii) because there is currently no book series that is a focus for multidisciplinary research in the field.

Specifically, the series will

- provide a range of key introductory and advanced texts that map out ethical positions on animals;
- publish pioneering work written by new, as well as accomplished, scholars, and
- produce texts from a variety of disciplines that are multidisciplinary in character or have multidisciplinary relevance.

The new Palgrave Macmillan Series on Animal Ethics is the result of a unique partnership between Palgrave Macmillan and the Ferrater Mora Oxford Centre for Animal Ethics. The series is an integral part of the mission of the Centre to put animals on the intellectual agenda by facilitating academic research and publication. The series is also a natural complement to one of the Centre's other major projects, the *Journal of Animal Ethics*. The Centre is an independent "think tank" for the advancement of progressive thought about animals, and is the first Centre of its kind in the world. It aims to demonstrate rigorous intellectual enquiry and the highest standards of scholarship. It strives to be a world-class centre of academic excellence in its field.

We invite academics to visit the Centre's website www.oxfordanimalethics.com and to contact us with new book proposals for the series.

Oxford, UK

Andrew Linzey
Clair Linzey

Contents

Part I Cultural and Historical Perspectives on Vegetarianism 1

1 **State of Meatlessness: Voluntary and Involuntary Vegetarianism in Early Twentieth-Century Italy** 3
Carol Helstosky

2 **Taking an Anti-Sacrificial Stance: The Essentializing Rhetoric and Affective Nature of Meat Consumption in Islam** 25
Nora Abdul-Aziz, Daniella Fedak-Lengel, and Lara Martin Lengel

3 **Because We Care: Veganism and Politics in Israel** 63
Sharon Avital

4 **Veg(etari)anism in Serbia: Attack on Traditional Values** 93
Mirjana Uzelac

5 **Ancient Text, Modern Context: Patanjali's *Yoga Sutras* and the Twenty-First Century Veg(etari)an** 119
Sharon Lauricella

Part II Veg(etari)anism as Embodied Practice 141

6 **The Accidental Vegetarian: Object-Oriented Ontology at the Intersection of Alpha-Gal Mammalian Meat Allergy** 143
Elizabeth Baddour

7 **"You Are What You Eat": Oprah, Amarillo, and Food Politics** 171
Callie F. Kostelich and Heidi Hakimi-Hood

8 **Queer Hunger: Human and Animal Bodies in Djuna Barnes' *Nightwood*** 195
Molly Mann

Part III Eco Versus Ego: The Transformative Potential of Veg(etari)anism 213

9 **Laying Down with the Lamb: Abolitionist Veganism, the Rhetoric of Human Exceptionalism, and the End of Creation** 215
David P. Stubblefield and Dynestee Fields

10 **Feeling Bad? Veganism, Climate Change, and the Rhetoric of *Cowspiracy*** 245
Alexa Weik von Mossner

11 **Constituting Vegetarian Audiences: Orchestrations of Egocentric, Anthropocentric, Ecocentric Exigencies in Jonathan Safran Foer's *Eating Animals*** 271
Oren Abeles and Emma Lozon

12 **Beyond Diet: Veganism as Liberatory Praxis** 291
Tara Roeder

Index 319

Notes on Contributors

Nora Abdul-Aziz is a Syrian-French-American pre-medicine undergraduate student at the University of Toledo and aims to specialize in surgery in medical school and improve the quality of care for underserved populations. She has published in the area of community mental health advocacy for Muslim communities within and outside the Middle East and North Africa.

Oren Abeles is Assistant Professor of Rhetoric and Composition at Michigan Technological University. His research focuses on the intersection of rhetorical theory, science, and agriculture, particularly in regard to biological evolution. His work has appeared in *The Quarterly Journal of Speech* and *Poroi: An Interdisciplinary Journal of Rhetorical Analysis and Invention*.

Sharon Avital holds a PhD in Rhetoric and Language from the University of Texas at Austin and is a lecturer at Tel-Aviv University. Her work explores the intersection between social movements, politics, and popular culture with a focus on affect, the body, and the role(s) of visuality and color. In addition to teaching classes in rhetoric and social movements, intercultural communication, and rhetoric and the body, Avital is a licensed therapist of holistic medicine and yoga, an activist on issues related primarily to the environment and geopolitical coexistence, and is a recently converted vegan after being a vegetarian for over 20 years.

Elizabeth Baddour is interested in the interconnected nature of language with race, class, and culture. Through historiography, Baddour's dissertation traced the interplay of social forces with the paradigm shift in the discipline of rhetoric and composition. Her most recent publication is found in the edited collection of *Rhetorics Change/Rhetoric's Change.*

Daniella Fedak-Lengel is in the Honors College of Bowling Green State University, studying toward a Bachelor of Science in Biology. Fedak-Lengel converted to a plant-based diet at the age of 12 and has been a committed vegan ever since. She was selected to participate in a veterinary program in Belize focusing on mitigating the illegal trade of exotic animals, wildlife medicine and conservation, population health assessment of endangered iguana, and contributions of veterinary public health in disaster situations.

Dynestee Fields is a graduate of Southern Wesleyan University where she studied English and Media Communication. She has spent the last 4 years writing, filming, and editing news packages and documentaries that center on animal-related topics, such as cold weather protection for dogs, the health benefits of a vegan diet, animal sanctuaries in South Carolina, and the need for conservation efforts in urban areas. Specializing in garnering attention for the causes of veganism and animal protection through the use of visual media, her most recent project is a documentary on the intricacies of chicken communication.

Heidi Hakimi-Hood studies rural representation in nineteenth-century British writings. Her research includes transatlantic literatures, hispanophone literatures, and historic culinary texts. Her dissertation, *Locating Rural Cosmopolitanism in Long Nineteenth-Century British Writings*, gives attention to the global awareness of rural populations and women writers such as Elizabeth Gaskell and Fanny Calderón de la Barca. She is a 2017–2018 recipient of an American Association of University Women (AAUW) Dissertation Fellowship. She is an associate coeditor for *An Anthology of Anglophone Transatlantic Literature, 1776–1920,* scheduled for 2021 publication by Edinburgh UP.

Cristina Hanganu-Bresch is Associate Professor of Writing and Rhetoric at the University of the Sciences, Philadelphia, USA. Her research on the rhetoric of psychiatry and scientific communication has appeared, among others, in *Written Communication, Literature and Medicine*, and *Medical Humanities*, and several edited collections. She co-authored *Diagnosing Madness* (2019) and *Effective Scientific Writing* (2020).

Carol Helstosky is an associate professor at the University of Denver, Colorado, where she teaches courses in modern European history, food history, and historical method and directs student research through the Veterans Legacy Program. She is author of *Garlic and Oil: Food and Politics in Italy* (2006); *Pizza: A Global History* (2008); *Food Culture in the Mediterranean* (2009); and the editor of the *Routledge History of Food* (2014). She is researching a global history of meatlessness.

Kristin Kondrlik is Assistant Professor of English specializing in Health Communication at West Chester University of Pennsylvania. Her research focuses on wellness movements in historical and contemporary medical journals. Her scholarship has appeared in *Victorian Periodicals Review*, *English Literature in Transition: 1880–1920*, and *Poroi* (*Project on Rhetoric of Inquiry*).

Callie F. Kostelich is Assistant Professor of Practice in First-Year Writing at Texas Tech University. Kostelich's research interests include rural literacies, agricultural literacy, feminist rhetorics, and first-year writing. Her book project, *Sponsoring Agricultural Literacy: Literacies, Ideologies, and the FFA,* explores how and for what purposes the National FFA Organization and supporting entities sponsor students' agricultural literacy acquisition. As an interdisciplinary, literacy-focused project, *Sponsoring Agricultural Literacy* offers insight into the state of corporate sponsorship in public education, prevailing agricultural narratives with deeply embedded rural literacies, and the connection between privilege and critical literacy acquisition.

Sharon Lauricella is an associate professor and Program Director in the Communication and Digital Media Studies program at Ontario Tech University in Oshawa, ON. She holds a PhD from the University of Cambridge (England). Her research focuses on feminist digital identities, food communication, and the use of mobile technologies in higher education pedagogy. Lauricella is a trained yoga instructor with Power Yoga Canada, and has been a practitioner of yoga for more than 20 years.

Lara Martin Lengel began her research program as a Fulbright Research Scholar in Tunisia and American Institute of Maghreb Studies Scholar in Morocco. Her refereed articles have appeared in *Journal of Health Communication*, *Journal of International and Intercultural Communication*, *French Journal for Media Research*, *Journal of Communication Inquiry*, *Feminist Media Studies*, *International Journal of Women's Studies*, *Text and Performance Quarterly*, and *Studies in Symbolic Interaction*. As a professor at Bowling Green State University, USA, she was awarded nearly $500,000 from Fulbright-Hayes, U.S. Department of State's Middle East Partnership Program and Bureau of Educational and Cultural Affairs to codirect international university and professional partnerships in the Middle East and North Africa.

Emma Lozon is a PhD student in the Rhetoric, Theory, and Culture program at Michigan Technological University where she teaches undergraduate writing courses. Her research interests include rhetoric and technical communication with a focus on critical food literacy and representations of healthy living.

Molly Mann serves as Assistant Dean in the Graduate Division of St. John's College of Liberal Arts and Sciences at St. John's University in Queens, NY. Mann's research interests include domestic fiction in the long nineteenth century, American modernism, women's labor and its literary representations, food studies, and gender and race in the digital humanities.

Tara Roeder holds a PhD in English from the CUNY Graduate Center, and is Associate Professor of Writing Studies at St. John's University in New York (Queens), NY. She is coeditor of the Parlor Press volume *Critical Expressivism: Theory and Practice in the Composition Classroom* and the author of multiple essays, poems, and chapbooks. Her recent work focuses on critical animal theory and veganism as they relate to pedagogy and liberation movements.

David P. Stubblefield is an instructor at the University of South Carolina at Union. He holds an MA in Philosophy and a PhD in Rhetoric and Composition from University of South Carolina. He specializes in rhetorical theory, affect theory, and continental philosophy and has published multiple scholarly articles, including articles on affect, performativity, and aesthetics. His most recent publications include "We Have Never Been Rational: A Genealogy of the Affective Turn" in *Affect, Emotion, and Rhetorical Persuasion in Mass Communication* and "Who Is Afraid of Neutrality: Performativity, Re-Signification, and the Jenna Six," coauthored with Chad Chisholm, *On Neutrality: Politics, Praxis, and Performativity* (forthcoming anthology).

Mirjana Uzelac is a doctoral candidate at the University of Alberta Department of Anthropology. A native of Belgrade, Serbia, Uzelac hold a BA in Archaeology and an MA in Anthropology from the University of Belgrade. Her research focuses on anthropology of science, gender, and post-socialism. The research examines the intersection of astrophysics and gender in post-Yugoslav, post-socialist Serbia. Through her conversations with Serbian scientists she seeks to learn what makes them tick, what makes them angry, and why so many STEM scientists in Serbia are women.

Alexa Weik von Mossner is Associate Professor of American Studies at the University of Klagenfurt in Austria. Her research explores contemporary environmental culture from a cognitive ecocritical perspective. She is the author of *Cosmopolitan Minds: Literature, Emotion, and the Transnational Imagination* (2014) and *Affective Ecologies: Empathy, Emotion, and Environmental Narrative* (2017), the editor of *Moving Environments: Affect, Emotion, Ecology, and Film* (2014), and the coeditor of *The Anticipation of Catastrophe: Environmental Risk in North American Literature and Culture* (with Sylvia Mayer, 2014).

Introduction: Legitimation Strategies in Veg(etari)an and Anti-Veg(etari)an Discourses

Vegetarianism (abstention from meat) and veganism (abstention from consuming or using any animal-derived products) are embodied practices of an ideology that is about much more than diet, blending social justice with moral philosophy and health consciousness, and individual protest with global activism. While vegetarianism and veganism used to be considered fringe (in her 2002 monograph on this topic, Donna Maurer was still pondering whether vegetarianism was a "movement or moment"), we have overwhelming evidence today that vegetarianism, and to a lesser extent, veganism, have gone mainstream, although not without generating perpetual conflictual ripples in the contemporary public sphere. Whether as a global consumer trend, as an ecological practice and philosophy, or as an aspiration toward a more peaceful and healthier world, vegetarianism has captured our collective imagination; it has, at the same time, garnered considerable backlash and ridicule, becoming one of the many fault lines that criss-cross our dappled social landscape.

Vegetarian practices have a long and illustrious history across Western and Eastern traditions alike. Recently, however, the moral quandaries of eating animals have garnered renewed interest and focus. In the past half-decade or so, veg(etari)anism has been intensely debated in philosophy and the social sciences—including history, critical geography, anthropology, and animal studies broadly construed; some of the stronger contributors to the conversation come from literary studies. This deeply interdisciplinary appeal may come from a conflation of factors that propelled veg(etari)anism to the fore of our cultural imagination. Among these factors are

- Extensive ethology research over the past hundred years or so, expanding our understanding of sentience among nonhuman animals; these studies showed that most if not all animal species, from invertebrates to primates, display levels of intelligence, emotion, sociability, and communicative abilities (including symbol usage) that chip away at the claims of human exceptionalism (e.g., human as the only tool-making animal) and make it much harder to justify our treatment of animals as "raw materials" or "objects";
- A slew of philosophical, sociological, and cultural-theoretical works questioning the inherent anthropocentric bias of our moral and political practices that has been growing steading since the 1970s (collectively adding to the foundation of Critical Animal Studies);
- Eye-opening investigations into factory farming practices (often covert and putting the activist and videographer at risk), showing mostly hidden but shockingly brutal practices, made even more barbaric by what we now know about sentience in nonhuman animals;
- Awareness of health benefits of plant-based diets, and clinical research correlating excessive consumption of animal products (in particular meat) with a variety of metabolic diseases, chronic vascular diseases, and cancers;
- The growth of the environmental movement and a new consciousness regarding our duty to maintain ecosystems and preserve animal species;
- Growing awareness of the impact of animal agriculture on the environment and our health—and in particular, awareness of the sizable contribution of animal farming to global warming;
- Increasing visibility and larger cultural footprints of countercultures predicated on nonviolence (e.g., hippie, organic movement, eastern philosophies, and meditative practices), as well as of animal rights groups (such as PeTA or ASPCA) on mainstream cultures;
- Wider, consumer-driven availability of varied vegan products and plant-based protein that successfully functions as a meat substitute

(note: this is not to say that all vegan products we consume or use are entirely ethical or cruelty free—farmworkers are often underpaid and exploited migrant labor, and too much of our produce comes from remote locations, significantly adding to fuel consumption and global warming).

This is not, of course, an exhaustive list; overall, however, it highlights the particular convergence of cultural, political, economic, scientific, and social factors that have brought veg(etari)anism into the mainstream. Our treatment and consumption of animals will continue to remain problematic, and keenly intertwined with human destiny, as emphasized by historical events such as the COVID-19 pandemic. Zoonotic diseases like the ones caused by the coronavirus originated with animal consumption, but, as some philosophers have warned, such occurrences are only the tip of an enormous iceberg, and will plague us as long as we continue to mistreat animals in the name of convenience, efficiency, tradition, or economy (Benatar 2020). Likewise, the effects of those pandemics are bound to be deepened by the relentless grip of the global neoliberal paradigms on nature cultures.

Veg(etari)an practices uncover tensions between individual dietary choices and social justice activism, between ego and eco, between human and animal, between capitalism and environmentalism, and within the larger universe of theoretical and practical ethics. This background makes for a tense, combative rhetoric in the pro- and anti-veg(etari)an arguments lobbed among a multiplicity of actors in what is by now a growing literature on this topic. In general, these arguments fall into three categories: (1) concern with animal welfare: in a nutshell, arguing that eating animals is ethically objectionable as it goes against universal ethical laws of compassion and fairness, which ought to be extended to nonhuman animals; (2) concern with human welfare, in which evidence is garnered that a "plant-based" diet is optimal for human health, whereas a meat/dairy-centric diet leads to a variety of ailments; (3) concern with the environment in general, which emphasizes reliance on animal-based diets in general, and factory farming in particular, are conducive to climate change and general environmental destruction. Thus, the topic of veg(etari)anism is intrinsically transdisciplinary, as it is built on a premise that argues with the status quo and seeks widespread, sustained change of habits and *habitus*. Of course, wherever anti-meat rhetoric emerges, discourses countering these arguments are quick to appear as well. This collection looks at some of these discursive strategies and their outcomes in a variety of contexts.

Strategies of Legitimation in Veg(etari)an-Adjacent Discourses

Vegetarianism and especially veganism have transcended their designation as dietary habits and can be thought of as philosophies; indeed, recently, Laura Wright proposed the field of "vegan studies" as a separate, multidisciplinary home of any and all research that not only takes veganism as its object, but uses it as a critical lens to be applied to other cultural artifacts, and, indeed, to a whole new theory of culture (Wright 2015). Without losing sight of that perspective, we are primarily concerned in this book with veg(etari)anism as an everyday social practice almost always at odds with ambient societal forces. Wherever and whenever it occurs, veg(etari)anism engenders discursive tensions related to its legitimacy.

In order to understand discourses for and against veg(etari)anism, Theo van Leewen's theory of discursive legitimation emerges as particularly useful. In his 2008 book *Discourse and Practice: New Tools for Critical Discourse Analysis*, van Leeuwen looks at discourse as a way to recontextualize, linguistically, social practices. Discourse, in his definition, is "a socially constructed knowledge of some social practice, developed in specific social contexts, and in ways appropriate to these contexts" (p. 6). Recontextualization, van Leeuwen argues, "involves not just the transformation of social practices into discourses about social practices, but also the addition of contextually specific legitimations of these social practices, answers to the spoken or unspoken questions 'Why should we do this?' or 'Why should we do this in this way?'" (p. 105). To that end, he identifies four major mechanisms for legitimation—each of which gets a lengthier treatment in his book:

1. ***Authorization***, that is, legitimation by reference to the authority of tradition, custom, law, and/or persons in whom institutional authority of some kind is vested.
2. ***Moral evaluation***, that is, legitimation by (often very oblique) reference to value systems.
3. ***Rationalization***, that is, legitimation by reference to the goals and uses of institutionalized social action and to the knowledge that society has constructed to endow them with cognitive validity.
4. ***Mythopoesis***, that is, legitimation conveyed through narratives whose outcomes reward legitimate actions and punish nonlegitimate actions. (vanLeeuwen, pp. 105–106).

These mechanisms (each of them with their own internal structures and hierarchies) may be combined to legitimize or delegitimize (p. 106) the social practice(s) in question. The choice of legitimation strategies is often telling of the political orientation and purposes of the individual or group using them, and, I would add (although, curiously, van Leeuwen does not discuss it), of the type of audience targeted.

Let's take, for example, moral evaluation in arguments related to ve(getari)anism. Reference to moral values can be made in absolute terms, and veg(etari)an discourses are known to make those appeal, using a vast armamentarium of arguments that have been honed in moral philosophy and critical animal studies since the publication Peter Singer's *Animal Liberation* in 1975. But such judgments can also come more obliquely through reference to values that are socially and historically contingent, whose genealogy would need to be unearthed in order to fully comprehend the degree to which they have been coopted as undisputed ethical values. "Natural," "pure," "organic," "clean," "detoxifying," and many other attributes that saturate contemporary discourses of "wellness," for example, are cultural constructs that went through a historical and rhetorical evolution to embody their positive moral valences, especially in reference to diet and lifestyle (see also Helstolsky's history of "natural diets" in early twentieth-century Italy, this volume). A subset of vegan/vegetarian discourses often resort to these oblique values in their arguments as well; but so do meat-centric discourses: meat is natural, primal/primordial, the caveman's diet (amusingly, never the cave woman's!), and we are born to eat meat, since it was protein that allegedly allowed our brains to grow and *homo sapiens* to evolve. Of course, words like "natural" or "pure" often function as signals of nostalgia, a longing for a return to a mythical golden age and status quo, and as such are very fraught concepts in general when applied to diet, as several historians have demonstrated (see, for example, Corinna Treitel's work on the concept of natural food, vegetarianism, and Nazism in the Third Reich, or Bobrow-Strain's work on purity and white bread).

Authority legitimation stemming from conformity and tradition is, unsurprisingly, frequently invoked in anti-veg(etari)an arguments. Most of the cultural traditions analyzed in the first section of this book delegitimize veg(etari)anism on account of its break with tradition, where tradition is essentialized as innate and indeed congruent with ethnicity, culture, religion, and, of course, personal and collective identity. Meat-eating is conservative, traditional; veg(etari)anism—disruptive and liberal (as our chapters on meat-eating in Islamic cultures and in a typical south-east European orthodox nation such as Serbia indicate—see Abdul-Aziz, Fedak-Lengel, and Lengel; Uzelac). This distinction does not, however, always hold up, as veg(etari)an proponents within these cultures appropriate and interpret parts of their own tradition to legitimate their practices. And in Israel, one of the countries where veganism has made great strides, pro-vegan arguments are somewhat paradoxically related to the political right, as one of our chapters discusses (Avital). To further nuance this strategy, as shown by another chapter on Patanjali's Yoga Sutras (Lauricella), traditional authority as a way of legitimating veg(etari)anism takes on a new meaning in the case of yoga, an ancient practice whose philosophy involves abstention from meat and whose import by the West reinterprets that dietary dictum for its new practitioners.

Mythopoetic legitimation works via storytelling—for example, via moral tales or cautionary tales, which offer substance and meaning to a social practice by outlining the consequences of conforming or not conforming to it. Stories are powerful and vivid and connected to our earliest learning experiences in the world. They are meant to make sense of a system of beliefs, or outline how or why they don't work. In the mythopoetic repertoire of veg(etari)an conversions we find many individual accounts of experiences ending roughly with the same resolution (adopting a new diet and a new belief system); most of these are moral tales. There are also the cautionary tales of "ex-vegans" who, unable for whatever reason to maintain their vegan lifestyle, turn their negative experience into a warning for anyone who might want to give it a try. A lot veg(etari)an legitimation occurs through a subtype of mythopoesis, the multimodal legitimation strategy (e.g., footage of abattoirs, feel-good animal videos, memes, Instagram accounts, lush or sensuous imagery of vegan food, documentaries). As some of the chapters in our book emphasize, these stories are always complicated, troubling the placid luster of abstract notions with the lived, embodied experience of meat abstention. One chapter explores what happens when one acquires a meat allergy through a tick bite in the midst of a Southern culture where socialization can be heavily reliant on barbecue (Baddour). Into what category do we push this "accidental vegetarian"? How does she make sense of her new identity? Another chapter (Mann) focuses on queerness and the animal body, and the blurred boundaries between the two in one novel: where does the human start and the animal end? How useful is that distinction? What are the implications queerness and hunger for meat eating? Another chapter (Kostelich and Hakimi-Hood) examines the story Oprah Winfrey tells the city of Armadillo when Texas Beef sues her after she declares her intention never to eat a hamburger again in the wake of the mad cow disease scare—a cautionary tale meant to show, among other things, what could happen when one angers the meat lobby. These stories of embodied experiences work strategically to foreground personal truths that legitimate veg(etari)an practices.

Rationalization legitimation strategies are present in both pro and anti-veg(etari)an arguments. A subtype distinguished by van Leeuwen (who follows Habermas's lead) is teleological: an action is judged in terms of its effectiveness or success—the ends justify the means. For example, if the goal is to get more protein (and iron, and vitamin B12), meat proponents appear to have found their means. (The fetishization of protein in the current dietary climate deserves, perhaps, its own book.) On the other hand, if the goal is to reduce our carbon footprint and slow down or reverse climate change, vegan environmentalists have the higher ground. Another subtype of rationalization strategies, theoretical legitimation, relies on whether the action is based on some sort of truth—on the way things *are*; this category includes scientific rationalization. Scientific studies have, in fact, been used to prop up both practices, both directly (as in studies researching the effect of meat on cardiovascular disease, or of animal farming on climate change), or indirectly (as in studies of Bovine Spongiform Encephalopathy, or "mad cow's disease," which may scare people into reducing or eliminating their meat consumption as a personal risk-reduction strategy). Importantly, van Leeuwen notes, scientific rationalizations "not only include modern science but also other systematic bodies of knowledge that are used to legitimize institutional practices, for instance, religions" (p. 116). As the final chapters in this collection show, pro-veg(etari)an arguments rely heavily on rationalization legitimation, particularly by emphasizing the social justice aspects of veg(etari)anism (Stubblefield and Fields; Roeder), and making the connection between the exploitation and consumption of animal bodies and a bleak climate change scenario entailing worldwide devastation (Lozon and Abeles; Roeder; Weik von Mossner). Intimately intertwined with such instrumental justifications are, of course, moral judgments about what ought to motivate humanity, collectively, when contemplating our duties to self, each other, and the Earth.

A Definitional Note: Veganism, Vegetarianism, Veg(etari)anism

In the past couple of decades, veganism has gained traction over vegetarianism and has come to supplant it, especially in critical animal studies and other contemporary theoretical discourses. We (and, we expect, our readership) are keenly aware of the definitional difference between vegetarianism and veganism: where the former means abstention from meat in one's diet, the latter means elimination of all animal products from one's lifestyle—diet, wardrobe, décor, and so on. This book uses "veganism," "vegetarianism" and the composite "veg(etari)anism" somewhat interchangeably, though not because we wish to conflate the terms; on the contrary, it is because we wish to keep our focus on what they have in common, which is, abstention from consuming animal flesh, stemming, in general, from a desire to at the very least minimize and, in an ideal scenario, abolish harming animals. Furthermore, historically, "vegan" has a much shorter career as a term than "vegetarian," which was used in the past to denote a wide range of plant-based diets, some of which were indeed, vegan. (Donald Watson and his wife, Dorothy, are credited with coining the word "vegan" in 1944, though the term did not take off until a few good decades later.) Even Carol Adams's groundbreaking *Sexual Politics of Meat* bore the subtitle, "Feminist-Vegetarian Critical Theory" (which she later acknowledged should be taken to mean "vegan").

We should, nevertheless, recognize that if vegetarians are abstaining from meat on account of the animal suffering it entails, they should also abstain from consuming dairy or eggs, given what we now know about the horrors of industrial farming. There is no less suffering involved in the production of nonmeat animal products than it is in the production of meat per se. (And to be clear, industrial farming is singled out here merely for its unprecedented scope of its cruelty, but even before its advent, in what we would call "traditional" or family farming, animals who may have suffered less and on a lesser scale invariably end their lives in the same abject space of suffering and death; David Nibert makes fairly decisive arguments about the cruelty and speciesism of domestication, which he re-dubs "domesecration"). Thus, those who are vegetarian for the sake of animals should really be vegan, or their position is a priori contradictory. The same inference may not apply to those who practice vegetarianism for other reasons, such as personal health or perhaps religious motives.

Josh Milburn convincingly parses out the philosophical arguments for vegetarianism (as opposed to veganism) and finds out that most contemporary scholarship argues in fact for veganism (not vegetarianism), with one notable exception (i.e., Tzachi Zamir, whom he critiques at length). Moreover, none of the usual arguments for vegetarianism holds under scrutiny—namely, that it is "easier" than, or a "less extreme" middle-of-the road approach compared to veganism; it is hard to dispute that the production of any animal protein—in the form of egg, dairy, or meat—remains a violent, fraught process forever at odds with our desire to lead moral lives. Vegetarianism, then, would be the type of compromise that tolerates the suffering and death of millions of some animals (destined for dairy and egg production) while drawing the line at the suffering and death of some animals destined for meat production. In utilitarian terms (i.e., adding up and comparing the amounts of suffering involved), the distinction does not hold and reveals vegetarians as hypocritical. Of course, by the same token, vegans who understand their activism strictly in terms of exclusionary consumption (i.e., nonanimal products) may also fail in their professed duties to the environment (e.g., by using plastic or other nonsustainable materials) and to fellow human beings (e.g., by buying cheap vegan products made in sweatshops, imported produce, berries picked by exploited labor, or even, to take the case to the extreme, produce obtained through non-vegan agriculture [i.e., using bone and blood meal fertilizers, pesticides, and harvesting means harmful to local fauna]—which is to say, roughly all of it). The politics of purity is a slippery downward spiral.

Milburn also offers a tentative sketch of an argument in support of vegetarianism (as opposed to veganism), revolving on the conceptualization of flesh as the *body* of animals, which may be said to be of a different nature than that of milk or eggs, which can be classified as secretions rather than bodily matter. (More can be added under this latter category, of course: honey, wool, leather, silk, and so on, all substances avoided by vegans.) While Milburn does not go so far as to offer a fully blown defense of vegetarianism, I find there is some ontological merit to this line of thought, predicated on the idea that what animates vegetarianism is the belief that consuming the bodies of others (=other sentient beings) ought to have no place in human sustenance. The term veg(etari)anism is a recognition of this avoidance of meat consumption as the common core of both vegan and vegetarian practices.

Going back to the issue of terminology: without denying the differences between veganism and vegetarianism, we wish to acknowledge the motivation that unites them, as well as the historical and cultural vagaries that have made one term more prominent than the other at a particular point in time. All three terms (veganism, vegetarianism, and veg(etari)anism) are used by the authors in this collection as they document the histories, cultural and embodied practices, and arguments related to meat consumption in a variety of contexts; some of the cultural histories presented by some authors necessarily need to use the term "vegetarian" where "vegan" would be anachronistic (especially in Part I).

Structure of the Book

The chapters maintain a sustained focus on the rhetorical arguments at the core of veg(etari)anism as they play out in a variety of media and are thus focused on the public sphere and grounded in veg(etari)an *discursive* practices. Furthermore, the book addresses a variety of transnational contexts and incorporates a variety of perspectives, including feminist, queer, religious, and ecological. Vegetarianism is inherently conflictual, as it threatens a carnist status-quo; hence, many of the essays in this book parse out the contentious aspects of veg(etari)an and anti-veg(etari)an discourse. Since this is an extremely complex and multifaceted topic, we are necessarily not exhaustive in its treatment. For example, we have decided against including topics such as PeTA advertisements for veganism, which have been extensively covered in a number of essays (see Pendergrast 2018, among others).

In Part I, chapters focus on how veg(etari)anism fares in specific geocultural traditions and contexts: Pre-WWI Italy (Helstolsky), Islam (Lengel et al.), Serbia (Uzelac), Israel (Avital), and Patanjali'sYoga Sutras (Lauricella). We believe that more work needs to be done on transcultural veg(etari)anism, which, as it can be clearly seen from this selection, has complex, context-dependent factors that may not align with Western veg(etari)an practices; to understand them, a deep dive into the history and cultural motivations of these regions is a must. Part II focuses on embodied experiences and legitimation strategies—in particular, the politics, identities, and ontological consequences coming from consumption of or abstention from meat: what discursive tactics work when abstention from beef enrages the beef lobby (Kostellich and Hakimi-Hood)? What sort of identities are forged at the boundary between veg(etari)an and meat-eater (Baddour), between human and nonhuman animal (Mann)? Finally, in Part III, chapters look at the motives, purposes, and implications of veg(etari)anism as a transformative practice that should revolutionize our value systems and, by extension, our future(s). How do we grapple with human exceptionalism, a concept that undermines animal welfare arguments, and replace it with a more inclusive, liberating Judeo-Christian interpretation that makes veganism a moral imperative

(Stubblefield and Fields)? How do we get others to think about meat in ecological terms (via eco-documentaries—von Mossner; or books—Lozon and Abeles)? And what sort of liberating, transformative potential do veg(etari)an practices have in the context of global challenges such as climate change, overpopulation, inequality, and so on (Roeder)?

We hope that this collection will add to the growing body of veg(etari)an studies and offer much-needed insights into the legitimation strategies of both veg(etari)an and carnist practices. More than a diet, veg(etari)anism invites us to reflect on our identity and relationship with our animal selves, and our entanglement with nonhuman animals on this earth; reexamine our moral grounding; rethink tradition and authority and the complexities of adopting veg(etari)anism in different cultures; and contemplate our deontological duties to our planet and fellow earthlings.

Philadelphia, PA, USA Cristina Hanganu-Bresch

Bibliography

Adams, Carol. 1990. *The Sexual Politics of Meat: A Feminist-Vegetarian Critical Theory*. New York: Continuum.

Benatar, David. 2020. Our Cruel Treatment of Animals Led to the Coronavirus. *The New York Times*, April 13. https://www.nytimes.com/2020/04/13/opinion/animal-cruelty-coronavirus.html

Bobrow-Strain, Aaron. 2008. White Bread Bio-politics: Purity, Health, and the Triumph of Industrial Baking. *Cultural Geographies* 15 (1): 19–40.

Maurer, Donna. 2002. *Vegetarianism: Movement or Moment?* Philadelphia: Temple University Press.

Milburn, Josh. 2019. Vegetarian Eating. In *Handbook of Eating and Drinking*, ed. Herb Meiselman. Cham: Springer.

Nibert, David. 2013. *Animal Oppression and Human Violence: Domesecration, Capitalism, and Global Conflict*. New York: Columbia University Press.

Pendergrast, Nick P. 2018. Patriarchy and Intersectionality. *Animal Studies Journal* 7 (1): 59–79. Available at: https://ro.uow.edu.au/asj/vol7/iss1/4

Treitel, Corinna. 2017. *Eating Nature in Modern Germany: Food, Agriculture, and Environment, 1870–2000*. Cambridge: Cambridge University Press.
Van Leeuwen, Theo. 2008. *Discourse and Practice: New Tools for Critical Discourse Analysis*. Oxford: Oxford University Press.
Wright, Laura. 2015. *The Vegan Studies Project. Food, Animals, and Gender in the Age of Terror*. Athens/London: The Georgia University Press.

Part I

Cultural and Historical Perspectives on Vegetarianism

1

State of Meatlessness: Voluntary and Involuntary Vegetarianism in Early Twentieth-Century Italy

Carol Helstosky

For much of the nineteenth century and the first half of the twentieth century, Italians found themselves in a state of meatlessness. The young nation, unified in 1861, confronted economic hardships, regional disparities, and social divisions. The majority of Italy's population lacked the money to purchase meat or animal products; agricultural production focused on grains, olives, and grapes; the domestic production and market for foods was limited; and a faltering economy hindered the development of sophisticated food retail sector. Meat was a rarity, consumed by Italians on holidays or special occasions like funerals. Most Italians were vegetarians by necessity, not choice.

Scientific professionals in nineteenth-century Italy viewed the Italian state of meatlessness unfavorably. In many of the scientific and social scientific writings published in Italy, the absence of meat in the Italian diet signaled monotony, malnutrition, and backwardness, especially in

C. Helstosky (✉)
University of Denver, Denver, CO, USA
e-mail: chelstos@du.edu

C. Hanganu-Bresch, K. Kondrlik (eds.), *Veg(etari)an Arguments in Culture, History, and Practice*, The Palgrave Macmillan Animal Ethics Series,
https://doi.org/10.1007/978-3-030-53280-2_1

the context of the dietary habits of other European nations (Britain, Germany) and the United States. Because of technological advancements like the railroad and refrigeration, Western European and American workers had access to meat, the protein source recommended by a growing number of medical professionals. At the end of the nineteenth century, physiological chemists established dietary norms and published daily requirements of calories and protein. American, German, and British physiologists recommended a daily intake of between 80 and 120 grams of protein a day (along with 2500–3000 calories) for an adult male engaged in moderate labor (Rabinbach 1990, 128–133). Although some physiologists and food reformers (such as Horace Fletcher and Russell Henry Chittenden) contested high levels of protein and calorie intake, scientific consensus formed around the idea that meat was an excellent and efficient source of protein, to build muscle, endurance, and strength (Offer 1989, Chapter 2). In Italy, scientific professionals knew that most Italians were unlikely to reach the recommended protein levels set by the international scientific community, primarily because individual Italians could not afford to consume meat on a daily basis. Urban and rural Italian workers consumed a monotonous diet based on wheat (bread) or corn (polenta), depending on where they lived. Popular diet was a source of great professional concern and diseases of malnutrition like pellagra plagued Italian regions well into the twentieth century. The Italian government conducted social scientific inquests into the nature and quality of popular diets across Italy, but the government possessed few resources to help Italians afford a more varied diet (Helstosky 2006, Chapter 1; Scarpellini 2015, Chapter 2).

The lack of animal products in popular diet or what I refer to as meatlessness was viewed negatively by many Italians. Those who did not consume meat or animal products for the sake of economy cannot be called voluntary vegetarians; they might have enthusiastically consumed meat if they could have afforded to do so. However, a number of Italians chose to evaluate popular consumption levels, not by what Italians could not eat, but what they did eat. Rather than viewing existing eating habits as monotonous or inadequate, some medical professionals and a small number of proclaimed vegetarians promoted a diet consisting mostly of grains and fresh produce as a way to maintain good health. Moreover, vegetarians reasoned that a simple diet based on grains, legumes, fruits, and vegetables was the most appropriate diet for Italians, who lived off the land and were therefore closer to nature than other European and American populations. For those who thought positively about vegetarianism, meatlessness was not a sign of Italian inferiority; meatlessness was a sign of individual and community well-being (Capatti 2016).

This chapter will examine the history of vegetarianism in nineteenth and early twentieth-century Italy, examining the publications of vegetarian organizations and published works by individuals in order to understand the ideas of Italian vegetarians regarding diet and its connection to individual, community, and national health. Rising to prominence in the early twentieth century, Italian vegetarians maintained organizations in Florence and Milan. These societies were loosely affiliated with other vegetarian societies across Europe, drawing inspiration from British, German and Swiss organizations. Italian vegetarian philosophy was influenced by naturism and Theosophism. Although Italian organizations cross-pollinated ideas and strategies with other groups, Italian vegetarianism remained distinctive from its European and American counterparts. With an emphasis on local produce grown in Italy and prepared according to varied regional cooking styles, vegetarian cuisine followed closely the general precepts of Italian cuisine: a minimal number of locally produced ingredients, simple preparation techniques, and a tendency not to mask or hide the authentic flavors of the foods used in dishes. Even when suggesting ways to create substitute meats, vegetarian cookbook authors remained true to local ingredients. During the early twentieth century and through the years of the First World War in Italy (1915–18), the small but vocal group of vegetarians pushed for dietary reform because they thought a simple and healthful vegetarian diet would be well suited for the alimentary needs of individuals as well as larger populations of Italians.

After the First World War and four years of economic and political dislocation, Benito Mussolini was given the reins of power, ostensibly to reform Italy and Italians. From 1922 until the outbreak of the Second World War in 1939, the fascist regime actively sought to promote Italian nationalism in all aspects of life, including agricultural production and food consumption, both of which were subject to the national goal or autarky or self-sufficiency. Given the contours of national agricultural production, the fascist regime promoted a dietary regime that was not unlike that proposed by Italian vegetarians: grains enhanced by an abundance of fresh produce and an emphasis on locally grown or produced foods. The two alimentary regimes were not identical, however. Fascists enthusiastically recommended the consumption of wine, given that grapes were abundant and easy to raise in Italy, and thus fit into the autarkic aims of the regime. Italian vegetarians seldom mentioned wine in their publications and considered alcohol an overly stimulating substance, with effects similar to those of meat. Although historians have noted ideological and culinary connections between early twentieth-century vegetarians and fascist agricultural and medical professionals (Buscemi 2019, 137–147; Capatti 2016), one must be careful not to draw too many conclusions about vegetarianism or fascism based on similar attitudes towards food. Italy's state of meatlessness originated with inequities in land use and a poor economy; both conditions ensured that Italians consumed a monotonous diet composed mostly of carbohydrates and lacking in animal-based proteins as well as fresh fruits and vegetables. The fascist regime understood that boosting protein consumption among Italians would be costly. Although the regime spoke openly about maintaining and improving racial health (through various initiatives), the government was more interested in controlling consumption patterns so that they accommodated national agricultural and economic imperatives through autarky. Despite similarities in thinking about and through Italy's condition of meatlessness, vegetarians and fascists differed sharply in terms of how they viewed the purpose of rejecting meat. For vegetarians, the rejection of meat was a choice to benefit individual health whereas for fascists, rejecting meat was one of several responses to the demands of economic nationalism.

The Case of Italy

Today, when we define Italian cuisine, we tend to think about uncomplicated dishes based on local ingredients that have been minimally processed, with grains comprising the foundation of many dishes. Two of the most globally recognized Italian dishes, pasta, and pizza, are made with little or no meat (unless one considers the fast-food pizza, laden with various meats, produced in the United States). In many Italian dishes, meat does not comprise the foundation of the dish but is treated more like a condiment, added sparingly for flavor. Historians have referred to Italy's cuisine as a cuisine of scarcity, based as it was on limited access to expensive ingredients (Montanari 1994; Dickie 2008; Helstosky 2006; Parasecoli 2014; Scarpellini 2015). There were topographical reasons for the nature of Italian cuisine, as the rocky quality of the Italian peninsula is good for growing grapes and olives more so than wheat and pastures for cattle. There were also historical reasons for Italy's simple cuisine; the nation lacked the economic and political resources to vigorously expand food imports, either through trade or imperialism. An abstemious culinary nationalism ruled the day for much of Italy's history because there were no alternatives.

Italy's limited food supply had a lengthy history: in 1614, Giacomo Castelvetro observed in his treatise *Brief Account of all the Roots, Greens and Fruits that are Eaten in Italy either Raw or Cooked* that the large number of Italians confined to a small space meant that everyone ate less meat (Montanari 1994, 113). Land use was only one factor of many; climate, poverty, and the Catholic religion (which emphasized Lenten practices) all influenced the meatless habits of Italians for centuries. While populations in northern European countries became more carnivorous after the Reformation, southern and eastern European countries did not. Meatlessness was not so much of a choice as it was a fact of life. When citizens of Italy, Eastern European nations, and Russia migrated to North America, their dietary habits changed from consuming mostly carbohydrates to consuming meat several times a day (Diner 2003). In the eighteenth century, vegetarianism became more of a choice, at least for well-off Europeans who defined their dietary practices as being distinct from those of the poor. European vegetarianism by choice implied a particular world view, especially during and after the Enlightenment, when consuming a vegetarian diet represented the revival of "well-tested Christian images and motivations: vegetable food represented the food of peace and non-violence, the choice for a 'natural,' simple and frugal life; vegetable food insured against bodily heaviness and so allowed the mind to work more freely" (Montanari 1994, 150). Prior to the nineteenth century, vegetarian food stood in sharp contrast to old ways of eating, especially the meat-centered diets of the European elite.

With industrialization, the progressive democratization of diet brought with it a change in the ideology of food. Over time, workers could afford foods previously deemed luxuries, such as sugar and meat. In industrialized Britain, both workers and factory owners could afford to consume meat, though they likely consumed different cuts of meat (Montanari 1994; Laudan 2015, Chapter 7). Individuals and communities who consciously chose to become vegetarians rejected the greater availability of meat as well as other consumption practices deemed unhealthy, like caffeinated beverages, tobacco, and alcohol. By the late nineteenth century, vegetarian societies thrived in the British cities of London and Manchester. In the early twentieth century, Germany and Switzerland were home to vegetarian organizations and doctors proclaiming the benefits of a natural, meatless diet. In addition to forming organizations, vegetarians across Europe published newspapers and opened restaurants, cooperatives, and food supply businesses, maintaining a loose network of committed and curious consumers. The diffuse quality of continental European vegetarianism meant that Italian vegetarianism borrowed health and hygiene concepts from Germany and Switzerland (Capatti 2016, 17). The Società Vegetariana d'Italia, founded at the end of 1905, was one of the most prominent organizations, with active vegetarians in Florence and Milan, cities with vegetarian restaurants and consumer cooperatives. Prior to the First World War, official members of vegetarian societies and associations likely numbered in the hundreds. In publications, the Society focused on the maintenance of good health with a vegetarian diet. Like their German and Swiss counterparts, Italian vegetarians endorsed natural diets: vegetables, fruits, grains, nuts, and legumes, prepared simply with minimal cooking time. German vegetarians espoused a natural diet for two reasons, according to Corinna Treitel. Natural diets enabled the German nation to use nutritional resources efficiently during periods of shortage and optimize public health in periods of abundance (Treitel 2017, 3). Ideologically, natural diets comprised an alternative to the consumer rhythms of industrial, modern life. As a result of industrialization, food had become toxic, German vegetarians argued. Thus, vegetarianism was the most obvious solution to counter the toxicity of modern, industrial life.[1]

Italian vegetarians agreed that modern consumption habits had become unhealthful, even toxic, though the cause of this trend in Italy was less clear, given that Italian industrialization was uneven and based largely in the northern regions. Italians were influenced by German thinking about food, but the overlapping of German vegetarian ideas onto the terrain of Italian consumption was awkward at best. Italian vegetarians understood that vegetarianism provided an alimentary counterbalance to materialistic desires for luxury and excess (Hoffman 1905, 10–14), but given Italy's lagging economy, there was little evidence of widespread overconsumption. It was clear that Germans consumed more protein, given that citizens had access to a greater variety of foods, including meat. Italian nutritional experts frequently grappled with German recommendations for protein, questioning how Italians could even approach German standards. Physiologist Carl Voit of the University of Munich recommended in 1870 that German workers consume 118 grams of protein per day, as protein was the most important nutrient to build muscle and enable the body to perform physical labor (Treitel 2017, 96–97). Italian vegetarians argued against the idea that meat consumption resulted in good health, criticizing the high protein recommendations of German physiologists (Rinonapoli 1902, 5).

While Italian vegetarians agreed with German and Swiss vegetarians on the healthful properties of a natural vegetable diet, they acknowledged other reasons for rejecting meat that originated with a concern for animals. Although Italy had a fledgling animal welfare movement, they were also influenced by British vegetarians, who emphasized the moral and ethical benefits of vegetarianism. Italians were curious about Theosophy, Confucianism, and Buddhism, all of which considered the lives of animals when making choices about diet. For some Italian vegetarians, eating meat was immoral and therefore strongly indicative of a regression on the evolutionary scale (Rinonapoli 1902, 5). Although health and proper nutrition were significant reasons for vegetarianism, a fundamental belief of several prominent vegetarians was that animals did not exist to serve humans as food, as beasts of burden or as objects of vivisection. Indeed, the realities of the slaughterhouse and of the vivisection laboratories, offering "such images of fear, terror and horror," were sadly commonplace in the unhappy lives and deaths of so many living creatures (Rinonapoli 1902, 13).

Meat and the Body

Starting in the 1870s, scientific experts throughout Europe recommended increased levels of protein consumption, echoing their American counterparts. Recommendations for 80–120 grams of protein per day were not impossible to meet in Germany or Britain, where improved wages and living standards allowed middle class and even working-class families to be able to afford meat. In Italy, wages and living standards did not rise to accommodate improved consumption habits, at least not in the nineteenth and early twentieth centuries. Thus, Italians consumed far less than what was recommended for workers elsewhere. This gap in protein consumption split the Italian medical community. Some experts treated high protein recommendations as a goal to be achieved and not a reflection of living conditions. Pietro Albertoni and Felice Rossi, two distinguished Italian physiologists, reached positive conclusions after introducing meat to vegetarian diets in a clinical setting. Meat aided health by increasing the body's absorption of vitamins and minerals, they argued. Thus, Albertoni and Rossi concluded, with increased meat consumption, the health of individuals and therefore the nation would steadily improve (Albertoni and Rossi 1910). A few experts disputed high protein recommendations. Doctor Alessandro Clerici, who was sympathetic to vegetarianism, advocated a protein intake half that of the German and American recommendations, 50 or 60 grams per day. Favoring a diet that did not exclude meat, Clerici reasoned that a vegetarian diet could provide adequate protein intake and contribute to health and longevity (Clerici 1909, 128–9).

Doctor Ettore Piccoli, who published extensively on the merits of a vegetarian diet, vehemently opposed the protein recommendations of German and American experts. Piccoli warned that if Italians were to consume 120 grams of protein per day, the cost to the nation would be enormous. Recommending 40 grams of protein per day, Piccoli argued that vegetarianism could become the basis for a "new alimentary theory," so that individuals and the nation could prosper within the economic restrictions Italy faced (Piccoli 1911). Piccoli's new alimentary theory differed from existing nutritional theories, which broke food down into the basic elements of proteins, fats, and carbohydrates. Instead, Piccoli advocated thinking about foods according to their essential traits or qualities. Thus there were plastic foods (eggs, dairy, legumes), energy-giving foods (carbohydrates like pasta and potatoes), fats (butter and oil), and mineral-rich foods (fruits and vegetables). Piccoli recommended that a simple plate of pasta, seasoned with cheese and butter and accompanied by fresh fruit for dessert, provided the foundation for a wholesome Italian diet.

Piccoli and Clerici stood out in the Italian medical profession, perhaps, in advocating for or at least accepting a vegetarian diet as favorable for Italians. Most medical professionals reluctantly acknowledged meatlessness as a specifically Italian condition, yet held out hope that in the future, Italians would consume more meat. Italian vegetarians strenuously objected to the way meat was regarded by the medical community, arguing instead that meat actually had a toxic effect on digestion and therefore individual health. In particular, vegetarians argued that meat had a toxic effect on the nervous system, producing the condition of intestinal tuberculosis, a build-up of excess bacteria in the digestive system ("Proprietà curative delle frutta e degli erbaggi"1911, 2).

Legumes, vegetarians argued, were an acceptable meat alternative because they were more digestible with nutritional value equal to that of meat (Rinonapoli 1902, 6–7; Hoffman 1907, 11–14). In vegetarian publications, meat was regarded as a toxic substance that frequently overloaded the human body because it was difficult to digest, exciting the nervous system, and possibly leading to all kinds of health complications, including gout, diabetes, neurasthenia, hysteria, apoplexy, and heart problems. Equally important, perhaps, the stimulating effects of meat consumption left the body craving more stimulating substances like sugar or alcohol, compounding poor nutrition, and creating other illnesses (Rinonapoli 1902, 6–7). By contrast, simple foods like legumes, fruits, and vegetables calmed the body's digestive and nervous system because they were easily digested. The defense of vegetarianism over meat consumption was based on a theory of digestibility, with foods that were easily digested, like vegetables or rice, taking precedence over more difficult to digest substances like meat, alcohol and tobacco. Fittingly, perhaps, a simple and easy-to-digest diet was recommended for the sober lifestyle of most Italians. Vegetarians pointed to the Italian peasant, who consumed a primarily vegetarian diet. Similar to laborers in Ireland, Russia, and China, the Italian peasant ate little to no meat, yet had an adequate level of energy. The peasant from southern Italy "lives on legumes, black bread and fruit; they are always healthy and is everywhere in demand for his qualities of being an excellent laborer" (Rinonapoli 1902, 7). The vegetarian diet was linked to greater health and sufficient labor power, vegetarians reasoned, because it taxed the body less in terms of digestion.

What to Cook

Information about early twentieth-century Italian vegetarianism can be gleaned from organizational publications (such as the Associazione Vegetariana d'Italia's *Propaganda della riforma alimentare e della vita igienica*), as well as national newspapers (such as the *Corriere della Sera*), nutritional education books, and vegetarian pamphlets. Few vegetarian cookbooks were published in the early twentieth century because few non-vegetarian cookbooks were published at that time. Unlike the United States and Britain, which had longer histories of cookbook publication, Italian cookbooks did not become popular until the early fascist period, when domestic economy books, culinary periodicals, and cheap recipe collections advised women on how to cook frugally under fascism. The few vegetarian cookbooks published in Italy prior to 1922 (*Ricette di cucina vegetariana* 1907; *100 Nuove ricette di cucina vegetariana* 1907) attempted to show how simple it was to prepare vegetarian dishes. In fact, these books stressed how vegetarianism could be incorporated into one's life with a minimum of inconvenience or disruption. Vegetarian cookbooks adopted the same format of other Italian cookbooks, separating, and organizing dishes by when and how they are served during the meal. Thus, vegetables were integrated into a meal structure suited for middle- and upper-class readers: a first course (soup or pasta) followed by a main course of a main dish with side dishes and a dessert.

Although Italian vegetarians acknowledged that carbohydrates like bread, rice, and polenta comprised the basis of daily consumption, vegetarian cookbooks contained only a few recipes for starchy dishes. Presumably, readers already knew how to make polenta or pasta. There were some creative recipes for rice, however, involving mixing rice with sugar, milk, and butter for a satisfying meal or mixing rice with egg, sugar, and breadcrumbs, then shaping the mixture into balls in order to boil or fry them. Vegetarian cookbooks advised readers on how to turn vegetables into hearty and satisfying dishes. This was usually achieved by covering vegetables in sauce and baking them off in a casserole dish or coating them with breadcrumbs and frying them. Both techniques mimicked cooking preparation techniques for meats. In addition to finding different ways to prepare fresh vegetables, readers were also advised to prepare dishes with legumes and nuts, which could be ground up and used as meat substitutes in dishes like croquettes, timballo, and pasticcio. An Italian pasticcio was a type of pie made of varied ingredients, sometimes layered, and baked in an oven. Vegetarian pasticcios featured legumes that were cooked and mixed with other ingredients to create a hearty baked dish. Dishes to imitate well-known meat dishes were also featured. A fake meat consisted of eggplant fried in oil until soft, then mixed with ground nuts, parmesan cheese, béchamel sauce, parsley, pepper, and egg yolks. This mixture could be layered with other ingredients and baked, or coated with breadcrumbs and sautéed to emulate the Italian dish Cotoletto alla Milanese (a veal cutlet). In addition to vegetables and legumes being cooked and combined with other ingredients to imitate meat, vegetables themselves were stuffed with ingredients and seasonings then baked, to provide a hearty baked main dish, again, in emulation of a main meat dish. The most popular vegetables for stuffing were onions, eggplants, and artichokes. Readers were advised that if vegetables were served as main course dishes, they should be served with potatoes, mimicking the structure of a meat-based meal (*Ricette di cucina vegetariana* 1907; *100 Nuove ricette di cucina vegetariana* 1907).

Readers of the vegetarian periodical *La nuova scienza*, launched during the First World War, were given cooking advice with articles about meat substitutions in recipes. Chestnuts, nuts, milk, eggs, peas, chickpeas, and fava beans all provided adequate protein and these foods could be manipulated in such a way as to substitute for meat in favorite dishes (*La Nuova Scienza* 1915, 216–17). In addition to providing creative ways to substitute for meat, vegetarian cookbooks highlighted the versatility of vegetables in soups and salads, mainstays of the Italian diet. Elaborate salads like the Insalata Regina featured green beans, lettuce, artichokes, potatoes, carrots, cucumbers, and asparagus, each individual prepared then mixed with capers, parsley, olive oil, and mayonnaise. Soups featured vegetables and legumes and sometimes featured fake or pseudo meats or in some cases, pseudo fish and seafood, presumably to imitate Lenten dishes, consumed on Fridays and Catholic holy days. Shredded or diced vegetables like mushrooms or pumpkin and even chestnuts, properly seasoned and cooked, took the place of fish or seafood (*100 Maniere di preparare le vivande di magro 1906*). Both *Ricette di cucina vegetariana* and *100 Nuove ricette di cucina vegetariana* contained numerous recipes for desserts, including puddings, cakes, tortes, and fruit dishes. Recipes for sweets dominated *Ricette di cucina vegetariana*, consisting of 34 pages, longer than any other recipe category. Vegetarian cookbooks like *Ricette di cucina vegetariana* also tried to impart nutritional information to readers, stressing the dynamic qualities of vegetables. Onions and garlic activated gastric secretions and saliva; asparagus purified blood; spinach improved kidney function; and celery calmed the nerves and eased pain associated with rheumatism.

The First World War began in 1914 but Italy entered the conflict in May of 1915 on the side of the allies. Two months before Italy entered the war, the *Società Vegetariana d'Italia* launched a monthly periodical titled *La Nuova Scienza. Vegetarianismo e perfezionamento umano* (*The New Science. Vegetarianism and Human Improvement*). The timing of the launch of the periodical seems ill-fated, perhaps, but the Italian Vegetarian Society felt a pressing need to give a greater voice to advocacy. The editors wrote that the time was propitious for launching a magazine that taught people how to be physically and morally healthy through the practice of vegetarianism. At the height of a terrible war that caused such pain and misery, vegetarianism could extend food supplies and provide a healthy diet for individuals as well as nations (*La Nuova Scienza* 1915, 1–6).

Indeed, vegetarians were prescient in understanding how individual health was tied to national health. During the war, the Italian government diverted meat to the military, leaving less for civilians, who would have to substitute legumes, vegetables and fruits. Meat substitutions were lauded by government officials for aiding the economy, improving public health and increasing national resistance (Capatti 2016, 101). Italy's position during the First World War was unique among European nations, in that loans of money and food from the allies facilitated adequate military provisioning and increased availability of food for civilians. In particular, the subsidized price of wheat and price controls for bread allowed many Italian families to spend less on the staple of their diet. Ironically, perhaps, working families had money left over to purchase foods previously considered too expensive, including meat, during the war (Helstosky 2006, Chapter 2). While nations like Germany and Austria had to tighten their belts, enduring severe privations during the last two years of the conflict, Italians ate better than ever, at least in comparison to prewar standards of consumption. Although more Italians were able to afford meat during the war, they were unable to purchase much of it. Calls for meat substitutions, whether issued by the government or by vegetarians, were likely heeded by most civilians because Italy continued to remain in a state of meatlessness.

Vegetarianism Institutionalized Under Fascism

When Benito Mussolini assumed power in 1922, there were few dramatic economic changes affecting popular consumption habits in Italy. By the mid-1920s and through the rest of the fascist period, popular consumption, including food habits, had to adjust to fascist economic policies and foreign policy decisions (like the invasion of Ethiopia in 1935 and the alliance with Nazi Germany in 1939). Fascist propaganda advised citizens not to waste food, not to purchase foods from foreign countries, and to restrict food intake whenever it interfered with national economic health. Officially, this policy was known as autarky and restrictions grew more severe as Italy inched closer to war. Domestically produced foods were highlighted in fascist propaganda (film, posters, newspapers, women's magazines, cookbooks) and officially sponsored food holidays reminded citizens that their dietary choices had national significance. Not surprisingly, many of the foods touted by the fascist regime fell under the category of vegetarian: grapes, bread, lemons and citrus fruits, and fresh vegetables. Fascist era cookbooks promoted simple fare while scientific publications touted the superiority of the Italian peasant's mostly vegetarian diet. While the fascist government did not publicly endorse or mandate a vegetarian diet for citizens, it certainly did much to promote meatlessness for reasons of political prestige. The health of the nation and the viability of fascism as a system of governance were at stake (Helstosky 2004). In this political and culinary atmosphere, vegetarian societies thrived. In the 1930s, the Associazione Naturista Italiana formed with branches in Milan, Trieste, and Rome; ultimately the organization spread to several other cities on the peninsula and published the periodical *L'Idea naturista*. The Association, along with the Unione Naturista Italiana, promoted vegetarianism and a natural lifestyle as the healthiest choice for Italians. Although Italy was not as advanced at the United States or Germany in terms of vegetarian restaurants and businesses, vegetarian restaurants, stores, and clubs opened up throughout the 1930s in northern and central cities in the nation (Capatti 2016, 107–125).

Historians writing about Italian vegetarianism under fascism have linked vegetarian beliefs to the fascist *strapaese* (literally super-village) movement, which focused on and promoted the simple life of Italian peasants in a conscious effort to promote or return to local rural traditions. Like some vegetarians, the *strapaesani* also integrated spiritualism into their advocacy of a simple life (Buscemi 2019, Capatti 2016). Francesco Buscemi, in tracing the history of vegetarian beliefs from the postwar era through fascism, argues that followers of radical nationalism who sought to purify the nation rejected meat, which threatened to corrupt Italian purity. Viewing meat as an essentially foreign culinary habit, nationalists who formed the core of the Italian Regency of Fiume and the early fascist movement looked to vegetarianism as a way to cleanse the body and soul from corruptions and intoxications of daily life (Buscemi 2019, 141). The threads between radical nationalism, fascism and vegetarianism are intertwined but not the same; while all three communities promoted simple, wholesome vegetarian fare, most vegetarians embraced a kind of culinary internationalism. Whereas fascists rejected any type of foreign influence as debilitating or decadent, vegetarians sought to maintain ties to international organizations, or were at least aware of trends and movements outside of Italy. Lastly, while nationalists and fascists championed all domestic crops, including wine grapes, many vegetarians abstained from alcohol, including wine. When it comes to specific intents, fascism, and vegetarianism map imperfectly on to each other, although they shared a similar outlook in regards to preparing food.

The fascist era is a revealing one for understanding the political implications of controlling consumption. In prefascist Italy, vegetarians promoted a simple diet for the sake of individual and community health. They were fully aware of the nation's agricultural limitations, but their advocacy of a vegetarian lifestyle had little to do with Italy's inability to feed its citizens meat. Rather, early twentieth-century vegetarians supported the maintenance of a simple lifestyle and drew strength from spiritualism, international associations and scientific expertise. Under fascism, however, a simple lifestyle became a political imperative, as the regime attempted to rein in consumption for the sake of domestic and foreign policies. In Nazi Germany, as Corinna Treitel has argued, "turning to nature" in dietary terms was critical for pursuing racial health and ultimately, racial war (Treitel 2017, 193). Thus, eating a natural, vegetarian diet would enable performance and productivity within the limits set by Germany's agricultural production. Pre-Nazi Germany confronted a situation in which land played a key role in supporting a burgeoning population; meat consumption was not the most efficient way to use this land, a critical consideration for a population hemmed in by the post-First World War treaties. Cutting back or eliminating meat constituted an important plank in German and later Nazi biopolitics; vegetarianism not only paved the way for individual emancipation through better health, it also enabled national self-determination in the face of external, sometimes uncontrollable, forces (Treitel 2017, 32).

Italians were also hemmed in geographically by post-war treaties and, lacking a vast empire, did not possess the agricultural resources to adequately feed the population. Prior to fascism, however, naturalism and vegetarianism were not as popular in Italy as they were in Germany and the scientific community was deeply divided about the role meat would play in national life. The Italian fascist regime was perhaps bold in its efforts to link what people ate with the destiny of the nation, but these associations had a history that stretched back to prefascist vegetarian rhetoric, specifically calls for a simple life and diet (Hoffman 1905; Piccoli 1911; Rinonapoli 1902). Whereas prefascist era vegetarians were thinking about the health of individuals, the involvement of the government during the war and under fascism linked individual health to a national community; both individual and nation could be strengthened together through the practices of consuming food.

Early twentieth-century medical experts were conflicted about protein recommendations and the consumption of meat by Italians. While experts prior to the First World War understood that meat was a convenient source of protein, they also recognized that most Italians would be unable to reach European and American recommendations for protein consumption. Under fascism, medical experts lowered their expectations considerably, advising between 40 and 100 grams of protein per day and urging Italians to consume less meat or no meat (Helstosky 2004, Chapter 4). These experts were not vegetarians. Rather, they championed the simple diet of the Italian peasantry, one that rejected meat and tobacco and selectively rejected alcohol (since grapes were such an important crop, the regime did much to promote wine consumption). In a shift from earlier nutritional advice manuals, which appeared to be written for a gender-neutral reader, fascist advice manuals were aimed at female readers, especially when it came to maintaining health for reproduction and feeding children properly. Experts bemoaned women who consumed too much meat. Doctors had differing opinions as to how much meat was too much; one expert advised eating meat only at one meal (Crovetto 1933, 83) while another suggested that Italians consume only herbivorous or omnivorous animals (Vesporina 1930, 53).

Calls to limit consumption of meat were commonplace in the nutritional literature, suggesting that the basic tenets of Italian vegetarianism, to live and eat simply, had found widespread acceptance in the medical community under fascism. Calls to reduce and sometimes eliminate meat consumption appeared regularly in fascist periodicals as well; vegetarian and non-vegetarian experts agreed that consumers should substitute eggs, milk, and honey for meat, to consume less, to eat slowly and not to waste anything (Manci 1931; Piccoli 1921). Not every medical expert agreed that meat should be eliminated or even reduced. There was, for example, ongoing scientific concern with monotonous diets, particularly corn-based diets that led to pellagra and several publications urged consumers to take up backyard farming of pigeons and rabbits in order to increase meat consumption with little expense. Yet an increasing number of medical experts counseled a healthy balance, a "perfect and harmonious equilibrium between food from the animal and vegetable kingdoms" (Bellotti 1926, 6).

The most well-known and successful cookbook published in the fascist era was the Duca Enrico Alliata di Salaparuta, *Cucina vegetariana: Manuale di gastrosofia naturista: con raccolta di 758 formule scelte d'ogni paese* (1930), a weighty tome of 432 pages that underwent multiple editions (the third edition, published in 1935, contained over 1000 recipes). As a naturalist, Salaparuta recommended simple Italian vegetarian fare: a plate of pasta seasoned with butter and cheese, accompanied by fruit, furnished all the necessary nutrients, vitamins, and minerals for the human body (Salaparuta 1930, xxxxv–xxxxvi). For readers desiring more than a simple plate of pasta, Salaparuta offered multiple recipes for fake or pseudo meats, including a pasticcio of pseudo hare, which consisted of boiled black beans, combined with egg yolks and seasonings. Caramelized sugar was added to give the mixture a brown gloss. Soups with fake clams consisted of shredded pumpkin and onions while soup with fake frog meat was made with chopped and shredded mushrooms. Bouillabaisse consisted of mushrooms, onions, and tomatoes, stewed with cloves and orange peel (di Salaparuta 1930, 91–94). Salaparuta's recipes for meat substitutions were very much in keeping with the time. Meat substitutes were commercially produced in Germany, the United States, and Switzerland. Many were produced during the First World War; continued sales after the war fluctuated and the products were marketed as foods to combat nervous disorders, given the food's high concentration of vitamin B. Meat-like spreads were quite popular with vegetarians in Germany, Austria, and Switzerland (Brodard 2010). Meat substitutions were one way to combat increasing austerity. As fascist economic and foreign policies put the squeeze on domestic budgets, housewives were instructed through cookbooks and domestic economy manuals on how to make do with less. Substitutions were standard recipes in many cookbooks of the 1930s, especially as economic conditions worsened in the wake of international sanctions against Italy for the invasion of Ethiopia. The fascist regime vigorously supported such restrictions on consumption and sought to educate women about their obligations to the regime through consumption (Helstosky 2006, Scarpellini 2015, Garvin 2017).

Humanitarian reasons for vegetarianism played a less significant role under fascism. Niccolò Grillo, the preferred name for Licò Nigro, wrote extensively on issues of animal rights and the degree of animal intelligence, as well as works on occultism and natural therapies for illness. His book, *L'alimentazione razionale: norme e ricette vegetariane* (1929) argued that Italians should reject meat for reasons of animal welfare as well as for health. Equally important, perhaps, was his advocacy of vegetarianism for simplifying food preparation. Vegetarian dishes, Grillo argued, took less time, required fewer complicated and expensive tools, and were easier to learn than recipes based on meat. In an era when the household rationalization movement took root in Italy, Grillo's support for uncomplicated cuisine would have resonated with fascist-era housewives, who were told to adopt a simpler, more streamlined approach to cooking (Grillo 1929, 60–61). Less clear is how a pro-animal welfare message would have affected readers in fascist Italy. Unlike Nazi Germany, where naturalism and animal welfare advocacy went hand in hand, the animal welfare movement in fascist Italy was not a strong presence in the minds of citizens.

Fascist calls for consumers to reduce consumption levels of meat and other foods grew more strident in the aftermath of the invasion of Ethiopia and assumed panicked tones at the start of the Second World War. Unaware that Hitler intended to invade Poland in the fall of 1939, Mussolini entered into a diplomatic and military alliance months earlier. The war was by all accounts a complete disaster for Italy, a nation unprepared for another military conflict and exhausted financially. Moreover, Italy's military ties to Nazi Germany turned on the delivery of food from Italy to Germans. Facing wartime shortages, Italians had few choices but to ship food to Germans. Conditions became worse as the war dragged on, with citizens forced to forage in the countryside for food or resort to the black market to buy staples for their families. Unlike German civilians, who were adequately fed until the very end of the war (thanks in part to Italians), Italians faced hunger and malnutrition for the last two years of the conflict. The scarcity, then absence, of food turned citizens against a regime that so desperately tried to control their everyday habits. When the allied forces landed in 1943, first in Sicily then on the peninsula, they found an exhausted and hungry population that offered little resistance to allied occupation. Italians asked or begged for food to the extent that the allied advance was hampered by provisioning shortages and the army had to request more supplies to feed hungry citizens. Gifts of chocolate bars and impromptu soup kitchens became the basis for new political alliances and affiliations.

This chapter has examined the history of Italian vegetarianism in the context of the larger history of meatlessness, an ongoing condition or predicament faced by many Italians in the nineteenth and early twentieth centuries. Meatlessness could be interpreted as involuntary vegetarianism. When Italians were able to afford meat after 1945, they purchased and consumed meat, and thus vegetarianism was defined more by choice than necessity. However, for much of Italy's history, vegetarianism was a necessity. Thinking about meatlessness allows us to understand the history of vegetarianism as something broader and more complex than the existence of organized societies. Meatlessness explains a great deal, not only about the particular trajectory of Italian vegetarianism, but the larger complexities of Italian food history and Italian history. The ideas of a very small number of voluntary vegetarians became mainstream during the First World War and during the fascist regime, which sought to prolong Italy's condition of meatlessness for reasons of political prestige. Because of meatlessness, it was not a major sacrifice for Italians to reduce or even eliminate meat consumption under fascism as they had been doing so for well over a century.

The history of Italian vegetarianism also reminds us to pay attention to the role that politics plays in affecting our food consumption habits. As food historian Rachel Laudan and anthropologist Sidney Mintz have observed, our ability to create different types of food turns on our other freedoms, to determine what tastes good and what we like to eat. The history of Italian vegetarianism serves as a cautionary tale, perhaps, that food choices can become food mandates under authoritarian regimes, with mixed results when it comes to the health and well-being of consumer-citizens and nations.

Note

1. This was meant figuratively and literally. Vegetarian cookbooks and periodicals argued that meat contained unhealthy levels of mineral salts, which could poison the human nervous system and tax the digestive system. See, for example, *Ricette di cucina vegetariana*, Biblioteca Casalinga (Milan: Soc. Ed. Sonzogno, 1907) and P. Hoffman, *Le tendenze ideali del vegetarismo: conferenza fatta alla Società vegetariana belga il 9aprile 1900* (Udine: Tip. del Patronato, 1905).

Bibliography

100 maniere di preparare le vivande di magro. 1906. Milan: Soc. Ed. Sonzogno.

100 nuove ricette di cucina vegetariana. 1907. Milan: Soc. Ed. Sonzogno.

Albertoni, Pietro, and Felice Rossi. 1910. *Nuove ricerche sulla influenza delle proteine animali nei vegetariani*. Bologna: Tip. Gamberini e Parmeggiani.

Associazione Vegetariana d'Italia, *Propaganda della riforma alimentare e della vita igienica* (Anno 1, No. 2, 1911).

Bellotti, Silvio. 1926. *Come devo alimentarmi igienicamente*. Milan: U. Hoepli.

Brodard, Pierre. 2010. From the Austrian Sausage Carnavegetale to the Swiss Spread Le Parfait: Meat Substitutes in Peace and in Wartime (1908–1959). *Food & History* 8 (2): 25–44.

Buscemi, Francesco. 2019. The Sin of Eating Meat: Fascism, Nazism and the Construction of Sacred Vegetarianism. In *Proteins, Pathologies and Politics: Dietary Innovation and Disease from the Nineteenth Century*, ed. David Gentilcore and Matthew Smith, 137–147. London: Bloomsbury.

Capatti, Alberto. 2016. *Vegetit. Le avanguardie vegetariane in Italia*. Lucca: Cinquesensi.

Clerici, Alessandro. 1909. *Come si deve mangiare*. Turin: Soc. Tip. Ed. Nazionale.

Crovetto, Angelo. 1933. *Cucina edigiene*. 2nd ed. Genoa: Soc. An. Coop. Fascista Poligrafici.

Dickie, John. 2008. *Delizia! The Epic History of Italians and Their Food.* New York: Atria Books.

Diner, Hasia. 2003. *Hungering for America: Italian, Irish and Jewish Foodways in the Age of Migration.* Cambridge, MA: Harvard University Press.

Garvin, Diana. 2017. Feeding Fascism: Tabletop Politics in Italy, 1922–1945. PhD diss., Cornell University.

Grillo, Niccolò (Licò Nigro). 1929. *L'alimentazione razionale: norme e ricette vegetariane.* Catania: F. Battiato.

Helstosky, Carol. 2004. Fascist Food Politics: Mussolini's Policy of Alimentary Sovereignty. *Journal of Modern Italian Studies* 9 (1): 1–26.

———. 2006. *Garlic and Oil: Politics and Food in Italy.* Oxford: Berg.

Hoffman, P. 1905. *Le tendenze ideali del vegetarismo: conferenza fatta alla Società vegetariana belga il 9 aprile 1900.* Udine: Tip. del Patronato.

Laudan, Rachel. 2015. *Cuisine and Empire: Cooking in World History.* Berkeley: University of California Press.

Manci, Vezio. 1931. *Nutrizione ed Assicurazioni (Appunti).* Castelplanio: Prem. Tip. Romagnoli.

Mintz, Sidney. 1997. *Tasting Food, Tasting Freedom.* Boston: Beacon Press.

Montanari, Massimo. 1994. *The Culture of Food.* Trans. Carl Ipsen. Oxford: Basil Blackwell.

Offer, Avner. 1989. *The First World War: An Agrarian Interpretation.* Oxford: Clarendon Press.

Parasecoli, Fabio. 2014. *Al Dente: A History of Food in Italy.* London: Reaktion Books.

Piccoli, Ettore. 1911. *Norme di igiene nuova (vegetarismo e fisiatria).* Milan: La Compositrice.

———. 1921. *L'alimentazione dell'uomo.* Milan: R. Quintieri.

"Proprietà curative delle frutta e degli erbaggi." 1911. Associazione Vegetariana d'Italia. *Propaganda della riforma alimentare e della vita igienica.* Anno 1, No. 2.

Rabinbach, Anson. 1990. *The Human Motor. Energy, Fatigue and the Origins of Modernity.* New York: Basic Books.

Ricette di cucina vegetariana. 1907. Milan: Soc. Ed. Sonzogno.

Rinonapoli, G. 1902. *Il vegetarianismo dal punti di vista igienico e Teosofico.* Naples: Francesco Perella.

Salaparuta, Duca Enrico Alliata di. 1930. *Cucina vegetariana: Manuale di gastrosofia naturista: con raccolta di 758 formule scelte d'ogni paese.* Milan: U. Hoepli.

Scarpellini, Emanuela. 2015. *Food and Foodways in Italy from 1861 to the Present.* New York: Palgrave Macmillan.

Società Vegetariana d'Italia. 1915. *La nuova scienza. Vegetarianismo e perfezionamento umano. Cura e vita naturale per guarire e prevenire le malattie*. Anno 1, Nos. 1–10.

Treitel, Corinna. 2017. *Eating Nature in Modern Germany. Food, Agriculture and Environment, c. 1870 to 2000*. Cambridge, MA: Cambridge University Press.

Vesporina, O. 1930. *L'alimentazione razionale. In base ai più recenti studi e alle nuove orientazioni*. Milan: Casa Editrice "Varietas."

2

Taking an Anti-Sacrificial Stance: The Essentializing Rhetoric and Affective Nature of Meat Consumption in Islam

Nora Abdul-Aziz, Daniella Fedak-Lengel, and Lara Martin Lengel

There is not an animal on earth,
nor a two-winged flying creature,
but they are communities like you.
—*Qur'an*, Surah Al-Anam 6:38

N. Abdul-Aziz
University of Toledo, Toledo, OH, USA
e-mail: nora.abdul-aziz@rockets.toledo.edu

D. Fedak-Lengel • L. M. Lengel (✉)
Bowling Green State University, Bowling Green, OH, USA
e-mail: dfedakl@bgsu.edu; lengell@bgsu.edu

C. Hanganu-Bresch, K. Kondrlik (eds.), *Veg(etari)an Arguments in Culture, History, and Practice*, The Palgrave Macmillan Animal Ethics Series,
https://doi.org/10.1007/978-3-030-53280-2_2

Dressed in a green headscarf, to highlight the importance of environmental sustainability and vegetarianism for Muslims, and *al-Khaḍrá* [Arabic: حضرة; green][1] being color of Islam, Suraiyya Benazeer came in peace. Walking toward the Taj-ul-Masajid one of Asia's largest mosques, with her fellow women PETA-India volunteers, all was well, at first. However, once Bhopal's Muslims, who were attending Taj-ul-Masajid, understood what Benazeer and her colleagues were doing—holding signs that read "Make Eid Happy for All. Try Vegan" in Urdu and English—some reacted far from favorably. As Benazeer and the other volunteers approached Taj-ul-Masajid, a crowd gathered and started yelling at the volunteers, then tried to hit them. The police helped them escape being beaten by the crowd. After receiving death threats, Benazeer went into hiding and restricted her social media from public access (Ayub 2014; Fatah 2015; People for the Ethical Treatment of Animals 2014; Srivastava 2016).

While the attack on Benazeer and her fellow PETA-India volunteers illuminates the contentious nature of meat consumption as it relates to Islam, it also touches upon more subtle considerations of the identity politics of consumption, including nationalism and xenophobia,[2] challenges to the masculinist resistance to women's advocacy and activism in the public sphere (Hong Tschalaer 2015; Kirmani 2009; Schneider 2009), and the cultural framing of meat consumption. This cultural framing has been called "meat culture" by Annie Potts (2016). Her work on veganism and vegetarianism, and that of Potts and Philip Armstrong (2016) on vegan intersectionality, situates these practices as markers of difference in relation to the dominance of carnism.

In this work, which has been informed by lived experience and field research in the Middle East and North Africa (MENA),[3]we interrogate rhetorics of anti-veg(etari) an discourses in transnational contexts. Building on the scholarship of Potts, Armstrong, Timothy Pachirat, Yamini Narayanan, and others, most notably feminist Sara Ahmed and her theorizing about emotion, we address the essentializing rhetoric of carnism and the affective nature of animal slaughter in Muslim religious practice, most notably for *Eid al-Adha* [Arabic: عيد الأضحى; Feast of the Sacrifice]. We critique specific discursive constructions and enactments of the emerging Muslim vegan movement, and how the movement's discourses re/imagine or resist the practices of and perceptions about Islam.

In our own efforts to reflect on our world views and practices, it is important to situate ourselves in this project. Our self-reflexivity, positionality, lived experience, and our respective pre-medical and pre-veterinary science studies and critical ethnographic research backgrounds[4] inform our study. Nora is a Syrian-French-American pre-med undergraduate student who aims to specialize in surgery and serve in *Médecins Sans Frontières* in the MENA.[5] She grew up on a variety of foods from both European and Middle Eastern cultures and, thus, knows there are alternatives to meat-centric diets. Nora notes that vegan eating in the MENA is nothing new. A wide variety of popular traditional dishes, such as falafel, lentil soup, hummus, and grape leaves, are vegan. After moving to her university house-share Nora found she was not eating much meat after moving out of her family home. She fully transitioned to vegetarian within her first year at university, shopping at the Middle East Market near her university.

Daniella is a biology student who plans on studying veterinary surgery. Animal malpractice is the primary reason she decided to abstain from animal products, as well as one of the many reasons she has entered the pathway to veterinary practice. She was born in London, one of the most progressive centers of veganism (Larsson 2019), where, this year, ethical veganism was ruled as a "philosophical belief," and covered as one of nine "protected characteristics" by the UK Equality Act 2010 (Catholic Concern for Animals 2019). Daniella decided to convert to a plant-based diet at the young age of 12 and has been a committed vegan ever since. Her most recent international veterinary study focused on the illegal trade of exotic animals during a veterinary practicum in Belize (Fedak-Lengel 2019).

Lara grew up as a lacto-vegetarian in the 1970s never having heard that label—her primarily vegetable diet emerged from effect and aesthetics rather than politics—she just avoided eating things that looked, as she then called it, "gross."[6] Only later in her youth did she connect her disgust reactions to the grotesque nature of industrial animal agriculture. Peers, neighbors, and extended family were pronounced in their collective opinion that her diet was odd; one of her elementary school peers went so far as to say she would die before she was 30. By that age she was not dead; rather she was living in the Mediterranean North African nation of Tunisia on a Fulbright Research Scholarship where she began, albeit in modest ways, advocating for plant-based diets and for environmental sustainability in the MENA. While her dissertation field research did not focus on either topic, rather on Arab Islamic women's performativity, what she observed and experienced in-field, such as the use of ancient Roman cisterns in Carthage as make-shift garbage dumps, the overly enthusiastic serving of meat-centric meals by her various hosts, and her delight in discovering a restaurant that would serve her *couksi bil barsha sfinaria* [couscous with a large amount of carrots], inspired her commitment to MENA environmental sustainability and veganism. Since her field research she has been involved in university partnership building initiatives on women and the media and on environmental journalism in Tunisia and Algeria.

We bring together our own commitment to animal welfare, environmental sustainability, community-based initiatives, our specific scholarly areas of veterinary science, biological science, MENA gender and women's studies, and work on Islamophobia and culturally sensitive healthcare approaches[7] to analyze the discourses and perceptions of veganism and vegetarianism, particularly as they intersect with identity, culture, and power. In order to do so, we apply a methodological approach that blends, first, our own aforementioned field research, second, critical discourse analysis, and, third, a medical and health sciences case study approach. We are guided by researchers (see, for instance, Shirazi, cited in Animals in Islam 2013; Taylor and Robichaud 2004), who have conducted critical discourse analysis in Muslim and Middle Eastern contexts. For example, in his work on social movements in the Middle East, Farid Shirazi (2013) uses critical discourse analysis to analyze meanings, contexts, and processes of sense-making in discursive spaces "where agency and text, symbols, speech and other communicative objects are generated to better understand the meaning of discourse" (30). We draw from medical and health sciences research on vegan and vegetarian practice among Muslims because "the case study approach allows in-depth, multi-faceted explorations of complex issues in their real-life settings" (Crowe et al. 2011, para 1). Units of analysis in our study include *Qu'ranic* verses pertaining to animals, animal welfare, *ḥalāl* practice; the entire corpus of media and news reports of Muslim veg(etari)anism in English and French languages, the entire corpus of comments to these reports, articles by Muslim veg(etari)an organizations, articles in individual blogs, and *khutbah* [خطبة; sermons or lectures presented by Imams] that have addressed meat consumption.

Our analysis of the discourse about Islam and consumption reveals several themes: Religious, historical, economic, contemporary cultural considerations; shifting conceptualizations of *ḥalāl* and *harām* [Arabic: حَرَام; that which is forbidden under Islam]; intersectional Othering and affect-driven resistance to Muslim veg(etari)anism; and differing politics of resistance.

Religio-Historical Considerations

> And the earth has He spread out for all living beings. (*Qur'an,* Surah Al-An'ām 6)

Although numerous reasons for converting to a plant-based lifestyle exist, the most prevalent ones are those pertaining to animal welfare and environmental impacts. Religion, faith, and spirituality also impact the choice to be vegan or vegetarian. In "Drivers for Animal Welfare Policies in the Middle East," Aidaros (2014) notes "One of the major components of Islamic thinking is consideration for non-human animals. As humans are considered to be speaking animals, we can understand how animals are highly esteemed in Islam" (86).

The *Qur'an* states that all living beings—humans, animals, even insects—are worthy of respect and care (Masri 2007; Nasr 2010; Tlili 2012) as indicated in verse 12 of Al-An'ām [الأنعام], the sixth chapter of the *Qur'an*, that announces "the earth was not created for humans alone" (cited in Aidaros 2014, 3). Despite the *Qur'anic* requirement for respect for all living beings, there remains a strong meat-eating culture in Islam. While not essentially opposed to veganism or vegetarianism, many Muslims consider meat to be a vital element in Muslim foodways and, more broadly, of Muslim culture. Beef and lamb are at the center of a tremendous amount of Muslim entrees; without meat, many consider meals as incomplete.

Historically, however, the centrality of animal-based entrees was not the norm. There are some Islamic religious leaders, such as Sheikh Hamza Yusuf, who note that "historically Muslims ate so little meat they were almost vegetarian. Meat is not a necessity in *sharia* [Islamic law] and, in the old days, most Muslims used to eat meat—if they were wealthy, like middle class—once a week on Friday. If they were poor—on the *Eids*" [high religious holidays]. Sheikh Yusuf reports the Prophet Mohammad did not advocate daily meat-eating as it could become addictive. The earliest Islamic religious leaders and scholars continually emphasized to their followers that "animals were to be cherished and treated in a humane manner, but many Muslims nowadays view animals as the dominion of people" (para. 4). Sheikh Yusuf argues, "The idea that animals are merely slaves to humans is not only abhorrent to animal-rights advocates, but seems to be at odds with the prophet's teaching" (para. 6).

"A good deed done to an animal": Is "Ethical Killing" Ethical?

> A good deed done to an animal is as meritoriousas a good deed done to a human being, while an act of cruelty to an animal is as badas an act of cruelty to a human being. (*Hadith* (cited in Rahman 2017, para. 7))

The *Qur'an* condemns acts of cruelty to animals, as indicated in the above *hadith* [tradition] and provides considerable support for conscientious attention to animal welfare, broadly, and animals for the purpose of consumption, specifically (*Qur'an*, surah Al-Anaam, verse 11). In "Drivers for Animal Welfare Policies in the Middle East," Aidaros (2014) argues, "while non-human creation is subjugated to human needs, the proper human role is that of conscientious steward and not exploiter" (84). *Dhabīḥah* [or *zabiha*; Arabic: ذَبِيحَة] is the set of Islamic rules detailing the practice of ethical ritual slaughter of animals intended for human consumption and, thus, *ḥalāl* [that which is permissible under Islam].[8] The *Dhabīḥah* dictates that unnecessary suffering to these animals must be avoided. First, animals must be in one of the categories permitted for Muslims to eat. The most well-known *harām* animal meat is pork which is specifically forbidden in the *Qur'an* (sura 2:173 and 16:115).[9] However, most non-Muslims are unaware that the list of *harām* [not permissible] food includes, but is not limited to, dogs, predatory creatures which have sharp teeth and claws and talons, certain birds, and all reptiles and insects. Second, animals must be alive, healthy, and not hungry at the time of their slaughter. Then there is a specific set of procedures "intended to provide a quick, humane and relatively painless death" (84). The slaughterer, who must be Muslim, must first reflect on *niyya* [intent], offer water to the animal, and invoke God through a recitation known as the *tasmiyah* [تسمية; naming, designation or calling], most frequently, "*Bismillah, Allahu' akbar*" ["In the Name of God, God is most Great"], which dedicates the animal to Allah. Prior to slaughter, the sharp knife to be used must be hidden from the animal. The act of slaughtering must consist of a swift and deep incision to the jugular vein, carotid artery and windpipe (Taqi Usmani 2006). This should be done out of sight of other animals waiting to be slaughtered (Aidaros 84). An additional step in the process, in the Shi'i tradition, is to turn the animal to face Mecca before slaughter (Robinson 2014, 278). The blood of the slaughtered animal must be completely drained, because *ḥalāl* dietary restrictions specifically forbid a Muslim to consume carrion, blood, swine, and animals dedicated to other than Allah (*Qur'an*, Surah 5; al-Ma'idah, ayah 3).

Ḥalāl practice has been widely studied in numerous disciplines.[10] Most researchers and animal agricultural practitioners argue that *ḥalāl*s slaughter is one of the more humane methods in the meat industry and the only acceptable method for Muslim consumers.[11] However, Aidaros (2014) argues "[c]urrent practices that are not in accordance with these religious teachings may cause great suffering to animals" (85). "Unfortunately," he notes, "many Muslims and Islamic religious leaders are not aware of the suffering that is inflicted on animals during handling, transport, lairage and slaughter itself" (87).

Although there are numerous privately owned butcher shops which provide *ḥalāl* meat following all steps of the *dhabīḥah*, the rising Muslim population worldwide has led to increased demand. Similar to the industrial practices in the U.S. that have placed profit over animal welfare, some of the enterprises cut corners on *ḥalāl* and *tayyib* [permitted and wholesome] *dhabīḥah* practice. Of all small and medium enterprises in Malaysia, for example, 90% do not have *ḥalāl* certification (Aqidah Azizi 2019). This is disconcerting given that 62% of the population in Malaysia, or 19.5 million people, are Muslim adherents (Aqidah Azizi 2019). While all meat produced in and exported from India is advertised as *ḥalāl,* a team from PETA India observed widespread breaches of *dhabīḥah* (People for the Ethical Treatment of Animals 2020).

A particularly egregious breach of *ḥalāl* and *dhabīḥah* practice occurs in numerous self-identified *ḥalāl* slaughter houses that claim to properly engage in the *tasmiyah* prayer before slaughter. Slaughterhouse owners and managers claim is central to Islam's humanity when animals whey they are killed for food. "This may have been true historically, but in today's 'halal' slaughterhouses, a pre-recorded prayer often blares nonstop as the animals are lined up and killed. That is a cop-out from what Islam teaches about 'humane' slaughter" (Mayton 2010, para. 7). Sarah Robinson's field research and interviews with laborers and others in the industry revealed the additional questionable practices such as the "same machine and blade used to kill pigs and halal animals, potentially rendering the halal-certified meat haram (not permissible) if the knife was not properly cleaned" (278). Further, in a practice Robinson names "'drive-by' halal certification" a shaykh recites a blessing as he drives past a slaughterhouse, "which is far from slaughtering by hand with the name of God on one's lips" (278).

Far from ethical killing has been discovered by Animal Aid, whose investigation and undercover filming in a North Yorkshire, England *ḥalāl* abattoir, revealed "gratuitous violence and contempt" in the treatment of animals prior to and during *dhabīḥah,* lead to four slaughter men having their licenses suspended by the UK Food Standards Agency (Morris 2015). The film was released just days after a petition demanding the banning of slaughter without pre-stunning[12] exceeded 100,000 signatures, adding pressure on British politicians to tighten regulations that would mitigate inhumane practices in slaughterhouses, be they *ḥalāl,* kosher, or secular.

Consistent with Pachirat's (2011) "politics of sight," Muslim advocates argue that more awareness of the lack of *ḥalāl* practice would lead to a rise in vegetarianism and veganism. Pachirat argues the "organized, concerted attempts to make visible what is hidden and to breach, literally or figuratively, zones of confinement" can "bring about social and political transformation" (15). Given the lack of consistency in *ḥalāl* practice, it is difficult if not impossible to determine whether or not the meat that is being consumed is certified in the name of God. "That is a cop-out from what Islam teaches about 'humane' slaughter" (Mayton 2010, para. 7).

Questioning the Sacred

> Meat is almost sacred in the Arab world. (Nada Elbarshoumi, United Arab Emirates-based author of the blog, "One Arab Vegan")

In her study of anthropatriarchy and Hindu nationalism, Yamini Narayanan (2019) asks, "How can insights from feminist and animal geographies politicize spaces of 'sanctuary' and refuge for animals repatriated from incarceration, exploitation, and violence?" (195). We apply Narayanan's question, which she posed regarding Hindu religio-cultural contexts, to Muslim contexts in order to ask how can we politicize the process of *Eid al-Adha* "so that animals are not only saved from the gendered, sexual, and disposal (slaughter) violence" of slaughter, but also "from being coopted as political, sociocultural, and religious symbols?" (Narayanan 195).

We realize this is treacherous terrain to tread. The affective nature of religion, already powerful, intensifies during times of high religious celebrations. As evidenced by the attack on Suraiyya Benazeer and her PETA India counterparts, contesting or resisting normative practices during religious celebrations can result in vicious, even violent response by those who claim they are devout believers and those who weaponize and politicize religion for various reasons (Anderson 2009; De Matteis 2018). Sarah Ahmed's work (2010, 2014) on affect studies and the "emotionality of texts" is relevant to this analysis. Her work traces how texts move and generate impact, a phenomenon known as affective resonance.

Islamophobia has thrived on the strategic rhetorical acts of misinformation, disinformation, or merely highlighting something while ignoring broader religio-cultural contexts. The case of the *ḥalāl* "exclusive" alleged exposé by *The Sun* illustrates this. On May 7, 2014, the UK newspaper had blazoned on the front page of its tabloid a massive, meme-style headline superimposed on a color photo of a pizza—"Halal Secret of Pizza Express." *The Sun* claimed it was revealing that the well-known pizza chain served *ḥalāl* chicken. A wave of affective resonance materialized by a social media-driven call to #BoycottPizzaExpress immediately ensued. #Halal ascended to one of the top trending terms on UK social media, as did #HalalHysteria. Other media organizations, such as the BBC and *The Guardian*, pointed out that Pizza Express had not kept any such "Halal Secret"; the company included information about the use of *ḥalāl* chicken in their website FAQ page and had tweeted about it previously. It had also been mentioned in news articles. Nevertheless, *The Sun*'s attempt to serve up a slice of Islamophobia had its desired affective response.

Ahmed's critique of the "emotionality of texts" is also useful to analyze one of the most affect-laden Muslim practices—the slaughter during *Eid al-Adha*. This Muslim holiday, which occurs on the tenth day of the last month of the Muslim year, commemorates one of the central stories of the *Qu'ran* which illustrates the willingness of the Prophet Ibrahim to sacrifice his son as an act of obedience to God.[13] According to scripture, seconds before Ibrahim engaged in the sacrifice God replaced the boy with a lamb. Muslims reenact this story by sacrificing a lamb or goat or another mammal such as a cow, or less commonly a camel,[14] on *Eid al-Adha*. The meat of the offering is divided into parts—the largest portion is given to the poor, the second largest to the person's relatives, and the smallest portion to the person's own family.

Eid al-Adha is often the turning point for Muslims who choose veganism and, in particular, for Muslims who commit to "affective feminist practice that views animal others as grievable, vulnerable, and valuable" (Jenkins 2012). Reflecting on her childhood experiences of her family's *Eid al-Adha*, Nada (2014) writes:

> When I was growing up, my family performed this almost every year. My brother and I would cower in the back of the garden watching a butcher

> read a blessing over a sheep he had hauled over in the back of his rickety truck. As per Halal requisites, he would feed it some water, say Bismillah (in the name of God) and swiftly slit it's [*sic*] throat. I sometimes find it odd that I never really questioned the practice while growing up—or that I continued to eat meat for years after, but something about seeing the ritual year-on-year from such a young age left me desensitized in a sense. It wasn't till I started eschewing meat altogether that I started to question this unnecessarily brutal ritual, which stood out to me like a sore thumb in a religion that preached compassion, love and understanding for all living beings. (para. 4)

Nada and many others, recalling the blood on the streets, their fathers slicing the neck of an animal that they considered a pet, advocate for alternatives and highlight a central argument: "Despite the fact that it is a ritual occurrence in our religion and culture; many don't realise that sacrifice is NOT a pillar of Islam" (para. 5). Nada (2014) goes on to note "Historically, pre-Islamic Pagan Arabs, Jews and Christians all offered some sacrifice in the hopes of attaining protection, acceptance or material gain from God—but the notion of 'vicarious atonement of sin' (that is absolving one's sins through the blood of another) is not mentioned at all in the Qur'an. Neither is the idea of gaining favor by offering the life of another to God—all that is demanded is one's personal willingness to submit one's ego and will" (para. 6). Rather than vicarious blood atonement through ritualized slaughter, *Eid al-Adha* is "the act of thanking God for one's sustenance and the personal sacrifice of sharing one's possessions and valuable food with fellow less fortunate people. The ritual itself is NOT about the sacrifice—it was about sharing the best of what you had" (para. 8). Nada argues:

> There are several Qur'anic verses that highlight the true purpose and objective behind sacrifice—but of all of them allude to the same thing. *The act of animal sacrifice is pertinent to the role animals played in Arabian society at the place and time.* Humans were commanded to give thanks to God and praise Him for the sustenance provided by him by **sacrificing something of value to themselves to demonstrate their appreciation for what they have been given. (para. 9, our emphasis)**

Similarly, the Regional Animal Welfare Strategy (RAWS) for the Middle East, developed jointly by the l'Organisation Mondiale de la Santé Animale (OIE) Regional Representation and Member Countries in the region, supports the development of national programs for "raising awareness of animal welfare and educating key players, particularly veterinarians, livestock owners, animal handlers, religious and community leaders and other key groups, such as schoolchildren and women" (87). These organizations raise awareness that Muslims, and practitioners of other religions,[15] do not need to kill animals, even during *Eid al-Adha*. Hakim and others in the Initiative, instead, focus on the ethical mission underlying *Eid al-Adha*—to help those in need.

Re-Examining the Contradictory Cultures of Meat, Blood, and Othering

> To show your sincerity
> on how much you want to help the poor,
> there are many other ways;
> you can give them plant-based foods,
> you can give them money.
> I think this practice (sacrificing an animal) has to be re-examined. (Sammar Hakim, The Vegan Muslim Initiative[16])

Recall Pachirat's (2011) "politics of sight," in which contemporary society hides the horrific from view. With rare exceptions, most notably Pachirat's ethnography of slaughterhouse in Iowa, large-scale industrial animal slaughter is conveniently invisible. Indeed, Lever (2019) argues, "[a]cross the secular West, the slaughter of animals for food has become an almost clandestine activity" (889). Compared to secular or Christian practice, *Eid al-Adha* provides scope for a clear politics of sight.[17] Many Muslim vegans and vegetarians recall the horror of the slaughters they witnessed as children. What is seen, hidden from view, or separated from others was evident in historical practices of *Eid al-Adha*. For instance, the Prophet *Umm'Atiyyah* Al-Ansaariyyah shared with followers a command from the Messenger of Allah that women, adolescent girls, menstruating women, and virgins be "brought out" on *Eid al-Fitr* and *Eid al-Adha.* However, menstruating women were to stay away from prayer; instead they "were to witness goodness and the gathering of the Muslims". Despite the preponderance of slaughtered animals' blood on *Eid al-Adha,*[18] women who, too, were flowing blood "were to stay away from the prayer" pushed, quite literally, to the margins.

Such contradictions contribute to the skewed view of*Eid al-Adha* by non-Muslims which, arguably, has fueled Islamophobia and the Othering of Muslims. It is not surprising that center- to far-right media organizations such as *The Sun* (UK), of *ḥalāl* expose fame, report on *Eid al-Adha* with affect-driven headlines like "BLOOD ON THE STREETS: Eid Al-Adha animal sacrifice festival sees roads turn red with blood as cows are beheaded" (Sullivan 2018). However, there are a myriad of non-Muslim religious practices and religio-cultural norms worthy of interrogation. There is evidence of Christian animal sacrifice in the village of Taybeh, near Jerusalem, for instance. The care of animals by those who self-identify as Christian certainly warrants critique. In his essay, "An Islamic Perspective Against Animal Sacrifice," Shahid 'Ali Muttaqi (2020) notes the hypocrisy of "Western 'Christian' countries that malign the Muslim world for sacrificing animals, yet have institutionalized factory farming and worldwide environmental destruction" (para. 5). The politics of sight, again, is relevant here. Rarely, if ever, is there discussion of "mass slaughter" of turkeys at Thanksgiving and Christmas. There is no such affective rhetoric. The "processing"[19] is kept from view, even codified through "ag-gag" laws not only silence animal welfare advocates and whistleblowers who attempt to call out inhumane animal agricultural practices, but open them to be prosecuted as "terrorists" (See Sanders 2019; Strong 2019).[20]

Eid al-Adha is increasingly divisive within Islam, sparking debates about continuing the practice of slaughter as it is "at odds with the prophet's teaching" ('Ali Muttaqi 2020, para. 5). Key points in the debate include, first, the argument that the story of Ibrahim/Abraham's willingness to sacrifice Isaac is interpreted in differing ways, most notably that the message was delivered to Ibrahim in a dream. Further, there is no definitive indication that the dream came directly from God. Those contesting this argument claim that dreams are a part of revelation. A second argument indicates that animal sacrifice on *Eid al-Adha* does not enhance Muslims' spiritual development. Third, to make the case against sacrificing animals, Muslims cite the Tafsir al-Jalalayn [تفسير] which states, *"Their meat will not reach Allah, nor will their blood, but what reaches Him is piety from you"* [al-Hajj 22:37] (Royal Aal al-Bayt Institute for Islamic Thought 2017). Finally, Muslims who are opposed to animal sacrifice, like The Vegan Muslim Initiative's Sammar Hakim, argue there are a myriad of strategies to help those in need, from giving people plant-based meals to investing in long-term projects to enhance economic justice. Similarly, Zain Syed Ahmed (cited in Fida 2019b) donates funds to the International Development and Relief Fund, an Islamic not-for-profit organization that connects donors from Canada with Muslim communities abroad to provide long-term and emergency aid. All of these endeavors adhere to the scriptural call "to change the condition of a people" (*Quran* 13:11).

One of the most reasoned discussions arguing to not abandon the practice of slaughter at *Eid al-Adha* is from Hira Amin (2013), a British Muslimah of Pakistani descent. She acknowledges the likelihood that the demand placed on *ḥalāl* butchers during *Eid al-Adha* results in breaches of *ḥalāl* practice. She suggests "working together and putting measures in place to allow for the humane treatment of animals around the world all times of year, particularly in the high-demand seasons." Specifically for *Eid al-Adha* the time period for *dhabīḥah* could be extended to allow those engaging in the sacrifice adequate time to carry out the procedures to fully comply with *ḥalāl* practice. Amin argues, "Muslims should be at the forefront of this issue and call for animals to be humanely treated according to the rules of Islam. Our efforts should not lie in abandoning the practice, but in reforming it in order to correctly fulfil the obligations set out by Allah to the best of our abilities" (para. 25).

Questioning the Centrality and Gendered Construction of Meat in Islam

> Vegetarianism is halal.
> Meat is not compulsory.
> Any food is permissible provided it is not harmful.
> Muslims are free to eat whatever they want provided it is halal.
> It is like wanting to eat a certain fruit and not the other. (Honorable Sayyid Fadhlullah[21])

Given that some Muslims equate eating meat with close adherence to Islam, Muslim vegans report that their family and friends tell them "you can't be vegan and Muslim." Those opposed to veganism claim God created animals for, among other things, food for humans. Many such Muslims, like anti-vegans more broadly, question the sustainability of a plant-based diet. Such questioning is consistent with the often ill-informed notions about sustainability and nutrition more broadly.[22]

Despite these qualms, veganism is gaining increased attention among Muslims within and outside the MENA. Support from prominent persons, such as Prince Khaled bin Alwaleed bin Talal of the Kingdom of Saudi Arabia, has helped. Prince bin Alwaleed bin Talal has been publicly vocal about his choice to be vegan, encourages others in the Kingdom and throughout the MENA to embrace plant-based diets, and is expanding the San Diego-based Plant Power Fast Food vegan restaurants throughout the MENA region (Starostinetskaya 2019). The Prince also has critiqued the animal agriculture industry, particularly in its contribution to the climate crisis, and is investing in companies who are finding a more sustainable ways to source protein for people (cited in CNBC International 2020).

While Prince bin Alwaleed bin Talal's leadership in shifting the culture toward veganism is encouraging, it tends to obscure the decades' long work of Muslim vegan women advocates and activists. Nevertheless, as veganism continues to carry emasculating cultural codes, the Prince's identity as the "leading man" of Muslim veganism will likely be influential to other men. Sammar Hakim notes that when people learn he is vegan, "[t]hey say, 'Oh, you are doing some feminine, New Age movement'" (cited in Fida 2019b para. 19). Hakim reports that people ask him, "Why do you have to do this to yourself?" although, he notes, "[b]ut that's a very small percentage of people. Most people take it at face value." However, when resisting the animal sacrifice of *Eid al-Adha* men, in particular, report strong negative responses. Hakim (cited in Fida 2019b) hears that "I'm an apostate, (I'm) trying to change Islam, (I'm) misguided" (para 19).

Several Islamic theologians have provided religio-scholarly opinion suggesting those who claim that Muslim vegans are apostate are the ones who are misguided. There are several *fatāwa*[23] on vegetarianism and veganism, such as that of Sayyid Fadhlullah above, by *ulamā'* [scholars trained in Islam and Islamic law], *mutakallimun* [theologians], *muftis* [group of theologians or canon lawyers], *qadis* [judges], professors, and *Sheikh al Islam* [the highest-level state religious official]. Mufti Ebrahim Desai, for instance, issued the following *fatwā*: "A Muslim may be a vegetarian. However, he should not regard eating meat as prohibited." Muzammil Siddiqi's *fatwā* states, "Muslims do recognize animal rights, and animal rights means that we should not abuse them, torture them, and *when we have to use them for meat*, we should slaughter them with a sharp knife, mentioning the name of Allah" (our emphasis). The Prophet (SAAWS) said, "Allah has prescribed goodness (ihsan) in everything. When you sacrifice, sacrifice well. Let you sharpen your knife and make it easy for the animal to be slaughtered. So, Muslims are not vegetarianists. However, if someone prefers to eat vegetables, then they are allowed to do so. Allah has given us permission to eat meat of slaughtered animals, but He has not made it obligatory upon us." Sheikh M. S. Al-Munajjid announces, "there is nothing wrong with being a vegetarian or not eating animal products, but you need to be aware of the following: You should not think that these things are Haraam (forbidden), because Almighty Allah says: *'O ye who believe! Make not unlawful the good things, which Allah hath made lawful for you, but commit no excess: for Allah loveth not those given to excess.'* (Al-Maa'idah 87)." Sheikh Al-Munajjid also shares the following scriptural passage: "'Say: who hath forbidden the beautiful (gifts) of Allah, which He hath produced for his servants, and the things, clean and pure (which He hath provided) for sustenance? Say they are, in the life of this world, for those who believe, (and) purely for them on the Day of Judgment thus do we explain the Signs in detail for those who understand.' (Al-'Araaf 32)." Finally, Muslim jurist Ayatullah Shirazi, concludes, "Being vegetarian is OK and halal, and in fact we have hadith in Islam that encourages us to eat less meat" (cited in Ali Muttaqi, Shahid 2020).

Difference and resistance are tremendously important factors in veg(etari)anism for Muslims, as the "meat culture" within Islam remains extremely strong (see Ali 2015; Dahlan 2019; Dahlan-Taylor 2012). The affective nature of dominant consumption patterns fuels affective responses: "Because you are dealing with a lot of deep rooted traditions cultures and dogma in some cases, there has been strong opposition" (cited in Fida 2019a, para. 21). We see intersectional Othering as a marginalized group within a marginalized group. There are multiple layers of intersectional difference for Muslim women vegans who resist what Erika Cudworth (2016) calls a "carnist discourse of naturalization, necessity and normalization" (237; see, also, Cudworth 2008). Contemporary foodways are central to cultural identity, at the individual and group levels. Elbarshoumi (cited in Kavaler 2019) argues, "I think the barriers to veganism in the Arab world are largely cultural–there's a perception of plant-based diets as being inferior from a nutritional point of view, and a notion that not eating meat makes you inherently less 'Arab'" (para. 12).

Anti-vegetarian and anti-vegan discourses abound. In a comment on a May 2019 *Jerusalem Post* article on veganism in the Arab world, one reader wrote, "Veganism is a new religion in the West. Some fanatics combine veganism with climate change and try to enforce their obscure views on others…It seems that the Arab world is happy to embrace the craziest Western ideas, but reject the good ones" (Gabor Ujvari 2019, comment to Kavaler 2019). This response illuminates the resistance to a more progressive Islam, and a critique of its anti-Westernism, a rejection of what some believe to be Islamic principles. Such rejections of non-dominant foodways take various forms, for various reasons. It might begin with a polite expression indicating confusion or concern, or an emasculating chuckle,[24] but can quickly escalates to nefarious Othering and Islamophobic outbursts on social media.

Intersections of Veg(etari)anism and Environmental Sustainability in the MENA

> Have you not seen that it is God
> to whom all the beings in the earth
> bow themselves down—and so too the sun
> and the moon and the stars
> and the mountains and the trees and the beasts? (*Qur'an* 22:18)

Attention to public health and environmental sustainability concerns in the MENA are vital. While Muslim health, wellbeing, and dietary practices are growing areas of research, this research is still not abundant.[25] In Europe, the U.S., and other industrialized nations, goals of ecological sustainability usually focus on "the need to maintain better balances between economic growth and social needs, while protecting local ecologies and reducing the negative impact of growth on the global environment" (Elegendy 2011, 10). The balance between primary development goals of economic and infrastructural stability and higher-level goals of environmental and social justice is still a tremendous challenge for many areas in the MENA and other developing regions. Elegendy (2011) argues the need to address more "basic developmental challenges, such as eradicating poverty, means that "other environmental and social aspects are considered a luxury a developing nation cannot afford" (10). However, there are advocates who support vegan and vegetarian practices for development and environmental sustainability strategies, along with their commitment to and compassion for human and animal health and wellbeing for animals. The groundbreaking Vegan Muslim Initiative is at the forefront of this advocacy. Founded in 2016 The Vegan Muslim Initiative's mission is "to help educate and inform Muslims and their communities about Veganism and the immense impact it can have on their lives. We aspire to relay a better understanding about how food choices affect their health, the animals and the impact on the environment."[26] Co-founder Sammer Hakim (cited in Pevreall 2017) argues, "It's not a necessary thing to do, killing so many animals…at a time when this very act is directly contributing to so many critical environmental issues is highly irresponsible" (para. 4).

Additional sustainability problems in the MENA include high population densities in urban areas, a large youth population, non-"green" building methods, lack of infrastructure for renewable energy systems, deforestation, the stress of high heat for animals (Caulfield, Cambridge, Foster and McGreevy 2014), food and environmental safety (Ghareyazie 2012; Moosa 2009), and loss of habitat for wildlife. Recently there have been new problems. Advocates, activists, experts, nongovernmental organizations (NGOs), and civil society organizations in the MENA that are focused on ecological and environmental sustainability are increasingly under attack, with authorities threatening to shut down their activities and silencing them (Schwartzstein 2019). The UN Environment Programme (2019) reports this trend is similar in other parts of the world. Between 2002 and 2013 a total of 908 environmental advocates were killed across 35 countries. This has been escalating at an alarming rate: 197 environmental activists and advocates were murdered in 2017 alone, which is five times the rate of deaths of environmentalists from just a decade ago. This is a particularly dreadful concern in the areas south of the Sahara Desert where, for decades, environmental activists have been murdered for their efforts to restrict the poaching of endangered animals. Now, even campaigning for water security or climate change action can lead to advocates being imprisoned, albeit for unrelated, spurious charges (Schwartzstein 2019).

There are encouraging developments, however. NGOs including the Arab Group for the Protection of Nature, Arab Office of Youth and Environment, Middle East Network for Animal Welfare, Ras Al Khaimah Animal Welfare Center, Middle East Children's Alliance, Alliance for Water Justice in Palestine, Emergency Water, Sanitation and Hygiene Group (EWASH), have increased civil society efforts. EcoPeace Middle East (2017) includes Palestinians, Jordanians, and Israelis who advocate for cooperation for the sake of their shared environment, particularly water scarcity in the Jordan river.[27]

Future Directions

We have aimed here to critique religious discourses surrounding meat consumption by Islamic scholars and examine contemporary discourses about veg(etari)anism as they construct cultural, religious and ethno-national identities, practices, and ideologies. In our efforts to do so, we have found surprisingly little scholarship on Muslim veg(etari)anism, or on Muslim women's activism for animal welfare and environmental sustainability. We have tried to fill this gap by assessing the few, newly emerging Muslim advocacy efforts for limiting meat consumption, drawing upon religious discussions concerning meat consumption that highlight, social and economic justice, health, environmental, and animal welfare concerns. But there is much more work to be done to analyze Islamic scripture, *fatāwa*, and *hadiths* regarding veganism and vegetarianism, critique the rhetorical framing of *ḥalāl*, particularly in Islamophobic mediated and other popular discourses, and unpack the intersectional Othering and affect-driven resistance to Muslim veg(etari)anism.

The contentious and affective nature of dominant and resistant forms of consumption is central to essentializing and Othering processes. Recall the violent, gendered backlash of anti-veg(etari)an Muslims who attacked PETA-India activist Suraiyya Benazeer in Bhopal. Through analyzing these forms of Othering, the anti-vegan and anti-vegetarian rhetoric in Islam, and the few, newly emerging efforts to advocate for veg(etari)anism for Muslims, scholars can trace the nuanced religio-cultural dialogues concerning consumption and identity that underlie and inform animal welfare concerns, public health considerations, and social, environmental, gender, and economic justice.

It is vital to continue to analyze the differing politics of resistance to dominant cultural and consumption practices, not only of food, but all resources. In particular, it will be important to evaluate the impact of Muslim leaders, most notably Prince Khaled bin Alwaleed bin Talal, as they continue to advocate for plant-based diets and invest in vegan food system infrastructures in ways that highlight intersections of health, social, environmental, and economic justice.

More also needs to be done to protect the environment, people, animals, and biospheres in the MENA (EcoPeace Middle East 2017; O'Regan 2011). Highlighting environmental sustainability-centered approaches to veganism, vegetarianism, and nutritional health should be central to any analysis of vegan/vegetarianism in the MENA and in Islamic communities outside the MENA. Arguing for sustainability, MENA researchers and leaders like Prince bin Alwaleed bin Talal raise concerns that the growing demand for livestock for human consumption, coupled with the devastating impact of climate change, threatens food security and rural livelihood in the region. They also are concerned about how climate change, deforestation, water supply concerns, and various types of harmful human intervention are all threatening many wildlife species in the region. Such approaches, however, are challenging within strongly faith-based Muslim communities, as many Muslim religious communities' leaders and organizational members maintain conflicting attitudes and beliefs about veganism and vegetarianism, perceptions of the need to consume meat for both health and religious purposes. This is not surprising given that public knowledge about veganism and vegetarianism has received significantly less attention in Muslim communities than in other faith-based communities and cultures.

More knowledge about nondominant foodways can help to dismantle the hegemonic discourses of social, cultural, and religious norms. Knowledge can also dismantle the affective resonance that situates nondominant practices as Otherness. Tracing how the emotionality of texts concerning nondominant cultural practices move and generate impact, can illuminate the emotionality of Othering, whether the subject of analysis is resistant praxis, Islamophobia, or the gendered, classed and raced nature of consumption practices.[28] We call for others to continue analyzing the discursive, cultural, and socio-political nuances of Muslim health, wellbeing, and dietary practices, the anti-veganism and vegetarianism rhetoric in Islam and the newly emerging efforts to advocate for sustainable, ethical plant-based food systems for Muslims. The future of the Middle East and North Africa, other Muslim-majority nations, and the entire world will rest, in large part, on resisting hegemonic normative thought and unsustainable practice.

Notes

1. The color green is also known as *Akhḍar* in Modern Standard Arabic.
2. An analysis of ethnocentrism, xenophobia, and increasing Islamophobia in India, specifically, is beyond the scope of this study. For more on these topics, please see Anand (2007); Sinan Siyech and Narain (2018); Schneider (2009).
3. See Lengel (1998).
4. See Abdul-Aziz (2019); Abdul-Aziz, Smidi, and Lengel (in press); Fedak-Lengel (2019); Fedak-Lengel, and Abdul-Aziz (2019); Abdul-Aziz, Lengel, and Smidi (in press); Lengel, Atay, and Kluch (2020).
5. Alphabetical order of authorship indicates equal contribution from each author.
6. For work on "disgust responses" see Fessler, Arguello, Mekdara, and Macias (2003); Hamilton (2006); Twine (2010).
7. See Abdul-Aziz (2019); Abdul-Aziz, Smidi, and Lengel (in press); Fedak-Lengel (2019); Fedak-Lengel, and Abdul-Aziz (2019); Abdul-Aziz, Lengel, and Smidi (in press); Lengel, Atay, and Kluch (2020); Lengel and Smidi (2019); Smidi and Lengel (2017).
8. Arabic translation into English is conducted phonetically resulting in variances in spelling. See Ryding (2005). We have used diacritic marks in Arabic terms throughout, with the exception of Arabic terms in direct quotations which are as included here precisely as written in the original cited texts.
9. It is also widely known that pork is also forbidden in the Jewish faith, according to Jewish laws and traditions of eating *kosher* meat. There is an exception in the *Qur'an* indicating that if a Muslim is starving and there is nothing else to eat but pork, Allah will allow for this.
10. See, for instance, Belhaj (2018); Benkheira (1995, 1999, 2000); Bergeaud-Blackler (2004); Bonne and Verbeke (2008); Burt (2015); Campbell, Murcott and MacKenzie (2011); Dahlan (2019); Dahlan (2012); Fischer (2009, 2010, 2011); Franck, Gardin, and Givre (2015); Friedenreich (2011, 2014); Hashimi and Mohd Salleh (2012); Lever (2013); Lever and Miele (2012); Lockerbie (2017); Tieman (2014); Yusoff and Sarjoon (2017).
11. A comparison of *Dhabīḥah* to *Shechita* [the religious slaughter of mammals and birds done in accordance with the Jewish dietary law] or "con-

ventional slaughter technology," as Abdullah, Borilova and Steinhauserova (2019) call it, is beyond the scope of this study. See Aghwan and Reggenstein (2019).

12. Analysis of the nuances of *Dhabīḥah* and of the debates about and critiques of stunning of animals before slaughter is beyond the scope of this study. For analyses of *Dhabīḥah* without stunning, see Fuseini, Knowles, Hadley and Wotton (2016). For the recent ruling in Belgium requiring stunning before slaughter, which has incited a debate about religious freedom, see Zaragovia (2019).
13. While the *Qu'ran* does not name Ibrahim's son, Muslims interpret the text to indicate that God ordered Ibrahim to sacrifice his son, Ishmael. In Abrahamic religions, the prophet Abraham is an important patriarch for the spiritual development of Christianity, Islam, and Judaism. The story is also prominent in both Judaic and Christian traditions, with the father named Abraham and the son named Isaac. For more details, see Hughes (2012).
14. Given dromedary camels (*Camelus dromedarius*) are considered the most common source of the Middle East Respiratory Syndrome coronavirus (MERS-CoV), measures to control the spread of MERS-CoV in Saudi Arabia and elsewhere in the region have included herd screening and isolation of infected camels, regulation of camel movement, use of personal protective equipment by handlers and, in some cases, a ban on the slaughter and consumption of camel meat and unpasteurized camel milk (see Azhar et al. 2014; Hemida et al. 2014, 2017; Schnirring 2019; Wong et al. 2020; Zumla et al. 2016)
15. For historical analyses of animal sacrifice in Afro-Brazilian, ancient Greek, ancient Roman, Christian, Hindu, Jewish, Muslim, and Sámi religious beliefs and practices, please see, for instance, Arcari (2010); Cosmopoulos and Ruscillo (2014); Hutt (2019a, b); Oro (2006); Salmi, Äikäs, Spangen, Fjellström, and Mulk (2018). For analyses on contemporary animal sacrifice practices, including coverage on attempts by self-identified Jewish activists to sacrifice goats and sheep on the Temple Mount in the Old City of Jerusalem, please see Boaz (2019); Gilad (2016); Halperin (2019); Lazarus (2019); and *Van Der Schyff* (2014).
16. Sammar Hakim, The Vegan Muslim Initiative, cited in Fida (2019a para. 14).
17. See, also, Tulloch and Judge (2018, 1), for an application of Pachirat's "politics of sight" and "the pedagogy of conscientization" to the New Zealand dairy industry.

18. See Safi (2016) for a brief video on the blood in the streets in Dhaka, Bangladesh during Eid al-Adha. See Selby (2018) for responses of Muslim women and girls about the stigmatization of menstruation.
19. Like Pachirat (2011), Hamilton and Taylor (2013) for analysis on the discursive practices in slaughterhouses, for example, animals as "units" and their deaths as "processing" that aids "the ideological obfuscation of them [animals] as embodied subjects…Language is used here to sanitise and justify their deaths" (79).
20. In contrast to the U.S. animal agriculture companies who have pushed "ag-gag" laws to keep their work out of public view, Al Safa, a Canadian company that supplies frozen *ḥalāl* products to grocers, maintains an open-door policy allowing anyone to visit its facilities unannounced (ElBoghdady 2005).
21. The Honorable Sayyid Fadhlullah, December 1, 2001, cited in "Fatwas on Vegetarianism."
22. One example emerges from Lara's field research year in North Africa. On more than one occasion she was questioned for drinking water during a meal. At the home of a host family who invited her to dinner, the male head of household explained to her that drinking water while eating bread will expand the food in one's stomach, causing weight gain. See, also, our earlier discussion of the use of Roman cisterns as trash receptacles.
23. *Fatawā* is the plural of *fatwā*, an Islamic religious ruling and a scholarly opinion on a matter of Islamic law. See Aljahli (2017).
24. See Bruce Feirstein's *Real Men Don't Each Quiche*. Published in 1982, this book was one of the first, albeit satirical, analyses of the gendered politics about consumption.
25. See, for instance, Dehghani Bidgoli (2019); El-Sayed and Galea (2009); Fischer (2010, 2011); Padela, Killawi, Forman, DeMonner, and Heisler (2012); Padela and Zaidi (2018).
26. See, for instance, Abiad and Meho (2018); Bahn, El Labban, and Hwalia (2019); Burlingame and Dernini (2011); Dehghani Bidgoli (2019); El Labban (2017); Elmi, Alomirah, and Al-Zenki (2016); Seed (2015).
27. In addition to the efforts of these NGOs, new university programs increase possibilities for enhancing health, wellbeing, and ecological sustainability in the MENA. These programs include the dual degree program in Biology and Environmental Science at Qatar University where students focus on biology-centered topics in environmental science such

as biodiversity, remediation, environmental health, and toxicology (Potts 2011), and the first master's degree in environmental journalism at l'Institut de Presse et des Sciences de l'Information (IPSI) at l'Université de la Manouba in Tunis, Tunisia. The latter is directed by Dr. Hamida El Bour, a feminist journalist/scholar who has been at the forefront of environmental advocacy in the nation and the region. See Cassara, Brendlinger, and Lengel (2008).

28. See, also, Trein (2017). For studies on affect and veganism, see, for instance, Fuller (2016).

Bibliography

Abdul-Aziz, Nora. 2019. Overcoming the Stigma of Mental illness in Muslim Communities and Increasing the Need for Cultural Competence to Address Health Inequities. Presented at the Fourth Annual Undergraduate Symposium on Diversity and Inclusion. Bowling Green, Ohio: Center for Undergraduate Research and Scholarship, February 15.

Abiad, Mohammed G., and Lokman I. Meho. 2018. Food Loss and Food Waste Research in the Arab World: A Systematic Review. *Food Security* 10 (2): 311–322.

Adbul-Aziz, Nora, Lara Lengel, and Adam Smidi. In press. Transcending In/Visibility, Stigma, and Trauma: Community and Organizational Advocacy for Muslim Mental Health. In *The Borders of this Body: At the Intersection of Health and Intercultural Communication*, ed. Andrew R. Spieldenner and Satoshi Toyosaki. New York: Peter Lang.

Aghwan, Zeiad Amjad, and Joe Mac Reggenstein. 2019. Slaughter Practices of Different Faiths in Different Countries. *Journal of Animal Science and Technology* 61 (3): 111–121.

Ahmed, Sara. 2010. Killing Joy: Feminism and the History of Happiness. *Signs* 35: 571–594.

———. 2014. *The Cultural Politics of Emotion*. 2nd ed. Edinburgh: Edinburgh University Press.

Aidaros, Hassan. 2014. Drivers for Animal Welfare Policies in the Middle East. Revue Scientifique et technique de l'Organisation Mondiale de al Santé Animale (OIE) 33 (1): 85–89.

Ali Abdullah Abdullah, Gabriela Borilova, and Iva Steinhauserova. 2019. Halal Criteria Versus Conventional Slaughter Technology. *Animals* 9 (8). https://www.ncbi.nlm.nih.gov/pmc/articles/PMC6718994/

Ali, Kecia. 2015. Muslims and Meat-Eating: Vegetarianism, Gender, and Identity. *Journal of Religious Ethics* 43 (2): 268–288.

'Ali Muttaqi, Shahid. 2020. An Islamic Perspective Against Animal Sacrifice. *Animals in Islam*. https://www.animalsinislam.com/islam-animal-rights/sacrifice-eid-al-adha/

Aljahli, Abdulrahman. 2017. A Rhetorical Examination of the *Fatwā*: Religion as an Instrument of Power, Prestige, and Political Gains in the Islamic World. PhD dissertation, Bowling Green State University, August 2017. https://etd.ohiolink.edu/!etd.send_file?accession=bgsu1491483098012349&disposition=inline

Amin, Hira. 2013. A Rebuttal to 'Should Muslims Reconsider Animal Slaughter on Eid?' *Muslim Matters*, October 25. https://muslimmatters.org/2013/10/25/rebuttal-should-muslims-reconsider-eid-slaughter/

Anand, Dibyesh. 2007. The Violence of Security: Hindu Nationalism and the Politics of Representing 'the Muslim' as a Danger. *The Round Table* 94 (379): 203–215.

Anderson, Ben. 2009. Affective Atmospheres. Emotion, Space and Society 2: 77–81. https://doi.org/10.1016/j.emospa.2009.08.005.

Aqidah Azizi, Nur. 2019. 9 out of 10 SMEs Do Not Have Halal Certification. *New Straits Times* (Malaysia), August 1. https://www.nst.com.my/news/nation/2019/08/509122/9-out-10-smes-do-not-have-halal-certification

Arcari, Luca. 2010. Animal Sacrifice in Ancient Greek Religion, Judaism, and Christianity 100 BC to AD 200. *Henoch* 32 (1): 154–160.

Armstrong, Philip. 2008. *What Animals Mean in the Fiction of Modernity*. London: Routledge.

Atkins-Sayre, Wendy. 2010. Protection from Animal Rights Lunatics: The Center for Consumer Freedom and Animal Rights Rhetoric. In *Arguments about Animal Ethics*, ed. Greg Goodale and Jason Edward Black, 147–162. Lanham: Lexington Books.

Ayub, Jamal. 2014. PeTA Activist Benazir Goes Underground, Logs Out of FB & Twitter. *Times of India*, September 25. http://timesofindia.indiatimes.com/articleshow/43410358.cms?utm_source=contentofinterest&utm_medium=text&utm_campaign=cppst

Azhar Esam I., Sherif A. El-Kafrawy, Suha A. Farraj, Ahmed M. Hassan, Muneera S. Al-Saeed, Anwar M. Hashem, and Tariq A. Madani. 2014. Evidence for camel-to-human transmission of MERS coronavirus. *The New England Journal of Medicine* 37 (26): 2499–2505.

Bahn, Rachel, Sibelle El Labban, and Nahla Hwalia. 2019. Impacts of Shifting to Healthier Food Consumption Patterns on Environmental Sustainability in MENA Countries. *Sustainability Science* 14: 1131–1146.

Belhaj, Abdessamad. 2018. Halal Matters: Islam, Politics and Markets in Global Perspective. *Islam and Christian-Muslim Relations* 29 (3): 398–400.

Benkheira, Mohammed Hocine. 1995. La Nourriture Carnée Comme Frontière Rituelle: Les Boucheries Musulmanes en France. *Archives de Sciences Sociales des Religions* 92: 67–88.

———. 1999. Lier et Séparer: Les Fonctions Rituelles de la Viande Dans le Monde Islamisé. *L'Homme* 39 (152): 89–114.

———. 2000. *Islâm et Interdits Alimentaires: Juguler l'Animalité*. Paris: Presses Universitaires de Frances.

Bergeaud-Blackler, Florence. 2004. Le Chevillard et le Sacrificateur: La Viande Halal Entre Logiques Économiques, Légales et Religieuses. In *L'Islam en France*, ed. Yves Charles Zarka, Sylvie Taussig, and Cynthia Fleury, 537–545. Paris: Presses Universitaires de France.

Boaz, Danielle. 2019. The 'Abhorrent' Practice of Animal Sacrifice and Religious Discrimination in the Global South. *Religions* 10 (3): 1–20.

Bonne, Karijn, and Wim Verbeke. 2008. Religious Values Informing Halal Meat Production and the Control and Delivery of Halal Credence Quality. *Agriculture and Human Values* 25 (1): 35–47.

Burlingame, Barbara, and Sandro Dernini. 2011. Sustainable Diets: The Mediterranean Diet as an Example. *Public Health Nutrition* 14 (12A): 2285–2287.

Campbell, Hugh, Anne Murcott, and Angela MacKenzie. 2011. Kosher in New York City, Halal in Aquitaine: Challenging the Relationship between Neoliberalism and Food Auditing. *Agriculture and Human Values* 28 (1): 67–79.

Cassara, Catherine, Nancy Brendlinger, and Lara Lengel. 2008. Partners for a Sustainable Future: Aiding Future Practitioners in Tunisian and Algerian Environmental Journalism and Communication. Grant report on partnership program with l'Institut de Presse et des Sciences de l'Information (IPSI), Université de la Manouba, Tunis, and l'Institut de Communication et des Sciences de l'Information (ISIC), University of Algiers submitted to the U.S. Department of State Bureau of Educational and Cultural Affairs, Washington, DC.

Catholic Concern for Animals. 2019. CCA Welcomes 'Landmark' Ruling that Veganism is Philosophical Belief. https://catholic-animals.com/uncategorized/cca-welcomes-landmark-ruling-that-veganism-is-philosophical-belief/

Caulfield, Malcom P., Heather Cambridge, Susan F. Foster, and Paul D. McGreevy. 2014. Heat Stress: A Major Contributor to Poor Animal Welfare Associated with Long-Haul Live Export Voyages. *The Veterinary Journal* 199 (2): 223–228.

CNBC International TV. 2020. Prince Khaled bin Alwaleed: 'We Need to Find a Better Way to Source Protein for People'. *CNBC International TV*, February 12. https://www.youtube.com/watch?v=8N4RgDM6zNY

Cosmopoulos, Michael B., and Deborah Ruscillo. 2014. Mycenaean Burnt Animal Sacrifice at Eleusis. *Oxford Journal of Archaeology* 33 (3): 257–273.

Crowe, Sarah, Kathrin Cresswell, Ann Robertson, Guro Huby, Anthony Avery, and Aziz Sheikh. 2011. The Case Study Approach. *BMC Medical Research Methodology* 11 (100). https://bmcmedresmethodol.biomedcentral.com/articles/10.1186/1471-2288-11-100

Cudworth, Erika. 2008. 'Most Farmers Prefer Blondes': The Dynamics of Anthroparchy in Animals Becoming Meat. *Journal for Critical Animal Studies* 6 (1): 32–45.

———. 2016. On Ambivalence and Resistance: Carnism and Diet in Multispecies Households. In *Meat Culture*, Annie Potts, ed. 222–242. Koninklijke *Brill* NV, Leiden; Boston: Brill.

Dahlan, Magfirah. 2019. *Sacred Rituals and Humane Death: Religion in the Ethics and Politics of Modern Meat*. Lanham: Lexington Books.

Dahlan-Taylor, Magfirah. 2012. Beyond Minority Identity Politics: Rethinking Progressive Islam Through Food. Ph.D. Dissertation, Virginia Tech University, May 2. http://hdl.handle.net/10919/37730

De Matteis, Federico. 2018. When Durga Strikes: The Affective Space of Kolkata's Holy Festival. *Ambiances: Revue Internationale sur l'Environnement Sensible, l'Architecture et l'Espace Urbain*. https://doi.org/10.4000/ambiances.1467.

Dehghani Bidgoli, Reza. 2019. Vegetarian Food Recommendation in the Traditional Islamic Teachings to Increase the Fertility of Men. *Journal of Ethnic Foods* 6 (1). https://journalofethnicfoods.biomedcentral.com/articles/10.1186/s42779-019-0014-6

EcoPeace Middle East. 2017. *Come Together at the River: The Sacred Sites of the Jordan River*. Amman: EcoPeace Middle East.

El Labban, Sibelle. 2017. Sustainable Consumption for Food and Nutrition Security in the MENA Region. Paper presented at the Conference: Youth and the Mediterranean: Exploring New Approaches to Dialogue and Cooperation. Rome: Instituto Affari Internazionali.

ElBoghdady, Dina. 2005. A Taste for Halal Meat. The Washington Post, May 2. https://www.washingtonpost.com/archive/business/2005/05/02/a-taste-for-halalmeat/8ec1e30a-b53a-4bc3-919a-ba651e7684be/

Elegendy, Karim. 2011. Sustainable Development and the Built Environment in the Middle East: Challenges and Opportunities. *The Environment and the Middle East: Pathways to Sustainability* 1: 10–12.

Elmi, Abdirashid, Husam Alomirah, and Sameer Al-Zenki. 2016. Framework for Sustainable Agro-Ecosystems in Support of Gulf Cooperation Council (GCC) Countries' Food Security. *Arab Gulf Journal of Scientific Research* 34 (3/4): 97–105.

El-Sayed, Abdulrahman M., and Sandro Galea. 2009. The Health of Arab-Americans Living in the United States: A Systematic Review of the Literature. *BMC Public Health* 9 (272). PMCID: PMC2728720. https://bmcpublichealth.biomedcentral.com/articles/10.1186/1471-2458-9-272

Fatah, Tarek. 2015. Indian Muslim Mob Beats a Muslim Woman. Her Crime? She Wants to End the Mass Slaughter of Cows and Sheep as Animal Sacrifice During the Islamic Eid Festival. October 20, 2015. https://tarekfatah.net/hundreds-of-indian-muslim-men-beat-a-muslim-woman/

Fedak-Lengel, Daniella. 2019. Neotropical Wildlife Rescue and Rehabilitation Field Research: Preliminary Findings. Presented at the Undergraduate and Graduate Conference on Embracing Global Engagement. Bowling Green: Center for Undergraduate Research and Scholarship, October 16.

Fedak-Lengel, Daniella, and Nora Abdul-Aziz. 2019. Diversity in Islam, Muslim Veg(etari)anism and Environmental Sustainability in the Middle East and North Africa. Presented at the Fourth Annual Undergraduate Symposium on Diversity and Inclusion. Bowling Green: Center for Undergraduate Research and Scholarship, February 15.

Feirstein, Bruce. 1982. *Real Men Don't Each Quiche*. New York: Pocket Books.

Fessler, Daniel M.T., Alexander P. Arguello, Jeannette M. Mekdara, and Ramon Macias. 2003. Disgust Sensitivity and Meat Consumption: A Test of the Emotivist Account of Moral Vegetarianism. *Appetite* 41: 31–41.

Fida, Kashmala. 2019a. One of Islam's Biggest Holidays Is All About Meat. So What's a Vegetarian Muslim To Do? *The Star* (Edmonton, Canada), August 11. https://www.thestar.com/edmonton/2019/08/11/one-of-islams-biggest-holidays-is-all-about-meat-so-whats-a-vegetarian-muslim-to-do.html

———. 2019b. How Vegetarian Muslims Celebrate Eid. *Toronto Star* (Canada). August 12. http://search.ebscohost.com.ezproxy.bgsu.edu/login.aspx?direct=true&db=nfh&AN=6FPTS2019081255318940&site=ehost-live&scope=site

Fischer, Johan. 2009. Halal, Haram, or What? Creating Muslim Space in London. In *Muslim Societies in the Age of Mass Consumption: Politics, Culture, and Identity Between the Local and the Global*, ed. Johanna Pink, 3–22. Cambridge, England: Cambridge Scholars Publishing.

———. 2010. Halal Sanitized: Health and Science in a Globalized Religious Market. *Tidsskrift for Islamforskning* 4 (1): 24–47.

———. 2011. *The Halal Frontier*. New York: Palgrave Macmillan.

Franck, Alice, Jean Gardin, and Oliver Givre. 2015. La Mort Animale Rituelle En Ville: Une Approche Comparée de La « fête du sacrifice » à Istanbul, Khartoum et Paris. *Histoire Urbaine* 44: 139–168.

Fredengren, Christina, and Camilla Löfqvist. 2019. Finitude: Human and Animal Sacrifice in a Norse Setting. In *Myth, Materiality, and Lived Religion: In Merovingian and Viking Scandinavia*, ed. Klas *Wikström* af Edholm, Peter Jackson Rova, Andreas Nordberg, Olof Sundqvist, and Torun Zachrisson, 225–262. Stockholm: Stockholm University Press.

Friedenreich, David M. 2011. The Implications of Unbelief: Tracing the Emergence of Distinctively Shiʿi Notions Regarding the Food and Impurity of Non-Muslims. *Islamic Law and Society* 18 (2011): 53–84.

———. 2014. *Foreigners and Their Food: Constructing Otherness in Jewish, Christian, and Islamic Law*. Berkeley: University of California Press.

Fuller, Jery. 2016. The Cultural Creation of Affect Amongst Vegan Consumers: An Ethnographic Analysis of Online Consumption Communities. Master's Thesis, Department of Service Management and Service Studies, Lund University, August 31. https://lup.lub.lu.se/student-papers/search/publication/8887555

Fuseini, Awal, Toby G. Knowles, Phil J. Hadley, and Steve B. Wotton. 2016. Halal Stunning and Slaughter: Criteria for the Assessment of Dead Animals. *Meat Science* 119: 132–137.

Gayle, Damien. 2020. Ethical Veganism is a Belief Protected by Law, Tribunal Rules. *The Guardian* (UK), January 3. https://www.theguardian.com/lifeandstyle/2020/jan/03/ethical-veganism-is-a-belief-protected-by-law-tribunal-rules

Ghareyazie, Behzad. 2012. Food and Environmental Safety of Biotech Crops: Islamic Perspective. In *International Workshop for Islamic Scholars on Agribiotechnology (IWISA): Shariah Compliance*, ed. Shaikh Mohd Saifudden

Shaikh Mohd Salleh, 29–42. Penang/Selangor: Malaysia Biotechnology Information Center (MABIC).

Gilad, Elon. 2016. Why Jews Stopped Sacrificing Lambs and Baby Goats for Passover and Why Some Have Started Trying to Perform it on the Temple Mount Again. *Haartez*, April 24. https://www.haaretz.com/jewish/.premium-why-jews-stopped-sacrificing-lambs-for-passover-1.5440120?v=13FB6E26AFD71798DD76E112197A4B37

Halperin, Ehud. 2019. Is the Goddess Haḍimbā Tantric? Negotiating Power in a Western Himalayan Sacrificial Arena. *International Journal of Hindu Studies* 23 (2): 195–212.

Hamilton, Malcolm. 2006. Disgust Reactions to Meat Among Ethically and Health Motivated Vegetarians. *Ecology of Food and Nutrition* 45 (2): 125–158.

Hamilton, Lindsay, and Nik Taylor. 2013. *Animals at Work: Identity, Politics and Culture in Work with Animals*. Leiden: Brill.

Hashimi, Darhim, and Shaikh Mohd Saifuddeen Shaikh Mohd Salleh. 2012. A Background on Halal Industry and Principles. In *International Workshop for Islamic Scholars on Agribiotechnology: Shariah Compliance*, 12–20. Selangor: Biotechnology Information Center: Selangor; International Service for the Acquisition of Agri-biotech Applications: Los Baños, Laguna, Philippines.

Hemida, Maged G., Daniel K.W. Chu, Leo L.M. Poon, Ranawaka A.P.M. Perera, Mohammad A. Alhammadi, Hoi-yee Ng, Lewis Y. Siu, Yi Guan, Abdelmohsen Alnaeem, and Malik Peiris. 2014. MERS Coronavirus in Dromedary Camel Herd, Saudi Arabia. *Emerging Infectious Diseases* 20 (7): 1231–1234.

Hemida, Maged G., Ahmed Elmoslemany, Fahad Al-Hizab, Abdelmohsen Alnaeem, Faisal Almathen, Bernard Faye, Daniel K.W. Chu, Ranawaka A.P.M. Perera, and Malik Peiris. 2017. Dromedary Camels and the Transmission of Middle East Respiratory Syndrome Coronavirus (MERS-CoV). *Transboundary and Emerging Diseases* 64 (2): 344–353.

Hong Tschalaer, Mengia. 2015. Muslim Women's Rights Activists' Visibility: Stretching the Gendered Boundaries of the Public Space in the City of Lucknow. *South Asia Multidisciplinary Academic Journal* 11. https://doi.org/10.4000/samaj.3928.

Hughes, Aaron W. 2012. *Abrahamic Religions: On the Uses and Abuses of History*. Oxford, England: Oxford University Press.

Hutt, Curtis. 2019a. A Threefold Heresy: Reassessing Jewish, Christian, and Islamic Animal Sacrifice in Late Antiquity. *History of Religions* 58 (3): 251–276.

———. 2019b. Off the Beaten Track: Animal Sacrifice and Christian Traditions. *History of Religions* 58 (3): 221–224.

Jakarta Post, The. 2015. Saudi Bans Hajj Camel Slaughter. *The Jakarta Post,* September 12. https://www.thejakartapost.com/news/2015/09/12/saudi-bans-hajj-camel-slaughter.html

Jenkins, Stephanie. 2012. Returning the Ethical and Political to Animal Studies. *Hypatia* 27 (3): 504–510.

Kavaler, Tara. 2019. Veganism in the Arab World. *The Jerusalem Post,* May 14. https://www.jpost.com/Middle-East/Veganism-in-the-Arab-world-589668

Kirmani, Nida. 2009. Claiming their Space: Muslim Women-Led Networks and the Women's Movement in India. *Journal of International Women's Studies* 11 (1): 72–85.

Larsson, Naomi. 2019. Which Is the World's Most Vegan City? *The Guardian* (UK), January 17. https://www.theguardian.com/cities/2019/jan/17/which-is-the-worlds-most-vegan-city

Lazarus, David. 2019. Priests Sacrifice Passover Lamb in Jerusalem. Israel Today, April 22. https://www.israeltoday.co.il/read/priests-sacrifice-passover-lamb-in-jerusalem/

Lengel, Lara. 1998. Researching the 'Other', Transforming Ourselves: Methodological Considerations of Feminist Ethnography. *Journal of Communication Inquiry* 22 (3): 229–250.

Lengel, Lara, and Adam Smidi. 2019. How Affect Overrides Fact: Anti-Muslim Politicized Rhetoric in the Post-Truth Era. In *Emotion, Affect, and Rhetorical Persuasion in Mass Communication: Theories and Case Studies*, ed. Lei Zhang and Carlton Clark, 115–130. New York: Routledge.

Lengel, Lara, Ahmet Atay, and Yannick Kluch. 2020. Decolonising Gender and Intercultural Communication in Transnational Contexts. In *The Cambridge Handbook of Intercultural Communication*, ed. Guido Rings and Sebastian Rasinger, 205–226. Cambridge, England: Cambridge University Press.

Lever, John. 2013. The Postliberal Politics of Halal: New Directions in the Civilizing Process? *Human Figurations* 2 (3). http://hdl.handle.net/2027/spo.11217607.0002.306

———. 2019. Halal Meat and Religious Slaughter: From Spatial Concealment to Social Controversy—Breaching the Boundaries of the Permissible. *Environment and Planning C: Politics and Space* 37 (5): 889–907.

Lever, John, and Mara Miele. 2012. The Growth of Halal Meat Markets in Europe: An Exploration of the Supply Side Theory of Religion. *Journal of Rural Studies* 28 (4): 528–537.

Lockerbie, Michael. 2017. Public Awareness of Halal and Kosher Slaughter. *The Veterinary Record* 181 (6): 149–150.

Masri, Basheer Ahmad. 2007. *Animal Welfare in Islam*. Leicester: The Islamic Foundation.

Masud, M. Khalid. 2014. Food and the Notion of Purity in the *Fatāwā* Literature. In *Alimentacion de las Culturas Islamicas*, ed. Manuela Marin and David Waines, 89–110. Madrid: Agencia Español de Cooperación Internacional.

Mayton, Joseph. 2010. Eating Less Meat Is More Islamic. *The Guardian* (UK), August 26. http://www.theguardian.com/commentisfree/belief/2010/aug/26/meat-islam-vegetarianism-ramadan

Moosa, Ebrahim. 2009. Genetically Modified Foods and Muslim Ethics. In *Acceptable Genes? Religious Traditions and Genetically Modified Foods*, ed. Conrad G. Brunk and Harold Coward, 135–157. Albany: State University of New York Press.

Morris, Nigel. 2015. Halal Slaughter: Outcry After Undercover Film Exposes Brutality of Industry. *The Independent* (UK), February 3. https://www.independent.co.uk/news/uk/home-news/outcry-after-undercover-film-exposes-brutality-of-halal-industry-10019467.html

Muzafar, Naroosha. 2019. The State in 'Vegetarian' India Where 98 Percent of People Eat Meat. *Ozy*, February 5. https://www.ozy.com/acumen/the-state-in-vegetarian-india-where-98-percent-of-people-eat-meat/92383

Nada. 2014. Sacrifice and Islam: Weighing from a Vegan Perspective. One Arab Vegan, September. http://www.onearabvegan.com/2014/09/sacrifice-and-islam-weighing-in-from-a-vegan-perspective/

Narayanan, Yamini. 2019. 'Cow Is a Mother, Mothers Can Do Anything for Their Children!' Gaushalas as Landscapes of Anthropatriarchy and Hindu Patriarchy. *Hypatia* 34 (2): 195–221.

Nasr, Seyyed Hossein. 2010. Ecology. In *A Companion to Muslim Ethics*, ed. Amyn B. Sajoo, 79–90. New York: I. B. Tauris.

O'Regan, Frederick M. 2011. *Raising Animal Welfare Awareness in the Middle East*. Yarmouth Port: International Fund for Animal Welfare (IFAW).

Oro, Ari Pedro. 2006. The Sacrifice of Animals in Afro-Brazilian Religions: Analysis of a Recent Controversy in the Brazilian State of Rio Grande Do Sul. *Religião e Sociedad* 1: 995–1014.

Pachirat, Timothy. 2011. *Every Twelve Seconds: Industrialized Slaughter and the Politics of Sight*. New Haven: Yale University Press.

Padela, Aasim I., and Danish Zaidi. 2018. The Islamic Tradition and Health Inequities: A Preliminary Conceptual Model Based on a Systematic Literature Review of Muslim Health-Care Disparities. *Avicenna Journal of Medicine* 8 (1): 1–13.

Padela, Aasim I., Amal Killawi, Jane Forman, Sonya DeMonner, and Michele Heisler. 2012. American Muslim Perceptions of Healing: Key Agents in Healing, and Their Roles. *Qualitative Health Research* 22 (6): 846–858.

People for the Ethical Treatment of Animals. 2014. Encouraging Vegan Eating Is Not a Crime, but Assaulting Women is. *All Creatures*. https://www.all-creatures.org/articles/ar-vegan-eating-abuse-women.html

———. 2020. Investigating Halal Meat. https://www.animalsinislam.com/halal-living/investigating-halal-meat/

———. n.d. Animals in Islam. https://www.animalsinislam.com/halal-living/fatwas/

Pevreall, Katie. 2017. Vegan Muslims Campaign for an End to Slaughter during Eid Al-Adha. *Live Kindly*. https://www.livekindly.co/vegan-muslims-end-to-slaughter-eid-al-adha/

Potts, Malcolm. 2011. Environmental Science at Qatar University: Realizing Qatar's 2030 Vision. *The Environment and the Middle East: Pathways to Sustainability* 1: 40–44.

Potts, Annie, and Philip Armstrong. 2016. Vegan. In *Critical Terms for Animal Studies*, ed. Lori Green, 395–409. Chicago: University of Chicago Press.

Rahman, Sira Abdul. 2017. Religion and Animal Welfare—An Islamic Perspective. *Animals* 7 (2). https://doi.org/10.3390/ani7020011.

Rana, Junaid. 2011. *Terrifying Muslims: Race and Labor in the South Asian Diaspora*. Durham: Duke University Press.

Robinson, Sarah E. 2014. Refreshing the Concept of Halal Meat: Resistance and Religiosity in Chicago's Taqwa Eco-Food Cooperative. In *Religion, Food, and Eating in North America*, ed. Benjamin E. Zeller, Marie W. Dallam, Reid L. Neilson, and Nora L. Rubel, 274–293. New York: Columbia University Press.

Royal Aal al-Bayt Institute for Islamic Thought. 2017. *Tafsir al-Jalalayn*. Trans. Feras Hamza. Amman: Royal Aal al-Bayt Institute for Islamic Thought. https://www.altafsir.com/Tafasir.asp?tMadhNo=1&tTafsirNo=74&tSoraNo=22&tAyahNo=37&tDisplay=yes&UserProfile=0&LanguageId=2

Ryding, Karin C. 2005. *A Reference Grammar of Modern Standard Arabic*. Cambridge, England: Cambridge University Press.

Safi, Michael. 2016. Rivers of Blood Flow on Streets of Dhaka after Eid Animal Sacrifices. *The Guardian* (UK), September 14. https://www.theguardian.com/world/2016/sep/14/rivers-of-blood-dhaka-bangladesh-eid-animal-sacrifices

Salmi, Anna-Kaisa, Tiina Äikäs, Marte Spangen, Markus Fjellström, and Inga-Maria Mulk. 2018. Tradition and Transformation in Sámi Animal-Offering Practices. *Antiquity* 92 (362): 472–489.

Sanders, Shaakirrah R. 2019. Ag-Gag Free Nation. *Wake Forest Law Review* 54 (2): 491–539.

Schneider, Nadja-Christina. 2009. Islamic Feminism and Muslim Women's Rights Activism in India: From Transnational Discourse to Local Movement—Or Vice Versa? *Journal of International Women's Studies* 11 (1): 56–71.

Schnirring, Lisa. 2019. With 3 new Saudi MERS cases, study notes antibodies in Pakistan camel workers. Center for Infectious Disease Research and Policy, University of Minnesota, November 6. https://www.cidrap.umn.edu/news-perspective/2019/11/3-new-saudi-mers-cases-study-notes-antibodies-pakistan-camel-workers

Schwartzstein, Peter. 2019. The Middle East's Authoritarians Have Come for Conservationists. *The Atlantic*, March 30. https://www.theatlantic.com/international/archive/2019/03/middle-east-north-africa-environmentalism-espionage/585973/

Seed, Barbara. 2015. Sustainability in the Qatar National Dietary Guidelines, among the First to Incorporate Sustainability Principles. *Public Health Nutrition* 18 (13): 2303–2310.

Selby, Daniele. 2018. Muslim Women Are Over Period Shaming During Ramadan. *Global Citizen*, May 30. https://www.globalcitizen.org/en/content/period-shame-ramadan-twitter-menstruation/

Sinan Siyech, Mohammed, and Akanksha Narain. 2018. Beef-related Violence in India: An Expression of Islamophobia. *Islamophobia Studies Journal* 4 (2): 181–194.

Smidi, Adam, and Lara Lengel. 2017. Freedom for Whom? The Contested Terrain of Religious Freedom for Muslims in the United States. In *The Rhetoric of Religious Freedom in the U.S.*, ed. Eric Miller, 85–99. Lanham: Lexington/Rowman & Littlefield.

Srivastava, Shruti. 2016. Muslim woman for PETA India campaigning for 'Vegan Eid' gets brutally attacked! Speaking Tree (India), July 1. https://www.speakingtree.in/blog/muslim-woman-working-for-peta-india-is-beaten-brutally-by-men-from-her-community-reason-will-disgust-you.

Starostinetskaya, Anna. 2019. Vegan Chain Plant Power Fast Food Sales Spike by 438 Percent. *VegNews*, July 16. https://vegnews.com/2019/7/vegan-chain-plant-power-fast-food-sales-spike-by-438-percent

———. 2020. Saudi Arabian Prince Predicts Beyond Meat Will be Cheaper Than Animal Flesh by 2025. *VegNews*, February 14. https://vegnews.com/2020/2/saudi-arabian-vegan-prince-predicts-beyond-meat-will-be-cheaper-than-animal-flesh-by-2025

Strong, Maggie. 2019. The Show-Me State's Hidden Cruelty: How Missouri's Ag-Gag Laws Unconstitutionally Silence Animal-Welfare Whistleblowers. *St. Louis University Law Journal* 63 (4): 611–639.

Sullivan, Harvey. 2018. BLOOD IN THE STREETS. Eid Al-Adha Animal Sacrifice Festival Sees Roads Turn Red with Blood as Cows are Beheaded. *The Sun* (UK), August 22. https://www.thesun.co.uk/news/7074173/eid-al-adha-2018-animal-sacrifice-cows-beheaded/

Taqi Usmani, Mufti Muhammad. 2006. *The Islamic Laws of Animal Slaughter: A Discussion on the Islamic Laws for Slaughtering Animals and a Survey of Modern-Day Slaughtering Methods*. Trans. Amir A. Toft. Santa Barbara: White Thread Publishers.

Taylor, James R., and Daniel Robichaud. 2004. Finding the Organization in the Communication: Discourse as Action and Sensemaking. *Organization* 11 (3): 395–413.

Tieman, Marco. 2014. Synergy in Halal Supply Chains. *Islam and Civilizational Renewal* 5 (3): 454–459.

Tlili, Sarra. 2012. *Animals in the Qur'an*. Cambridge, England: Cambridge University Press.

Trein, Lorenz. 2017. Islamophobia Reconsidered: Approaching Emotions, Affects, and Historical Layers of Orientalism in the Study of Religion. *Method and Theory in the Study of Religion* 2017: 1–16.

Tulloch, Lynley, and Paul Judge. 2018. Bring the Calf Back from the Dead: Video Activism, the Politics of Sight and the New Zealand Dairy Industry. *Video Journal of Education and Pedagogy* 3 (1): 1–20.

Twine, Richard. 2010. Intersectional Disgust? Animals and (Eco)feminism. *Feminism and Psychology* 20 (3): 397–406.

Ujvari, Gabor. 2019. Comment to Article: Kavaler, Tara. Veganism in the Arab World. *The Jerusalem Post*, May 14. https://www.jpost.com/Middle-East/Veganism-in-the-Arab-world-589668

UN Environment Programme. 2019. *Dramatic Growth in Laws to Protect Environment, but Widespread Failure to Enforce*. Geneva: UN Environment Programme.

Van Der Schyff, Gerhard. 2014. Ritual Slaughter and Religious Freedom in a Multilevel Europe: The Wider Importance of the Dutch Case. *Oxford Journal of Law and Religion* 3: 76–102.

Vegan Muslim Initiative, The. (2016). About. *The Vegan Muslim Initiative*. https://www.facebook.com/pg/theveganmusliminitiative/about/?ref=page_internal

Wong, Gary, Yu-Hai Bi, Qi-Hui Wang, Xin-Wen Chen, Zhi-Gang Zhang, and Yong-Gang Yao. 2020. Zoonotic Origins of Human Coronavirus 2019 (HCoV-19 / SARS-CoV-2): Why is this Work Important? *Zoological Research* 41 (3): 213–219.

Yusoff, Mohammed Agus, and Athambawa Sarjoon. 2017. Anti-Halal and Anti-Animal Slaughtering Campaigns and Their Impact in Post-War Sri Lanka. *Religions* 8 (4): 1–15.

Zaragovia, Veronica. 2019. Animal Slaughter Ruling in Belgium Stirs Religious Freedom Debate. *Al Jazeera*, September 8. https://www.aljazeera.com/indepth/features/animal-slaughter-ruling-belgium-stirs-debate-halal-meat-190906101608102.html

Zumla, Alimuddin, Abdulaziz N. Alagali, Matthew Cotton, and Esam I. Azhar. 2016. Infectious Diseases Epidemic Threats and Mass Gatherings: Refocusing Global Attention on the Continuing Spread of the Middle East Respiratory Syndrome Coronavirus (MERS-CoV). *BMC Medicine* 14 (132). https://bmcmedicine.biomedcentral.com/articles/10.1186/s12916-016-0686-3

3

Because We Care: Veganism and Politics in Israel

Sharon Avital

After months without a Chief of police and on the verge of possible military confrontations with Gaza on the south border and Hizaballah on the north border, the Prime Minister of Israel, Benjamin Netanyahu, posted a short video in which he announced his decision to treat the urgent problem of unchecked breeding of street cats and dogs. By his side proudly stood his "personal assistant on matters related to animals," *The Big Brother* reality television program winner, Tal Gilboa. This was one of a series of videos posted prior to the general elections in Israel in which Gilboa and members of the Netanyahu family discussed matters related to animal welfare and announced their deep devotion to the cause. The videos and other content on the topic were praised by some but mostly ridiculed by vegans and omnivores alike. They represent the ways in which veganism has become politically charged in the volatile Israeli political climate even as other environmental issues have been largely ignored by both the media and the political system.

S. Avital (✉)
Tel-Aviv University, Tel Aviv-Yafo, Israel

C. Hanganu-Bresch, K. Kondrlik (eds.), *Veg(etari)an Arguments in Culture, History, and Practice*, The Palgrave Macmillan Animal Ethics Series,
https://doi.org/10.1007/978-3-030-53280-2_3

According to the most current research, conducted between 2014 and 2016 by the Central Bureau of Statistics in Israel (BLMS(, 5% of the general population in Israel define themselves as vegetarians, with great differences across gender and religion: 5.3% of the Jews and 1.3% of Arabs avoid meat. The more extreme differences were found across the gender divide, with 7.15% of women and 1.9% of men defining themselves as vegetarians.[1] The organization Vegan Friendly also surveyed the population in 2016 and estimated higher figures for non-meat eaters, 5% vegetarians and 5% vegans.[2] While they only randomly surveyed 10,000 people, their findings nevertheless equate to an increase from their previous 2012 survey in which only 1% identified as vegans.[3] This reflects a high percentage of vegans and vegetarians per capita and a growing interest in the issue. Veganism has become mainstream in the Israeli culinary scene, popular culture, the army, and politics. Some high-profile vegetarians include the Israeli president, Rubi Rivlin, the Israeli Defense Forces' (IDF) Chief of Staff, Yair Cohavi, and about ten other candidates for the 2019 parliament. Veganism has even become one of the main platforms of the Green Party that ran for parliament in the second cycle of the 2019 elections under the Democratic Union.

This chapter examines the transition of veganism into the mainstream and the ways in which it intersects with other political issues in Israel, such as the conflicts with the Palestinians and with the economy. I believe that the vegan movement has succeeded due to a wide range of sociological factors, such as a wide variety of vegan options, vegans' presence in popular culture, and a focused rhetoric. The latter includes the identification of a clear enemy and a clear solution. Vegans were able to affect change by linking various struggles while still allowing the audience to decouple the vegan crusade from more controversial topics such as the dispute with the Palestinians or corporate responsibility. The links and separations were made possible by defining the battle as a binary, zero-sum argument that is based on posthuman logic and ethics of agency.

This research is important for various reasons. Studying the unusual success of a difficult political battle in a challenging political climate, coupled with the divided Israeli society, might be instrumental in understanding issues of social justice and consent. Specifically, this study can contribute to our understanding of the ways in which affect can be rallied in the service of activism. Lastly, this paper extends our understanding not only of vegan arguments but also of the use of arguments in an extremely dynamic online environment.

Vegan Rhetoric

I have been following the discourse on veganism in Israel for the past few years. For the purposes of this research, I rhetorically analyzed websites and talks dedicated to the issue in the summer of 2017 and in the summer of 2019, just prior to the general elections for the Israeli parliament. I also conducted in-depth interviews with over 40 vegans and used self-report surveys shared in vegan Facebook groups such as "Vegan Friendly," "Animals," "Vegans and Interested," and "Veganism in Israel."[4]

The vegan campaign is a very controversial and sensitive topic, with emotions running high.

When interviewing people and reading the comments sections, I often encountered the claim that the battle for animal rights is the most important battle to be currently fought in the world. This statement was made as early as 2017 not in Switzerland, but in Israel, long before the 2019 fires that decimated the Amazon rainforest were directly linked to the meat industry, thereby changing the environmental discourse.

The same activists also told me that this is not only the most important but also the most challenging battle to be fought, as it requires mental, emotional, and physical transformation that goes against the ways in which most Israelis were socialized.[5] For many, if not most, of the vegans I talked to, the argument for veganism must be framed as zero-sum. They explained that while in any other political battle, one can remain neutral, when it comes to animal rights not being vegan implies that one consumes animals and thus participates in the abuse of the animals and of the environment.

Not only is the argument framed as a clear-cut zero-sum game, so also is the culpability of the meat and dairy industry undeniable. While the political dispute in Israel is seen by most Israelis as complex, blurred by many factors and nuances, when it comes to animal abuse the magnitude of evil and the liability of the villains are seemingly crystal clear. The

Israeli left has been impotent since the Hamas has been controlling Gaza and the Arab Spring in Egypt and Syria faltered. Many doubt the possibility of collaboration and peace with the Palestinians and other Arab nations. This belief is summed up best by the then Israeli prime minister, Ehud Barak's stance that "there is no partner (for peace)" having been accepted universally while doubt and even hatred toward various groups in Israel have increased. The many interviews I conducted and comments I analyzed indeed reveal deep pessimism for the possibility of change in this regard. The information exposed to the meat and dairy industries, however, reveals cruelty and abuse on a massive scale that cannot be denied. The lies told by these industries and the inherent ties between them and other influential institutions, such as the media, expose them as manipulative global players trading our health and the planet's safety for profit.

Not only is the villain easily identified but the solution is also close at hand. The talks and the campaigns I analyzed in 2017 reveal a tendency to frame the solution mostly in terms of individual consumption. A clear example of this is a famous talk given by the American animal rights activist, Gary Yourofsky. Yourofsky's lecture, entitled "The most important lecture you will ever hear" was subtitled into Hebrew[6] by animal rights activists and was viewed by over three million people across many media platforms—not a small feat for a country with fewer than eight million Hebrew speakers. The video was especially influential between 2012 and 2016 and has indeed convinced many to stop consuming animals. The lecture was later banned by the Ministry of Education, a fact lauded on the first slide of the video as a sign of its ethical stance. Amidst data on factory farming abuse and graphic and heartbreaking videos, Yourofsky addresses the audience with a direct call for action, stating that "even if there is nothing good you can do in your lifetime, at least you can refrain from participating in the killing of animals. There is nothing more important you will do."[7]

A vegan interviewee told me that Yourofsky's message had deeply resonated with her: "There is so much suffering in the world, I don't even know if I believe in peace anymore, and after the failed protests of 2011 I realized that this corrupt government might never change. But if I don't eat meat, at least I don't contribute to this massive evil machine."[8] Another interviewee shared that following problematic lab tests, his physician suggested he might need to eat meat: "I felt horrible. Reflecting back on my life, I felt like my biggest contribution to this planet was not eating animals. To eat animals felt like a horrible regression, but luckily my test results improved."[9] A highly circulated ad from 2017 directly builds on this logic. It identifies and channels the guilt, anger, and motivation for action by visually translating "vegan time" into the number of animals' lives saved: "8+ in a month, 100+ in a year, and 300+ in 3 years." As another interviewee told me: "When I first tried a vegan diet I didn't know if I would be able to keep it up. I remember thinking after a day, 'Oh, one chicken is alive thanks to my veganism, oh, I saved another animal.' <laughs> I felt like I was doing something heroic or something."[10] It is not surprising then that vegans I spoke with two years later in 2019, after changes in the culinary scene and popular culture have become even more visible, expressed absolute faith in the ability of their personal choices alone to transform the market and even the world.

Mobilizing Affect

An important rhetorical tool that integrates the above elements is the use of the <Holocaust> as an ideograph. Many vegans I spoke to in 2017 described the meat industry as the greatest ongoing holocaust the world has ever known, far exceeding in numbers and duration the holocaust of Jews. Their repeated statements struck me as interesting as the Holocaust seemed to be on everyone's minds, yet it was not directly stated in most of the official articles I analyzed. The term "ideograph" was first coined by McGee to designate a term that captures and reinforces an ideological position and commitment for action in politics.[11] Condit and Lucaites explain that it is a general term that is able "to represent in condensed form the normative, collective commitments of the members of a public, and they typically appear in public argumentation as the necessary motivations or justifications for action performed in the name of the public."[12] Ideograph is an abstraction such as <liberty>,[13] <equality>,[14] or <terrorism>,[15] and it can unify and motivate for action when used strategically. <Holocaust> refers to a specific historical event from recent history and is not as abstract as other ideographs previously studied, yet it carries important symbolic value as the embodiment of absolute and organized evil.

In Israel, the Holocaust is a frequent reference point for discussions concerning morality and is commonly evoked and exploited in a variety of contexts. It often serves as the ultimate symbol of the dangers of helplessness, the necessity to fight, sometimes brutally, in order to defend a cause, and ultimately it stands forth as the justification for the existence of Israel. As such, the term <Holocaust> in Israeli society has stopped referencing only the historical event and has become a term that evokes identification with key values concerning the right to live, morality, and independence. It links together ideology and commitment for action and can therefore be understood as an ideograph.

Gary Yourofsky's famous talk was posted by a user named "The animal holocaust." In the talk, Yourofsky compares meat-eating practices to slavery, torture, and murder. He challenges the school of thought that allows for the concept of so-called humane slaughter, claiming this is a contradiction in terms akin to humane rape, humane slavery, or a humane Holocaust.[16] A similar message was shared in a viral Facebook post that addressed possible attempts to refute the fallibility of the analogy. Written in white font on a black background with a yellow Star of David adjacent to an image of caged chickens, it reads:

> How dare you compare the Holocaust of the Jewish people with the meat industry? That they both have wire fences, extermination by gas, a number branded to their skin, a yellow badge, imprisonment and isolation, physical violence, arduous labor, planned starvation, piles of corpses, separation of mothers from their children, transports, murder of the weak/old/sick/inefficient/, experiments on other living beings and that in both everything was legal and was done out of a sense of superiority of one versus the other–doesn't mean that they can be compared!!!

In another viral Facebook post from 2017, a vegan activist quoted a line from the book *Shosha,* written by Nobel prize recipient, Jewish author and son of Holocaust survivor Isaac Bashevis Zinger: "For animals, all humans are Nazis and every day is Treblinka."[17] The picture showed animals marching toward their deaths, the Nazi flag swaying above them. The post was uploaded on the memorial day for the Holocaust and was shared hundreds of times in spite of the user's relative anonymity. While she received support from the vegan community, she was also the target of expletives and threats by those offended by the comparison. The quote, attributed to the protagonist in Bashevis Zinger's story "The Letter Writer" receives traction and ethos since Zinger himself lost his mother and young brother in the historical Holocaust. The quote appears with more comparisons between the animal industry and the so-called <animal Holocaust> made by other authors on the home page of the first and largest vegan activist group "Animals."[18]

In an op-ed piece called "The Nazis were vegans too," writer Rogel Alpher adds another perspective to the debate regarding veganism and morality. When the Nazis rose to power in Germany in 1933, they quickly passed numerous animal rights restrictions, including bans on the fattening of goose liver for foie gras and the abuse of pets, as well as imposing strict sanctions on the cooking of live lobsters and crabs in restaurants, invasive animal research, and hunting. He notes that not only were a number of prominent Nazis, including Hitler, vegetarians, but they also held animals in higher esteem than they did many humans. The author quotes from Joseph Goebbels' diary, "Man must not feel superior to animals. Man thinks that only he has intelligence, a soul and the ability to speak. Don't animals have those qualities?"[19] The article attempts to expose the discrepancy between the Israeli trend to veganism and the Israelis seeming indifference to the weaker parts of society—the refugees, homeless, or the Palestinians. The public intellectual, Gadi Taub, took to Facebook to comment on the subject in another popular post from 2017:

> I find it really weird the way vegans are walking amongst us with a sense of moral superiority that is taken for granted. People who think a chicken's life and a child's are equal, who compare a cowshed to Auschwitz are not moral people. They are misanthropes who lost their conscience and traded it for melodramatic kitsch for children. This is human hatred masquerading as the love of the non-humans. So far, they haven't taken themselves seriously enough, but ultimately they will start believing in their rhetoric and then it will become violent because if the meat industry is equal to the

> Holocaust, then almost anything is justified. Laugh if you will but remember what I said: it will become violent.[20]

Aware of the sensitivity of the term but desperate to make people conscious of the gravity of the situation, vegans used the ideograph in more subversive ways before they stopped using it almost completely. One interviewee told me that "vegans care too much for other peoples' opinions because they are afraid of alienating them and risk ruining the cause."[21] Indeed, the administrators of the large Facebook group *Vegans and Interested* have declared that the use of the term will not be permitted in the group. With the exception of Yourofsky, who doesn't live in Israel, activists and official social media pages prefer to conjure associations to the <Holocaust> through images and slogans. Edward and Winkler[22] have argued in their study of cartoons that some iconic images, such as the flag-raising on Iwo Jima, are themselves ideographs that shift in meaning over time and are subject to appropriation and dispute. Essential to visual arguments is the audience's ability to interpret the images. Delicath and DeLuca's work on image events demonstrates the importance of the audience's interpretation.[23] Environmental groups such as Earth First! have conveyed arguments about environmental protection, such as tree-sitting and other staged events, that are mostly, if not entirely, visual. Delicath and DeLuca contend that these image events open arenas for public participation, employ images that claim and refute, and shift responsibility to the audience that can construct an argument based on what they see.

An image posted by Tal Giloba on her Facebook page from 2017 depicts a number printed on the bloody flesh of a cow with the caption "no words." The connotations of the number sunken in flesh are clear to any Israeli and are read metonymically as a notorious practice of a sign of stewardship and a murderous industry. This is also the main symbol of the Israeli group *269* that tattoo their skin with numbers "in solidarity with animals that are branded, oppressed, and slaughtered by the industry."[24] Another good example could be found in a massive campaign against the import of animals into Israel and the crowded and miserable conditions of their overseas journey. The successful campaign has become a battle flag for animal rights organizations. Following a huge campaign led by two animal rights groups, Anonymous, currently known as Animals, it was found in a 2018 survey that 86% of the general population object to the shipments.[25] The campaign was accompanied by big billboards, online commercials starring Israeli actors, and huge marches of tens of thousands of people. The export of animals was named in the campaign "Live Exports" with the slogan "Live Exports are Death Exports." Both name and slogan evoke connotations to the death journeys experienced by Jews in the Holocaust and as a couple of interviewees reminded me, these Jews were also taken in cattle wagons. The imagery chosen for the campaign, a cow miserably gazing from between the bars of a cage, similarly arouses identification through the familiar memory of concentration camps and the ghastly train journey.

For the analogy between the historical Holocaust and the animal industry to be accepted and the <Holocaust> to evoke commitment for action, the premise that an animal's life is meaningful has to be accepted. In her field study comparing earlier Israeli vegans with those that converted into veganism around 2014, Weiss[26] identifies a shift in the ethics used to justify veganism and in the vegans' characteristics. Israeli vegans used to adopt veganism as a part of a coherent ideology that focused on humanism and the commonality of suffering between all parts of society. Those who converted to veganism in the past decade, however, tend to come from all parts of the ideological spectrum with more single-issue activists among them. According to her study, this new generation of vegans tended to focus more on arguments related to the agency. Weiss explains that compassion and commonality of suffering that is at the base of humanist thought have been linked with ideas of stewardship. The Bible establishes human ownership over animals due to promise and command given to man to steward and protect the land and animals in their charge. European laws protecting animals, especially domestic ones, are also at the heart of humanist thought and, according to Redfield,[27] can be traced as far back as the seventeenth century. Dogfighting, cockfighting, bull-running, and other forms of the more violent sport were all targeted for humane intervention.[28] Paternalistic care for animals therefore emerged in tandem with ideas of empathy as a guiding force in morality. Weiss mentions in this regard Jeremy Bentham, who articulated the problem of animal protection in a way that has influenced thinking on the humane treatment of animals ever since: "The question is not, can they reason? nor can they talk? but, can they suffer?"[29] This goes hand in hand with the underlying assumption in Western humanist thought that ethics is based on the premise that humans are not meant to suffer.[30] Weiss explains that the ethical approach that relies on suffering has been critiqued by a number of philosophers and anthropologists who claim that its rationality produces and reproduces problematic politics and hierarchies as humans are seen as caregivers and responsible for the mute and suffering animals.[31] Weiss quotes in this regard Didier Fassin[32] who suggested that the focus on suffering displaces other, more vigorous, forms of social justice, such as those based on rights. While Palestinians are perceived as responsible for their actions and terrorist attacks, animals are considered innocent and therefore should not be abused. The transition to ethics of agency enabled, she claims, vegan activists to decouple the vegan battle from the Palestinian one.

In my own analysis I found humanistic arguments based on the commonality of suffering and arguments based on the agency to be intertwined and expanded. Even recognizing animals' suffering was based on the assumption that they can feel pain as well as a variety of other emotions such as motherly love, longing, joy, and sadness. At the same time, the uniqueness of their suffering is underscored by their helplessness and muteness. The official websites and Facebook pages I analyzed display more than information about animals' abuse and suffering. At least half the posts showcase cute animals doing cute things. In one video, animals lovingly hug humans and in another, animals display expressions associated with a range of emotions normally considered human. Tal Gilboa's lecture "Glass Walls"[33] was viewed 2.7 million times and, in contrast to common expectations, does not open with disturbing images of tormented animals, although these will be shown later. Rather, it opens with a close-up on a goat's face while Gilboa narrates the thoughts running through the goat's mind. "You can really see the goat's face," he narrates. He is asking, "why is this happening to me? What have I done to deserve this?" Gilboa does not focus on the goat's tormented death in a religious ritual described above, but on the goat's philosophical musings on the meaning of life. Throughout the lecture, Gilboa calls the animals by familiar Israeli names. At one point we see a chicken lying with her legs spread apart. Gilboa sarcastically explains that "Bracha[34] is not lying in the sun with her legs spread apart because she wants to get suntan, but because she is abused." In other cases, she emphasizes the animals' attempts to escape their imprisonment and torture, once again demonstrating the animals' awareness of their own impending end and their desperate attempts to gain sovereignty. One section is dedicated to what Gilboa calls: "a mother's pain." It begins with Gilboa making a direct comparison between her own experience of motherhood and the cows'. She screens videos of cows acting frantically, running in circles, and gazing in bewilderment and despair as their calves are brutally taken away from them at the moment of birth.[35]

The integration of arguments based on the commonality of suffering and arguments premised on agency was even more prevalent in 2019. Although the focus in most of the Facebook groups and websites has since moved to practical concerns with recipes, restaurants, and petitions, as well as with burning issues before the general elections, most pages are still peppered with the information provided about animal abuse, with images and studies underscoring the animals' affective and intellectual capacities. A recent campaign by the group *Animals* aimed at enforcing "humane treatment of animals in the fish industry."[36] On top of the page in large font the title reads, "Research shows: Fish feel pain." Detailed information is provided in separate tabs about the extremely difficult conditions fish face in the pools, but it is interspersed with "hooks," more amusing videos, and research about the fish's affective capacities. In one video, we follow the journey of a salmon set to the sounds of tango music. Rather than describing his voyage as led by animal instincts, the fish's story is described as being motivated by love and longing. The fish not only swims but is "chasing happiness." He does not simply return to his origin. Instead, much like Odysseus, he overcomes difficulties on his long trip back; "he will swim for 1400 kilometers even against the current, jumping four meters in the air above springs and streams, risking the threat of predators, because he has never forgotten his way home."[37]

When I asked vegans when and why they decided to become vegans, they recalled a moment in time in which they suddenly became aware of the animals' experience, and almost all of them added the following disclaimer to their explanation: "I have become aware of the suffering of animals *or I stopped repressing what I had actually always known*, depending on how you look at it." Many people I talked to wanted to stop consuming animals already from a young age but dropped the idea until it became normalized in the recent decade. Hila, a 25-year-old woman, talked about a gradual process that started at 17 after a visit to the concentration camps with her school:

> This trip and exposure to the holocaust and evil on that scale made me think about animal eating. Although I do not think that people who eat meat are bad, I think that they do bad things. I remember reading about calves and it shocked me. It was a very difficult period; I was exposed to many things that you feel that you need to expose yourself to. That's why in the first years of veganism people talk about it a lot. It just does not feel balanced this thing of taking life in comparison to my meal that ends ten minutes later. It was so unbalanced that it was very clear to me it had to stop.[38]

For Hila, as for others, the exposure to most of the content portraying suffering was deliberate and followed long reflections about the topic. When I asked interviewees about the moment when the veil of denial was lifted and the dots clicked—a few pointed to an episode shown in Yourofsky's lecture in which a cow is being slaughtered. One of the interviewees, a woman in her thirties recalled the most chilling scene: "It was like an assembly line…she (the cow) did not want to die, and tried to escape but they had skinned her even before she completely died."[39] Someone else commented on that same scene: "I realized that animals suffer, that they also want to live…I had animal pets when I was a child but…yes…we repress …we eat and live in the way we are brought up…it's very strong."[40]

The historical Holocaust was characterized by rational, systematized, and corporate planning often associated with utilitarian and efficient modernist thinking. A former hi-tech employee and a current vegan activist and chef told me that "once I heard the terms meat *industry and* dairy *industry* something clicked for me. I worked in tech and performance improvement for years. I knew how humans were being treated. I did not need more than that to finally realize what animals are going through."[41] Some recognized that animals were eaten in the past and they would have perhaps accepted it under different conditions but the scale of death and abuse, the "assembly line" shown in Yourofsky's video made the abstract suffering more concrete; the <Holocaust> as ideograph represented through other means motivated them for personal transformation. Vegan Facebook pages used to evoke in this regard a slogan from the 2011 protests against the high costs of living and the corrupt system called "Fortune-Governance-Mafia." By demonstrating lies and fallacies told by official dieticians, authoritative sources, and corporations, and by exposing links between sources in the industry, they demonstrated its corrupt, systemic corporate nature. In this way, the vegan struggle became metonymical to the myriad unsuccessful challenges in Israel in recent years against corporations and political corruption. However, while victory in other battles is dependent on increasingly corrupt gatekeepers and requires skilled manipulation of courts, the vegan battle, so we are told, will be won when we change our diets. "Only *we* can change the industry. It is about you and you and you!"[42] calls Gilboa in her talk, echoing Yourofsky's emphasis that even if one cannot change the world, they can still significantly contribute by changing themselves and not eating animals.

Veganism, Popular Culture, and Politics

Veganism, popular culture, and politics have become enmeshed in Israel in recent years. The vegan culinary scene in Israel was selected by Reuters as the best in the world,[43] vegan menus have become commonplace in most restaurants in the city and tour operators have started offering vegan culinary tours. Some high-profile individuals such as Yuval Noah Harrari and Tal Gilboa have also helped to normalize veganism. A public intellectual and the author of a few global best-selling books, including his 2011 *Sapiens: A Brief History of Human Kind*, Harari appeared on many popular shows[44] in which he explained the reasoning behind his transition to veganism. An interesting and controversial figure is Tal Gilboa, previously mentioned in this paper, who became a household name after her victory in the 2014 Israeli version of *The Big Brother*. She claimed to have participated in the show only to raise awareness of animals' suffering, and vegans eager to keep the topic in prime time voted her to victory. Her direct style, charisma, and the information she relentlessly brought to the table (pun intended) have been said to have caused a change of heart for many Israelis. In a survey conducted by the popular newspaper Mako, half of *The Big Brother* viewers reported that their perception of meat eating changed as a result of her appearance on the show. About 60% of viewers considered changing their diet, with 37% reporting that they consume less meat.[45] About 5 percent have converted to veganism and 10% to vegetarianism. Gilboa's success even led to an opinion column in *Haaretz*, a newspaper associated with the political left, which compared her surprising and pivoting success to that of Rosa Parks'.[46]

The "new Rosa Parks," however, has been promoting in 2019 the right wing and majority party: the Likud. In her new role as "animal rights' consultant," a job paid for by the Likud's elections department, Gilboa started sharing videos applauding the government's care for animals. In one video, entitled "The Prime Minister and Tal Gilboa Care for the Animals," Gilboa thanks Netanyahu for handling what she calls "the most urgent crisis we are currently facing: the proliferation of street cats and dogs." Netanyahu lowers his voice and asks her, "do you know why we do it? You said it, because we care. We REALLY care." "We really care," Tal echoes him. Netanyahu replies by gesticulating toward his chest, "it comes from the heart."[47] Earlier in the campaign, in a video announcing Gilboa's appointment, Netanyahu explains that he became convinced of the importance of animal care mostly thanks to his dog, Kaya. Netanyahu uses here the same rhetoric of care based on the familiar premise that animals have intellectual and affective capacities, noting that "the ability to see her consciousness, emotions and feelings influenced me very very much."[48]

In another video that became a source of ridicule and contention, Gilboa cooks spicy vegan "fish" (tofu) with the prime minister's wife, Sarah Netanyahu. Netanyahu, a highly controversial figure in her own right, was recently found guilty in court for mistreating her employees.[49] The cooking session begins with Netanyahu's confession about the suffering she went through as a child when she discovered a fish swimming in her mother's bathtub. To console her, the fish was taken back to the store by her mother.[50] The video has circulated with much amusement in the vegan Facebook groups and was mocked also by the popular satirical show, Gav Hauma aka "The Nation's Back."[51] "If Netanyahu really cared for the fish," snarls the host, Lior Schleien, "she would have returned it to nature! To the sea! Not to the store where it was probably bought and smashed in the head by someone else." Throughout the eleven-minute episode, Schleien points to the absurdities of the Netanyahu family's newfound veganism. "How come," he wonders, "was Gilboa appointed as the Prime Minister's personal adviser for animal rights when the Israeli police had been working for months without a chief of staff?" Schleien answers himself, "A chief of police is a profession of the past. In the same way that we stopped needing milkmen when refrigerators arrived, we now do not need police. We have a dictatorship instead." As for the "most urgent problem in Israel – the multiplication of street cats and dogs"—Schleien calls it the second most urgent problem after the recent proliferation of missiles which at the time were launched daily from Gaza. The show juxtaposes the Netanyahus' declarations of care for animals with evidence of their lack of care for humans, neatly exemplified by Sarah Netanyahu's mockery of the Darfur's refugees. Ultimately, Schleien reaches the conclusion that Netanyahu appointed Big Brother winner Gilboa instead of an expert for cynical political purposes. He concludes that "veganism is associated in the public's mind with care, compassion, with helping the weak and this is what the Netanyahu family wants us to think about them."

Schleien's video was popular but was not warmly received by many vegans, who commented on the host's Facebook page that at least the Netanyahu family cares for animals.[52] It is possible that Schleine's claim for "vegan wash" was not accepted due to Schleien's own derision of veganism and lack of compassion toward animals. The analogy he establishes between care for people and for animals was probably not persuasive enough. In most of my interviews and in many of the comments and posts on the topic, it has become clear that for many Israelis care for animals and concern for certain populations in the country are independent topics that should not be and cannot be intertwined. Instead of continuing the narrative that attempts to demolish Netanyahu's authenticity by showing his disrespect for humans, one Facebook post refutes the very claim that he cares for animals. It accomplishes this by listing the "horrible things done to animals during Netanyahu's ten years' reign as a prime minister which include, amongst others, the increase in live shipments and fish farming."[53]

Netanyahu may or may not care for animals, but he is undoubtedly interested in the vegan aura and votes. Appearing to care for the Arab population, however, is not as highly valued in Israel, largely thanks to the prime minister himself. During the 2019 election cycle alone, Netanyahu perpetuated rumors that Arabs were trying to steal the elections and threatened his voters that Israel's doom would arrive if the left won and formed a coalition with the Arabs. These threats influenced his voters because as surveys of the Israeli Center for Democracy revealed, the major breach between Jews and Arabs in Israel is mainly at the national level. Although most Arabs and Jews report good interpersonal experiences with one another, 59% of Jews believe that one cannot identify with the Palestinian nationality while simultaneously remaining faithful to Israel, while 70% of the Palestinians see no contradiction between the two things.[54] Professor Tamar Herman explains that this disparity between Jewish and Palestinian beliefs coupled with the Palestinian leadership's emphasis on Palestinian nationality leads to great distrust on the national level but not necessarily on the interpersonal level between the groups.[55] Former MP Oren Hazan channeled this suspicion and even hatred toward the Arab leadership in a video modeled on the theme of "the Good, the Bad and the Ugly."[56] In the video, Chazan converses with a superimposed real audio text by former MP Zechalka. Chazan mercilessly fires and supposedly kills Zechalka after the latter's assertion that Israelis stole Palestinian lands and that he would prefer to die rather than to sing the Israeli anthem. Indeed, dehumanization of Arabs and even calls for violence have unfortunately become commonplace in Israeli society. Arabs have been compared to a variety of animals, infamously even to "intoxicated roaches in a bottle."[57] Other populations have also experienced hate speech and dehumanization from the right wing and religious parties. Betzalel Smotrich, an MP from the religious party *The Jewish Home* organized in 2006 "The Beast Parade" in which protesters marched along with their donkeys and goats to express their discontent with the pride parade. It was the animals' rights group Let Animals Live that appealed to the supreme court against the event as it concerned the abuse of animals but their appeal was denied.[58] National Rabbi, Ovadia Yosef, called secular Jewish teachers "donkeys" to designate their stupidity,[59] and the former Minister of Culture, Miri Regev, referred to the refugees from Sudan as less than humans or animals by calling them in 2012: "cancer."[60]

Ahmed, an Arab-Israeli vegan activist expressed to me his deep disappointment with Tal Gilboa's collaboration with the Likud and contemplated ways to address his feelings with his many followers and fellow vegans. By the time of the writing of this chapter, he considered publicly stating that it is good to vote for candidates who care for animals, but this candidate should care for humans as well. A critic addressed this issue by creating a survey on another Facebook group entitled "Vegans who vote for a right-wing party do not understand veganism."[61] Three days later, 160 people had disagreed with the statement, while only 30 people agreed. Over a hundred comments reflected once again the complexity of this issue in Israeli society. Most people simply did not understand the connection between politics and veganism, noting the multiple dimensions comprising one's political position. Some expressed anger at the statement and "the left's attempt to own ethics," while others resorted to the rhetoric of agency with claims such as, "I do not hate Arabs, I do not like terror." Someone else commented that he does not think that right-wing voters who are vegans do not understand veganism, but that leftist activists who eat animals do not understand the left. Interestingly, many of the comments took issue with the economic agenda of the left and explained that only the free market's approach is leading the vegan revolution. In a larger Facebook group *Vegans and Interested*, self-report surveys on the parties that participants intended to vote for revealed support mostly to center and left groups with far smaller support to rightwing groups. The administrators for this group have been actively promoting another party, the Democratic Union, who boasted a record number of five vegans in its first fifteen places. The party has made veganism an official part of its green agenda but has not won the support of many other vegans in the group due to what they called the Union's "illusory" fantasies about security and their socialist agenda, which they identified as harming the vegan agenda. In the two weeks leading up to the elections, those Facebook groups were debating the question of veganism in regard to the right and left agendas, with left-leaning people unclear about the disconnect between the rhetoric of care toward animals and its lack thereof toward humans, and right-wing activists mad at the attempt to link the two issues while remaining unclear about how one can identify themselves as empathetic toward other people while remaining ignorant of the suffering inflicted on animals.

In her op-ed in the online left-wing magazine *972*, Shenhav Goldenberg claims that veganism serves as a fig leaf for many Israelis. Veganism, she claims, is the easier mainstream choice which enables Israelis to resolve their dissonance about the Palestinians' situation in the occupied territories.[62] Shenhav points to a study on the social and political implications of the occupation in Israel.[63] According to this research, to dull their guilt and maintain their perception of themselves as good people, Israelis use various mechanisms, such as avoiding exposure to information on the topic. When I asked right-wing vegans about politics, they were confused, and when I tried to explain the relevance of this question by pointing to the condition of the Palestinians in the occupied territories, they were even more confused and angry. Quite a few of them asked, "What occupation?" This answer should be read from the Israeli perspective in which the occupation has become legitimized and questioning it has become delegitimized to the point that it is rarely discussed in the media as such. When I asked these interviewees about their media consumption, they agreed that they do not believe the media or the politicians. Most of them said that they get the bulk of their information from Facebook. I did not see conscious guilt, but partisan self-exposure to media which can be read as information bias and a mechanism of avoidance.

In "States of Denial: Knowing about Atrocities and Suffering,"[64] the criminologist and sociologist Stanley Cohen examines different levels of denial, from internal denial to state-wide or official forms of denial. By providing examples of the many atrocities we refuse to think about, Cohen examines the processes by which we cope with this denial. The narrative structures of denial emerge from a cultural pool of explanations that "follow the same internal logic."[65] Ignoring information that conflicts with ours is one method that enables us to preserve our moral dignity. Cohen provides a rich analysis of appeals made by human rights groups to examine the ways in which people nevertheless change their minds and hearts and commit to action. While one might think that providing information about the extent of suffering would help change minds, this is not always the case. In fact, exposure to information that contradicts our previous values might lead us to ignore information presented all together.[66]

Acknowledgment of other humans who suffer requires, according to Cohen, that one resonates with ideals such as universality, solidarity, and empathy. Following Cohen's logic, it is understandable why it might be easier for some people to identify with animals because sensitivity to their suffering requires empathy but there is no need for universality and perception of equality. "Animals are not like humans," said Or, a 26-year-old vegan and a student of criminology. She continued:

> It is true that veganism means defending the weak, but animals have no voice. They cannot protect themselves. It's the same when we talk about women' rights. I am a feminist but I do not perceive of myself as a victim in the same way. If someone beats me, I can still go to the police. It is not like a puppy that was beaten. As for the Palestinians, I was raised during the Intifada and they have earned their stereotype… I do think that they need to be given more rights, but it's not the same as animals' helplessness… I do the best that I can to help the suffering I am aware of. I cannot help Africa or the Palestinians right now. That suffering and evil exist in one place does not mean that other places need to be ignored.[67]

Or's emphasis on agency echoes another answer I was given: "Animals taught me that we need to be strong against the Hamas or we will end up like them. Those who cannot protect themselves suffer."[68] The Israeli lesson of the <Holocaust> is once again evoked to emphasize the importance of strength. Perception bias is reflected here in the analogy one strikes: while left-wing people identify with the strong and want to protect the weak—be it animals, refugees or the Palestinians; right-wing people tend to identify with the strong when it comes to animals but with the weak and under threat when it comes to other populations.

It is in this light that we should also see the huge uproar in some vegan circles when it became known that a spokesperson for an animal rights organization protested against the IDF's actions in Gaza. When this information was shared by the extreme right-wing group *Im Tirzu*, responses on their page ranged from acceptance to anger at the "traitor who gives vegans a bad name and risks the IDF by demonstrating against their actions."[69] The stereotype that all vegans are politically left should be dropped, they claimed in the comments section. This same incident was resurrected from the internet's dark corners and used before the 2019 elections to demonstrate the subversive and dangerous links certain vegans make between veganism and politics.

The discourse about veganism in the IDF can demonstrate the complex Israeli approach to morality and politics. On the 2019 Vegan Day holiday, the IDF published on its Twitter page the following caption: "with over 10,000 vegan soldiers (one in every 18 soldiers) we are proud to be the most vegan army in the world."[70] The caption began circulating in mostly right-wing Jewish and Israeli outlets in English with headlines such as "The Most Vegan Army in the World"[71] or "The Army That Has the Most Moral Diet."[72] A *Guardian* article on vegans in the US army addressed the possible discrepancy between military life and veganism: "I'm living in a world of violence by being in the military but trying to live the most peaceful lifestyle that I can…. Choosing not to be violent in my everyday life when I don't have to be is something I wholeheartedly say falls in line with my religious beliefs *and* military values."[73] None of the articles covering the topic in Israeli media has directly or indirectly dealt with the implied paradox, probably because for many Israelis, there is no paradox. Service in the army is obligatory and the catchy title paraphrases the slogan that so many believe in: "the most moral army in the world." However, as in the case of Schleien, some of my left-leaning interviewees accused the IDF of using veganism for image embellishing purposes, noting that information on the topic is often provided in English and thus pandering to the non-Israeli audience.

The IDF began accommodating vegan soldiers in 2014 by offering faux leather boots and barrette hats. Vegan soldiers' conditions have greatly improved after a former vegan soldier, currently a student, founded a front to advance the vegan soldiers' rights in 2018. Some of their recent achievements include the lifting of bureaucratic hurdles for vegan soldiers,[74] a richer variety of vegan meals, and the rights of vegan chefs to refrain from using any animal products in their work. These milestones were shared across Israeli media in Hebrew and in English. The activists note the surprising help they have been receiving from the food service in the army, but when I asked one of the Front's activists about the possible discrepancy between veganism and the army, he acknowledged that most vegans serve in supporting roles such as the intelligence "although some of the warriors are vegans which is great." He paused for a second and continued:

> I learned that you cannot predict how someone would behave in different dimensions of one's life because we repress things all the time. Layers and layers of repression. Someone may choose to stop repressing one thing but may still not understand what he does with the Palestinian population in Gaza because other factors influence him to keep repressing it. This is why so many vegans are on the right (politically). But on the left people may fight for human rights and speak very passionately about harmless populations that require support, yet…you know (eat animals). This is also what I have learned from being exposed to soldiers' stories. They can be very compassionate towards animals but not necessarily very compassionate towards people in the checkpoints.[75]

In Israel, it might be easier to lift the veil of ignorance and repression when the culprits are clear and the solution is feasible and seemingly depends only on our individual actions. Even as the growing vegan community becomes diversified in its political orientations and even if veganism has been used for political purposes—the problem and solution are much clearer than in other fights.

Associating a clear culprit with a clear solution means not only that other human rights issues can be considered irrelevant but also that large scale economic questions are not perceived as a part of the problem. While I have not seen much use of the <Holocaust> ideograph in 2019, I did see the results of the simple problem and solution equation that was established by vegan groups in earlier stages of the battle. Almost all interviewees revealed deep faith in the power of their consumer choices.

While most people expressed distrust in the government and in big corporations, most people did not feel a need to boycott them. On the contrary, as we previously saw in the comment sections, many expressed beliefs that the free market and their personal choices exclusively move the wheels of the vegan revolution. In a recent article on the left website *Local Talk,* author Michael Sapir laments this narrow approach to veganism.[76] While vegan diet is important, he claims, vegans need to take an extra step and engage big corporations in their struggle. Sapir explains that although it appears that the consumption of dairy and meat had shrunk in some Western countries, it has actually increased around the world. This is due in part to the fact that consumer choices do not enforce meaningful change in the food industry. While big companies now offer vegan alternatives (e.g., Alpro's soy milk) they also increase their efforts in other countries that traditionally have not been great consumers of animal industries. "While we have cleaned our own hands from blood and the direct link to the killing, our money is being used for these purposes exactly," comments Sapir. He concludes that while individual consumer choices are important, they need to be leveraged by demands on behalf of the vegan and environmental movements to change policies.

Discussion

Veganism was normalized in Israel relatively early, in comparison to other countries, due to successful integration into the culinary scene and popular culture. Vegans have been able to focus on particular goals and rally opinion leaders and a great number of supporters for the cause. Their visible success was in turn taken by more established institutions such as political parties. Since most Israeli vegans seem to have adopted this lifestyle due to moral rather than dietary concerns, veganism has become a badge of compassion and was therefore used as a political currency by different parties across the spectrum, mostly the right-wing majority party The Likud. This rhetorical move was made possible due to the fact that veganism decoupled itself in recent years from political issues related to humanistic principles and care writ large toward other groups of people.

Vegan rhetoric has focused on animal rights and offers a crystal-clear problem and solution model according to which the meat and dairy industry abuse animals unnecessarily and on a large scale. The solution offered is based on individual action, in other words, dietary change. The model established a zero-sum argument as one must choose whether to participate in the abuse of animals or to refrain from it, there being no middle ground.

Vegans have mustered this argument through a focus on effect. The suffering inflicted on animals and the animals' affective capacities, consciousness, and agency established the premise according to which animals are as worthy of living as any humans. The singularity of animal suffering was marked by their muteness. Interviews revealed perception bias; where left-leaning vegans have seen commonalities between the suffering of animals and humans, right-wing individuals empathized with the suffering of animals but their helplessness reinforced for them the old lesson learned in the <Holocaust> in which one should stand strong in the face of injustice from others.

The use of the <Holocaust> as an ideograph and the use of labels such as "industry" further highlighted the cruelty targeted at animals and framed it as meaningless, systematic, and as metonymic for other social battles. Veganism stood out as a battle that is singular in its capacity to save lives and moreover as a battle in which the culprits are clearly evil while the solution is feasible. This simplicity stands in stark contrast to other political disputes in the region which are considered more complex and for which a solution is not perceived as realistic or possible.

For many vegans, veganism serves as a lighthouse of a battle that can be won through individual action alone. Many vegans expressed doubt in the importance of boycotts on big corporations and did not see veganism as in any way related to other political issues. By the time last revisions for this paper were made, fires in the Amazons and Australia have led to important changes in vegan rhetoric in Israel as veganism has become increasingly tied to environmental concerns. Social and political issues, however, have remained inconsequential as even left-leaning vegans expressed a desire to keep veganism outside of that game so that possible converts would not be dismayed. As one comment made in an online forum stated, "I don't agree with your right-wing agenda, but at least you don't eat animals." Although right-wing media and politicians embrace veganism as a badge of morality, I do not think that we should critique veganism merely as a foil for other ethical concerns (such as the conflict with the Palestinians.) Vegans I talked to were committed to save lives but regardless of their sincerity, vegan activism has become a place where different groups that often resent one another (right and left, Arabs, and Jews) can collaborate around a shared cause. I strongly believe that we should build bridges and continue to normalize veganism and other environmental concerns and solutions instead of defining them a bipartisan issue. However, once the importance of these matters has been accepted, the rhetoric and the lens through which we view them should be expanded. This might be the time to point to corporate and governmental responsibility to care for all beings and the environment instead of focusing only on individual action.

Notes

1. The Central Bureau statistics of Israel (2019, p. 93).
2. Refael (2018).
3. Ibid.
4. The groups' names have been translated from Hebrew to English for this paper.
5. Unlike Hinduism, for example, Judaism and Islam, the most prominent religions in Israel, are not vegetarian. However, in both traditions there is a rich variety of vegetarian and nondairy alternatives. In recent years, a growing number of people have also started interpreting Jewish texts as calling for the restriction of meat.
6. The lecture was also subtitled into Arabic and I have interviewed people who were greatly influenced by it as well.
7. Never Again (2011).
8. Dina, Interview with author (June 25th, 2017). Names of all interviewees in this chapter have been changed in order to respect their privacy.
9. Shira, Interview with author (June 30th, 2017).
10. Yoram, Interview with author (July 1, 2017).
11. McGee (1980).
12. Condit and Lucaites (1993, xii–xiii).

13. McGee (1980).
14. Condit and Lucaites (1993).
15. Long (2013).
16. Never Again (2011).
17. @Levy (April 23, 2017).
18. Animals (n.d).
19. Alpher (2014).
20. Taub (2017).
21. Yoram, Interview with author (July 1, 2017).
22. Janis and Winkler (1997).
23. Delicath and DeLuca (2003).
24. @269LifeISrael. n.d. "About," Facebook: 269Life Tel-Aviv, Israel. https://www.facebook.com/pg/269lifeIsrael/about/?ref=page_internal
25. Arlichman (2018)
26. Weiss (2016).
27. Redfield (2013).
28. Ritvo (1987).
29. Bentham (2007 (1789), 311).
30. Redfield (2013).
31. Ticktin (2015, 49–71).
32. Fassin (2011).
33. Glass Walls (2015).
34. A common Israeli name that means 'blessing.'
35. Glass Walls (2015).
36. Animals (2019a, b).
37. Ibid.
38. Hila, Interview with author (July 30, 2019).
39. Dana, Interview with author (July 12, 2019).
40. Ron, Interview with author (July 15, 2019).
41. Doron, Interview with author (July 16, 2019).
42. Glass Walls (2015).
43. Cohen (2015).
44. For example, Harari appeared on the very popular at the time "London and Kirshenbaum: Never Again," "Yuval Noah Harari on London and Kirshenbaum: A Fascinating Interview," Channel 10, Israel, Youtube Oct 22, 2013. https://www.youtube.com/watch?v=jpexDNZNsog
45. Nikolivski (2014).
46. Tal-Shir (2014).

47. Netanyahu (2019a).
48. Ibid.
49. Levi (2017).
50. Netanyahu (2019b).
51. Gav Hauma (2019).
52. Schleien (2019).
53. Storch, Ofer (2019). Facebook comment – Vegans and Interested group, Sep 7th, 2019, https://www.facebook.com/groups/veganism.israel/permalink/2444974018884349/?hc_location=ufi
54. Herman (2019).
55. Ibid.
56. Bender (2019).
57. This statement was made back in the 80's "by former IDF general, Refael Eytan who later claimed he had never said it.
58. Forsher and Hegin (2006).
59. Sela (2007).
60. Azoulai and Efraim (2012).
61. Abu Nimr, Mahmoud. 2019. "Survey – Vegans Who Vote," Facebook Survey – Vegans in Israel Group, Sep 4, 2019. https://www.facebook.com/groups/1498383493788874/permalink/2133622616931622/
62. Shenhav-Goldenberg (2019).
63. Halperin, Bar-Tal, Sharvit, Rosler, and Raviv (2010).
64. Cohen (2001).
65. Cohen (2001, 177).
66. Cohen (2001, 265).
67. Or, Personal interview with author (July 19 2019).
68. Yoram, Personal interview with author (August 10, 2019).
69. Im Tirzu (2018. Facebook Post, Dec 27th, 2018. https://m.facebook.com/imtirtzu/posts/2094165250641097).
70. @IDF, "With over 10,000 vegan soldiers we are proud to be the most vegan army in the world." Twitter, Nov 1st 2018.
71. Ahronheim (2018); Sones (2018); Channel 7 (2018).
72. Peretz (2014).
73. Krantz (2019).
74. Zeitun (2019).
75. Dan, personal interview with author (Aug 15, 2019).
76. Sapir (2019).

Bibliography

Ahronheim, Anna. 2018. The Most Vegan Army in the World. *Jerusalem Post* 3 Oct. https://www.jpost.com/Israel-News/The-most-vegan-army-in-the-world-568595

Alpher, Rogal. 2014. Vegans Can Be Compared Also to Nazis. *Haaretz*, June 9. https://www.haaretz.co.il/opinions/.premium-1.2341972

Animals. 2019a. Fish Story. Animals' Facebook page. https://www.facebook.com/animal.org.il/videos/2355812204526710/?t=50

———. 2019b. Join Those Who Signed the Petition. Facebook petition. https://animals-now.org/investigations/fish/?group=d17&fbclid=IwAR3QXNf90nk0p-ZOlEGYLvbvwcujeXEfQf6DUkoGsN2_i3DGICce2fkt908, Accessed 1 Sep 2019.

———. n.d. Everyday Is Treblinka. *Animals*. https://anonymous.org.il/art55.html

Arlichman, Aryeh. 2018. Survey: 86% of the Population Object to Live Shipments. *Ynet*, July 5. https://www.ynet.co.il/articles/0,7340,L-5304573,00.html

Azoulai, Moran, and Omri Efraim. 2012. Regev Called the Infiltrators Cancer: The Situation Encourages Violence. *Ynet*, May 24. https://www.ynet.co.il/articles/0,7340,L-4233655,00.html

Bender, Arik. 2019. Following Election Video—Zechalka file law suit against murder. *Maariv*, March 19. https://www.maariv.co.il/elections2019/news/Article-690253

Bentham, Jeremy. 2007 [1789]. *An introduction to the principles of morals and legislation*. Mineola, N.Y.: Dover.

Bot News. 2019. Tal Gilboa Supports the Likud, Would Advice the Prime Minister about Animals. Youtube, June 11. https://www.youtube.com/watch?v=5b2EtDV4Yu0.

Channel 7. 2018. The Army of Vegans for Israel. *Channel 7*, Oct 3. https://www.inn.co.il/News/News.aspx/383258

Cohen, Stanley. 2001. *States of Denial: Knowing about Atrocities and Suffering*. Malden: Polity Press and Blackwell Publishers.

Cohen, Tova. 2015. In the Land of Milk and Honey, Israelis Turn vegan. Reuters, Life Style, July 21. https://www.reuters.com/article/us-israel-food-vegan/in-the-land-of-milk-and-honey-israelis-turn-vegan-idUSKCN0PV1H020150721

Condit, Michelle, and John Louis Lucaites. 1993. *Crafting Equality: America's Anglo-African Word*. Chicago: University of Chicago Press.

Delicath, John W., and Kevin Michael DeLuca. 2003. Image Events, the Public Sphere, and Argumentative Practice: The Case of Radical Environmental Groups. *Argumentation* 17: 315–333.

Fassin, Didier. 2011. *Humanitarian Reason: A Moral History of the Present*. Berkeley: University of California Press.

Forsher, Efrat, and Adi Hegin. 2006. The Supreme Court Decided and the Beast March Will Take Place. *nrg*. Nov 9. https://www.makorrishon.co.il/nrg/online/1/ART1/503/334.html.

Glass Walls. 2015. Glass Walls – The Lecture. Youtube, May 22. https://www.youtube.com/watch?v=bV4-tnq1hZY. Retrieved on Sep 9 2019

Halperin, Eran, Daniel Bar-Tal, Keren Sharvit, Nimord Rosler, and Raviv Amiram. 2010. Socio-psychological implications for an occupying society: The case of Israel. *Journal of Peace Research* 47 (1): 59–70.

Gav Hauma. 2019. Netanyahu – The Vegan Family. Youtube, Sep 1. https://www.youtube.com/watch?v=d17T2iBlO_U

Herman, Tamar. 2019. Hate Speech? Reality Proves Differently. The Israeli Institute for Democracy website, Nov 7. https://www.idi.org.il/articles/28869

Janis, Edwards L., and K. Carol Winkler. 1997. Representative Form and the Visual Ideograph: The Iwo Jima in Editorial Cartoons. *Quarterly Journal of Speech* 83 (3): 289–310. https://www.tandfonline.com/doi/abs/10.1080/00335639709384187.

Krantz, Rachel. 2019. Green Zone: The Vegan Soldiers Are Fighting for Vegan Food. *The Guardian*, Feb 26. https://www.theguardian.com/us-news/2019/feb/26/us-military-vegan-plant-based-food-mres

Leichman, Abigail Klein. 2017. Israel Has the Largest Vegans Per Capita and the Trend Is Growing. *Israel 21ct,* March 27. https://www.israel21c.org/israel-has-most-vegans-per-capita-and-the-trend-is-growing/

Levi Zohar, Shachar. 2017. The Supreme Court Rejected Sarah Natanyahu's Appeal-Forced to Pay Court Expanses to Mani Naftali and Guy Eliyahu. *Calcalist*, July 7. https://www.calcalist.co.il/local/articles/0,7340,L-3716973,00.html.

Schleien Lior. 2019. Bibi Sarah and Yair Love Animals, Bibi Sarah and Yair Love Animals, Why Don't They Stop Talking About It? Here Is My Opinion. *Facebook*, Aug 29. https://www.facebook.com/lior.schleien/videos/394822178108596/?t=36

Long, Kelly. 2013. Terrorism in the Age of Obama: The Rhetorical Evolution of President Obama's Discourse on the "War on Terror." *Undergraduate Review* 9 (10): 90.

Matar, Hagai. 2013. Can Animal Rights Take Precedence Over Human Rights? *+972*, November 12.

McGee, Michael Calvin. 1980. The "Ideograph": A Link Between Rhetoric and Ideology. *Quarterly Journal of Speech* 66: 1–16.

Netanyahu. 2019a. Prime Minister Netanyahu and Tal Gilboa Care for the Animals. Youtube, Aug 21. https://www.youtube.com/watch?v=slyI57uGUxU

———. 2019b. What Do the Prime Minister's Wife, Sarah Netanyahu and Tal Gilboa Cook? Youtube, Aug 22. https://www.youtube.com/watch?v=-Zbnzrs-BqM.

Never Again. 2011. Most Important Lecture You Will Ever Hear. Youtube, April 3. https://www.youtube.com/watch?v=omweihtaYwI&t=5s

———. 2013. Yuval Noah Harari on London and Kirshenbaum: A Fascinating Interview. Channel 10, Israel. Youtube Oct 22. https://www.youtube.com/watch?v=jpexDNZNsog

Nikolivski, Yair. 2014. Tal's Big Victory, Half of *The Big Brother* Viewers Do Not Look at Meat in the Same Way. *Mako, The Big Brother,* Aug 27. https://www.mako.co.il/tv-bigbrother/season6-articles/Article-b9e18cc83431841006.html

Peretz, Osnat. 2014. The Army That Has the Most Moral Diet in the World. *Yediot Acharonot,* Sep 4. https://www.yediot.co.il/articles/0,7340,L-5321894,00.html

Redfield, Peter. 2013. *Life in Crisis: The Ethical Journey of Doctors Without Borders*. Berkeley: University of California Press.

Refael, Rina. 2018. Will 400,000 Vegans Be Added in Israel in the Next Seven Years? *Calcalist*, April 4. https://www.calcalist.co.il/articles/0,7340,L 3735187,00.html

Ritvo, Harriet. 1987. *The Animal Estate: The English and Other Creatures in the Victorian Age*. Cambridge, MA: Harvard University Press.

Sapir, Michael. 2019. Vegan Consumerism Would Not Stop Climate Change but It Certainly Would Help. *Local Talk*, July 27. https://www.mekomit.co.il

Sela, Neta. 2007. The Rabbi Ovadia: Secular Teachers are Donkeys. *Ynet*, Dec 2. https://www.ynet.co.il/articles/0,7340,L-3477837,00.html

Shenhav-Goldenberg, Rachel. 2019. By Going Vegan Israelis Can Avoid Talking About Human Rights. *972*, May 22. https://972mag.com/veganism-israel-occupation-denial/141572/.

Sones, Mordechai. 2018. The Most Vegan Army in the World. *Aroutz Sheva*, Oct 3. http://www.israelnationalnews.com/News/News.aspx/252681

Tal-Shir, Rachel. 2014. A vegan Revolution via *The Big Brother*. *Haaretz, Food Section*, Sep 4. https://www.haaretz.co.il/food/eatornot/1.2424543

Taub Gadi. 2017. I Find It Really Strange. Facebook, June 21. https://www.facebook.com/gadi.taub/posts/10154733780742151

The Central Bureau Statistics of Israel Health Survey. 2019. *Annual Report on Health and Nutrition on Adults 18–64 Years Old Between 2014–2016*, 93. https://www.health.gov.il/publicationsfiles/mabat_adults_2014_2016_383.pdf

Ticktin, Miriam. 2015. Non-Human suffering: A Humanitarian Project. In *The Clinic and the Court: Law, Medicine and Anthropology*, ed. Ian Harper, Tobias Kelly, and Akshay Khanna, 49–71. Cambridge: Studies in Law and Society.

Weiss, Erica. 2016. There Are No Chickens in Suicide Vets: The Decoupling of Human Rights and Animal Rights in Israel. *Journal of the Royal Anthropological Institute*. 22: 688–706.

Zeitun, Yoav. 2019. After a Struggle and Hundreds of Complaints: The IDF Will Provide Hundreds of Colorful Vegan Barrettes. *Ynet, Army and Security*, July 7. https://www.ynet.co.il/articles/0,7340,L-5552190,00.html

4

Veg(etari)anism in Serbia: Attack on Traditional Values

Mirjana Uzelac

Introduction: Vegetarians and Meat Eaters in Serbia

Meat is a staple of Serbian traditional cuisine. In many ways, meat is a synonym for Serbian food. It is seen as a necessary part of a "proper" meal and a tradition that needs to be protected (Baltic et al. 2018). This does not leave much room for alternative dietary choices, such as vegetarianism. Being a vegan or a vegetarian in contemporary Serbia is not easy, and people who practice this lifestyle encounter numerous challenges. These problems range from difficulties in finding appropriate food to social misunderstandings and conflicts with friends and family.

It is not easy to find statistics about percentages of people in Serbia who practice vegetarianism, but those in most Eastern European, postsocialist countries are below European averages (Leahy et al. 2010). Serbia is a postsocialist country that is still struggling with economic transition

M. Uzelac (✉)
University of Alberta, Edmonton, AB, Canada
e-mail: muzelac@ualberta.ca

C. Hanganu-Bresch, K. Kondrlik (eds.), *Veg(etari)an Arguments in Culture, History, and Practice*, The Palgrave Macmillan Animal Ethics Series,
https://doi.org/10.1007/978-3-030-53280-2_4

after the breakup of Yugoslavia (Greenberg 2011). This insecure economic situation, unemployment, and other problems make everyday life difficult for most citizens, which is a trend in many postsocialist countries (Chelcea and Druta 2016; Ghodsee and Mead 2018). After the breakup of Yugoslavia, more and more people adopted a nationalist attitude that praises Serbian traditional culture (Jansen 2000). Food, attitudes about hospitality, and celebrations are important aspects of this collective identity. The vast majority of Serbian population are meat-eaters, and they do not necessarily have positive attitudes toward those who adopt vegan and vegetarian lifestyle. The reasons for this antagonism are rooted in Serbian traditional culture (or what people in Serbia perceive as such), particularly in terms of relationship with the West. In this research, I explore those reasons: why people in Serbia dislike veg(etari)ans, and how they contextualize veg(etari)an diet and lifestyle as the antonym of "Serbianness."

This chapter is based on the analysis of online comments made in Serbian Internet forums, news websites and online magazines, as a response to articles that focus on vegetarianism. The research was performed through non-participant virtual ethnography in early to mid-2019. Virtual ethnography provided rich material for research; Internet forums and article commentaries proved to be a valuable opportunity to observe people who held different beliefs, attitudes, and values (Eisewicht and Kitschner 2015; Hallett and Barber 2014; Hine 2000; Kozinets 2010). This is important for the issue of vegetarianism because online forums and article commentaries attracted people from both sides of the vegetarianism debate. It was an opportunity for both the proponents and the opponents of this lifestyle to interact and give their opinions.

I chose to follow news and articles written in the last ten years, or those that generated interest during that period. I purposely aimed to examine forum posts and news comments that appeared highest in search engine results. The main search term was "*vegetarijanci u Srbiji*" ("vegetarians in Serbia"). Among the high-ranking results, I focused on those that generated at least 30 comments made in the last ten years. In this way, I chose: two newspaper articles about difficulties that Serbian vegetarians face; a topic "vegetarianism: yes or no?" from a travel forum; two news items[1] about parents struggling to make a kindergarten provide vegetarian meals for their children (one news item was a Facebook article); a news report about the number of vegetarian restaurants in various European countries (including Serbia); and an online "confession" page with comments. News articles were published by prominent Serbian online portals,[2] while forum discussions and confessions chosen were the ones that generated the most replies. This is not a complete corpus of articles and online discussions about vegetarianism in this period. Rather, I focused on comments and discussions that revolved around common themes about vegetarianism, particularly vegetarianism in the Serbian context. The exact articles were not of specific importance, nor did I aim to collect everything that was said online about vegetarianism in the last ten years. I focused on the dominant themes, opinions and assumptions that were commonly repeated over different articles, particularly those that revealed attitudes that non-vegetarians had about vegetarians in Serbia.

In citing Facebook comments, I included a username (or the first name) and the year the comment was made. I used only publicly available comments in my analysis. For the most part, I did not frame the analysis around specific newspaper articles or forum themes, since the subject itself did not particularly matter in terms of the type of comments it generated.[3] Regardless of the article or news item, there were "standard" types of comments, revolving around issues of health, reasons for choosing a vegetarian lifestyle, and the "strangeness" of such lifestyle. Many comments revolved around the idea that vegetarianism goes against "Serbianness" and local tradition, which constitutes the main focus of this chapter.

It is also important to note that many people used the terms "vegetarians" and "vegans" interchangeably, so I treated both together. Specific differences are noted when important, but for the most part, commenters, including vegans and vegetarians, did not insist on differences. This is why I treat all of these lifestyles under the umbrella term of "vegetarianism."

Unsurprisingly, the comments revealed that meat-eating is a norm in Serbia, with veg(etari)anism being an alternative lifestyle that is not fully accepted. Meat has an important role in Serbian cuisine, so those who avoid it often run into a risk of being labeled "not really Serbs." There was a notable traditionalist and nationalist element that prevailed in many comments made by the majority (meat-eaters). Because of this, I choose to focus more on those who positioned themselves as the "Serbian norm" (meat-eaters) than on vegetarians themselves. However, comments made by vegetarians were not ignored and were used to illustrate "the other side" and to complement themes and attitudes proposed by the majority.

"It's Their Choice, But..."

Comments on media articles about vegetarians revealed that, on the surface, Serbian people view vegetarianism as a "personal choice." Also, many commenters claim that they respect such personal choices and that people have a right to choose vegetarianism if it suits them. However, this seemingly neutral attitude was quickly made more complex by people's elaboration of their opinion. Many commenters were quick to point out reasons why vegetarianism is bad and vegetarians are strange or spoiled. The fact that vegetarianism might be someone's personal choice did not mean that such a choice would be respeted: it was often ridiculed and condemned.

At the same time, articles about vegetarianism and veganism also attracted proponents of the lifestyle, who shared their opinions and experiences. Typically, articles contained a mix of positive and negative comments, with at least one vegetarian participating in the discussion. These discussions often turned into arguments, typically over health benefits of vegetarianism versus meat-eating. Issues such as: which type of nutrition is more balanced, whether people are biologically made to be omnivores, carnivores or herbivores, physical strength of meat-eaters versus vegetarians, what kind of nutrients humans need, and so on, were common across all platforms. For the most part, motives for meat abstention voiced by Serbian vegetarians, such as refusal to eat animals, health benefits, dangers of food industry practices, and ethical issues, were very similar to motives chosen by vegans and vegetarians around the globe (Fox and Ward 2008; Rajaram and Sabaté 2000; Rozin 2004).

Here are some of the comments expressing a common "I respect vegetarians, but..." attitude:

> I respect vegetarianism (and any other choice made by other people in any way imaginable) as long as it doesn't affect me… it is one thing to be a vegetarian and mind your own business, and a different thing is to insist for others to be vegetarians and to try to persuade a gourmand that being a gourmand is bad. (Zeljo_bg 2011)

> OK, I respect when someone decides to do this, but I don't think there is much logic to it. Meat is to be eaten; not to mention all the nutrients and beneficial ingredients [...] I don't know why anyone would give it up. Nonsense. (Laza GNR 2011)

> The biggest problem is that they are forcing and pushing their lifestyle onto the others. I have nothing against vegans; everyone has the right to live a life they want and eat what they want, but if they have chosen that lifestyle they should not look at it as something special and to make victims out of themselves. (Segovia 2017)

> Nobody threatens them, nor are they forced to practice their "orientations" between their four walls and similar; but they have a problem that people make fun of them? Boohoo, I am so sorry they didn't win a right to be untouchable, something that doesn't belong (nor should belong) to any group in the world. (Vlada 2011)

Despite "live and let live" attitude that many people express, there is a tendency among meat-eaters to frame their own food choices as the norm, or the best way to enjoy food. Enjoyment of food and "being a gourmand" were associated with meat-eating, and often positioned as important factors surrounding the food debate. This is not surprising: in a recent (2014) study examining the motives for food choices among Serbian consumers, taste and sensory appeal of food were highlighted as the most important factors, while ethical concerns were ranked among the least important (Gagić et al. 2014). Furthermore, vegetarians were often accused of presenting their food choices as "something special" and themselves as "victims," which is in line with common "vegan killjoy" narrative (Twine 2014).

Vegetarians and vegans agree that there is a low level of acceptance for this lifestyle in Serbia. They also agree that it is not easy for them to find adequate food. Some vegetarians were enthusiastic about a few vegetarian/vegan restaurants popping up in larger cities,[4] but they agreed that the situation is still far from good. They reported difficulties in finding adequate foods in supermarkets and other stores; these foods exist but choices are extremely limited. Restaurants pose a particular challenge, since so many do not have any vegetarian or vegan items on the menus. Serbian national cuisine restaurants have a standard offer of dishes, most of which include animal products, such as *ćevapčići* (a minced meat dish similar to kebab) and *pljeskavica* (meat patty). Plant-based food options exist, but these are typically appetizers or side dishes (Stojanović and Čavić 2018), such as *prebranac* (baked beans), *proja* (cornbread), salads, and fried vegetables.[5] Furthermore, vegetarians complained about family and friends not showing understanding for vegetarian and vegan lifestyles, which created significant social problems. Vegans and vegetarians who participated in discussions agreed that their friends and family members often expressed negative and hostile opinions. It was not uncommon for friends and family to openly condemn a veg(etari)an lifestyle choice and to try to persuade the person to change:

> "You don't even eat fish?" "Chicken breast…?" and similar questions… I just answer those with a question: "Fish and chicken are not meat?" It is true about alcohol; it is fortunate when you are a driver, so you always have an excuse not to drink, but when they get a hold of someone who doesn't drive, heavy persuading [to drink] ensues. (andrea 2014)

> I hide the fact that I am a vegetarian and I feel, in a way, that the others would discriminate against me and make fun of me… Most of those who find out, try to "save" me or something. Even a doctor had the same attitude; he attacked me out of nowhere, insisted to tell him my reasons. […] In Serbia, being a vegetarian makes your life more difficult. (ashgdahdgah 2012)

> People used to look at me like I'm an alien and ask: "are you one of us (a Serb)?" "Poor guy, what do you eat?" But they got used to it by now. Unfortunately, restaurants did NOT improve at all… a shame and disgrace. (Macola 2012)

> If someone is lucky that the friends and family don't bother them about their choice, and if there is enough culture and politeness to respect other's opinions, everything else is more or less bearable. (Sandra 2012)

> I generally HIDE that I am a vegetarian. Of course, everybody finds out sooner or later, but I do my best not to emphasize this. I had so many bad situations in which someone tried to "save" me. People asked me so many questions: from what kind of illnesses I have, to cults and various disorders. (Mirjana 2012)

Shame, uneasiness, and being forced to hide their food choices are common themes for vegetarians in Serbia. This uneasiness mainly comes from expectations imposed by friends and family, and from general Serbian attitudes about "proper food." Vegans and vegetarians often feel that they have to hide their identity in fear of people's reactions.

If these comments and complaints sound like Serbs take other people's food choices personally, it is because they do. In part, it is because they feel judged by vegetarians or feel that their own food choices are questioned (Kluever Romo and Donovan-Kicken 2012). On a more direct level, someone's nutrition does have an influence on other people, because of traditions and social norms of Serbian hospitality.

Tradition of Hospitality

Hospitality is one of the things Serbs praise the most about Serbian culture. Eating is often considered a group activity, particularly during celebrations, and there are specific roles expected from a host and a guest. Serbs invite others into their homes often, and food has a central place. To feed a guest, and to feed them well, is imperative, particularly during important celebrations. The host's obligation is to provide varied foods in several courses and in large quantities. On the other hand, the guest's obligation is to enjoy such food and to eat it, preferably in large quantities (Čajkanović 1973:72).

Comments about vegetarianism are often contextualized through the lens of Serbian hospitality. Providing food for a guest in one's home becomes more complex if the host or a guest is a vegetarian:

> I had many vegetarians as guests; it was perfectly fine. I eat my own; they eat theirs :) A problem is if someone starts to lecture, and if they complain in your own home because they can smell meat, or if they force you to walk a couple of kilometers so they can find something that suits them. (Chivitli 2011)

> My opinion is that vegetarianism is perfectly OK. The only thing I dislike is a spoiled, fanatic vegetarianism. Like, you are hungry, there is nothing to eat without meat, and then […] the whole city needs to be searched until we find something for you. Or, you come to someone's home as a guest, he makes a pie and you say: oh no, I don't eat that. (Lazar 2011)

> It is bad etiquette on both sides. I also expect, if a vegan is my host, that he prepares something that I eat; meat, for example. Also, in my home, that I will prepare for him something that he can also eat. Those vegans who expect people to prepare things specially for them but give guests only food that they eat are not OK; same goes for people who give vegans meat. (Anonymus 2017)

> What kind of a twisted logic is that those who eat meat can eat your vegan food, but you can't eat theirs? And so they need to make an effort to be good hosts to you, and in return you give them whatever you prepare for yourself. What if someone doesn't want your food just like you don't want theirs? They will remain hungry? Well done. You are egoists. (Anonymus 2017)

> If someone invites you over to their home for a lunch, you either eat or don't eat what you are served. I don't know who raised you to whine and make demands in another person's home to prepare food for you? You are just demonstrating your impudence and lack of manners. (Anonymus 2017)

> They get offended when there is not a meal specially prepared for them. I have a Muslim friend, who believes that he should eat whatever a host offers to him, even if it's pork. So, who is crazy here? (Anonymus 2017)

> Seriously, anyone can eat whatever they want, but when a vegi comes to my home, he will remain hungry. I am sorry. On the other hand, I would never accept an invitation for a vege dinner, even if it meant I would stay hungry. Especially if they would bore me with "healthy" food conversation. (Gillan 2011)

> I really don't understand the source of irritation. This is called being a good HOST. Those who think that it's beneath them, or that it's too much trouble to prepare something for a guest that fits the guest's religious/ethical/health/etc. needs, should never invite anyone ever. Unbelievable; we lost the basic notion of hospitality. (Anonymus 2017)

Vegans and vegetarians who refuse to eat animal products pose a specific problem in the light of a highly valued traditional hospitality. The fact that these guests do not eat meat is perceived as a direct complication, or even hindrance, to hosts exerting proper Serbian hospitality. In Serbia, the host always prepares food; a guest bringing a dish is considered offensive. It is the host's responsibility to provide the key element of any Serbian celebration: the abundance of food. Feeding the guest goes beyond food preparation; hosts will often carefully monitor how much is eaten and offer more. The guest should make an effort to try as many dishes and to eat large quantities of food. Repeatedly asking whether a guest wants more and not taking "no" for an answer is a staple of Serbian celebrations. Being pushy and putting more food on a guest's plate despite objections is a common occurrence. Once it is on a plate, it is rude not to eat it, which both the host and guests know well. Eating small amounts of food, or only eating certain dishes, is not considered the best scenario; depending on the occasion and closeness between hosts and guests, it can be taken as a minor annoyance or a serious disrespect.

In these situations, vegetarian guests easily get labeled spoiled or disrespectful. While hosts might accept the fact that someone refuses to eat animal products in their personal life, this is not necessarily seen as a good enough reason to skip meat dishes when invited over for a meal. By refusing animal food at celebrations, veg(etari)an guests refuse the host's hospitality. This can be seen as a serious violation of social rules, particularly if the hosts view vegetarianism as a caprice or unreasonable demands made over a "whim." Of course, these situations can be avoided if the hosts are willing to prepare special meals for their vegetarian guests, or if they do not get offended if the guest skips certain dishes. These scenarios do happen, particularly when hosts and guests are close, and the meal/hospitality is less formal. However, as the above comments indicate, there are many of those who value traditional hospitality much higher than someone's dietary or lifestyle choices.

Kindergarten Vegetarians—A Child Abuse?

Negative attitudes about vegetarianism are particularly harsh when children are involved. In late 2018, media reported a case of vegetarian parents who complained that their children did not get vegetarian meals in the kindergarten.[6] Since rules and regulations nominally forbid any type of discrimination, and since children are (again, nominally) guaranteed balanced meals in the kindergartens, the parents complained. One of their main arguments was that the kindergarten forbade them from preparing and bringing meals to their children, which they were ready to do. Their plea reached the news and, after a lengthy persuasion, they managed to secure vegetarian meals for their children. While some parents praised this move, admitting that they, too, had a problem securing vegetarian meals for their children, many comments expressed strongly negative opinions. The parents were accused of neglect and child abuse, with many commenters highlighting the apparent hypocrisy of limiting children's freedom of choice:

> Those children didn't choose to be vegans. If they don't want to, parents don't have a right to force them into it. [...] Their parents are the ones who are discriminating against them by forcing them to do something they might not want to do. So... bullshit. (Katarina 2019)
>
> A child vegan? You make me puke. (Dragana 2019)

> Just like you talk about freedom of choice, those children are forced into vegan lifestyle by their parents, which means that those children don't have freedom of choice!!! (Stefan 2019)

> I don't understand this whole attack on the kindergarten. First, a child cannot decide not to eat meat, which means it is a parent's caprice. […] Plus, a kindergarten never offers only meat for lunch; there is always a side dish, so the child can eat a side dish and fruits and salads. And should eat at home what the parents prepare. (mica 2018)

Comments that viewed parents' vegetarianism as a "caprice" were common. These commenters often questioned whether a public institution (kindergarten) should fulfill such "whims":

> Who doesn't like what the state offers for a discounted price, should find a cook to care for his child and prepare meals a la carte. (Zoran Kg 2018)

> Kindergartens don't have enough funds, and now they are forced to fulfill everyone's caprices. (mica 2018)

> This is not about differences; nobody cares what someone eats. It is about someone demanding that the food is cooked specifically for their child. […] Kindergartens have (limited) funds for food, they sure won't fulfil parent caprices. (oca 2018)

However, some commenters pointed out that being a parent means choosing things for your children, such as religion or a meat-based diet:

> Parents represent their children until adulthood. Similarly, children are often forced into eating meat and other animal products, out of tradition, and not because of health reasons. (Урош 2018)

> Nobody asks children if they want to be religious. And we all know that children are often forced to eat all kinds of foods, even when they dislike it. This is what parents do; it's always been like that. But it's normal for you when parents make their children observe religious holidays or when they

> make them eat meat. But when parents make a "different" choice for their children, it's a problem for you. (Anonymus 2018)

To which a commenter replied:

> Religious choice cannot have health consequences, while lack of certain foods can. So, it should be forbidden by law for unreasonable parents to cause consequences for their children's health through forced vegetarianism. (Lucifer 2018)

"Think of the children" comments emphasize the importance of balanced nutrition; since vegetarianism is often perceived as unhealthy even for adults, commenters felt strongly that such nutrition is damaging and inadequate for children. Again, those sorts of complaints are hardly unique for Serbia (Cofnas 2019). However, in this case, these attitudes are heightened by the fact that vegetarianism goes against the tradition. As mentioned by some commenters, parents represent their children and (legally) make choices for their children. This is not something that is widely debated in Serbia when it comes to socially acceptable and traditional choices, but it is seen as harmful in the case of choices that go against Serbian traditional values or against accepted social norms. It is also interesting to note that many commenters framed this discussion as parents' caprice (vegetarianism as a fad) and parents' hypocrisy (discriminating against their children by forcing them to be vegetarians).

Vegetarians, Cults, and Gay Rights

Many comments made by those who are against vegetarianism contain references to cults (*sekte*). These comments reflect the attitudes about vegetarianism and veganism in which these lifestyles are perceived as being "sectarian" in nature. Veg(etari)ans are seen as a dangerous group that engages in harmful, obsessive behaviors:

> Who says that consuming meat is unnecessary??? Is this some new cult (*sekta*)? (bjuti 2011)

> (In reference to a vegetarian book that another commenter praised) That author totally brainwashed you; it's a cult (*sekta*). (Dragance 2011)

> Well said. There is nothing worse to me than those cult members (*sektaši*) who look at everyone from the above just because they eat differently… (Anonymous 2017)

> So, those of us who eat meat, are the enemies? And whose enemies, if I may ask?! What kind of chauvinism is this? […] If you want to eat only plants – eat, but don't annoy me. This story is too pushy and culty to me. (Svetozar Markovich 2016)

References to cults carry a specific weight in the Serbian context. Calling something "a cult" (*sekta*) is more than a reference to its alleged obsessive nature. It is a direct implication of something being against the proper religious beliefs (religious sect). Lifestyles and behaviors labeled cultists are seen as a direct opposite of the Serbian Orthodox beliefs (Kuburić 2008). As such, they are perceived to be particularly dangerous and damaging to everything that traditional Serbia stands for. The trend of comparing groups, lifestyles, and behaviors as a "cult" became common in the 1990s (Đorđević 1999), after the breakup of Yugoslavia and during the rapid rise in nationalist sentiments. Foreign, different, and the unknown remains to be labeled in such a way even 25 years later.

Vegetarianism and vegetarian rights are sometimes perceived in the same light as other marginalized groups who seek their rights, particularly LGBT people. Those who are critical of LGBT rights, voice similar complaints about vegetarians:

> It looks like we will soon have a vegan pride, too…. And since their lobby is getting stronger, soon there will be a prohibition for meat eating in public places… :D (Gillan 2017)

> What kind of a bullshit is this? What discrimination against vegetarians? Does anyone attack them on the streets because they eat plants? Do they get hate mail? Did a politician announce that they are second class citizens? Does anyone forbid them to eat plants? What, some of their friends make

> fun of them because they eat plants and it suddenly means they are discriminated? (mnogo vike nizašto 2011)

> Great! Now I can use this case for my needs. We only eat quail's eggs and organic tomato at home. We believe that is the healthiest choice. (Borac 2018)

> It seems that the understanding and acceptance of "vegetarian/vegan" nutrition is on the level of LGBT acceptance. (Ma hajde, BRE! 2012)

> If someone pushes something onto you, of course you will voice your opinion and attitude. Vegan activist? [...] I've never heard of a straight alliance or a "normal nutrition" activist. I don't care what anyone does in their life as long as I don't have to look at it, don't know about it and it's not forced at me at every corner! (ns 2011)

> I am sitting here and waiting to see what's next, which twisted head will fight for his "rights".... I don't think anything can surprise me anymore. (Stefan 2011)

> I have a feeling that this will become a condition to join the EU at one point... :) (Aleksandar 2016)

These opinions go beyond equating veg(etari)an rights with other human rights, or complaining about both. LGBT rights are hotly debated subject in Serbia, with a high degree of homophobia (Bilić 2016). While typically framed as being against family values or Serbian Orthodox beliefs (Valić-Nedeljković et al. 2017), the issue of LGBT rights is often rooted in "foreignness." This is seen as something imposed from the West and pushed against Serbian tradition. (Kuhar 2013) It is not necessarily about people choosing same-sex partners—it is perceived as a powerful culture (the West) pushing its ideas and values onto the powerless. In an act that is seen as imperialistic and colonizing, the West imposes its rules onto others. Gay pride parades (Mikuš 2011), control of locally produced alcoholic drinks[7] or vegetable shapes[8] are often seen in the same light: as a power game against those that lack power. It is not unusual for these ideas to be seen as hypocritical; many people in Serbia do not believe that Westerners "really" care about those issues. In this sense, the insistence on gay rights or vegetarian rights is seen as a dishonest power trip on the side of the West.

Eastern Orthodox Fasting and Vegetarianism

For many people, vegetarianism is seen as incompatible with Serbian tradition and values. This is particularly true due to high emphasis on meat dishes and food hospitality as a marker of Serbianness. At the same time, it is important to note that there is a segment of Serbian traditional cuisine that is compatible with the vegan lifestyle: Lenten fasting.

The vast majority of Serbs are Eastern Orthodox Christians; religion is often tightly related to national identity. By expressing themselves as Orthodox Christians, people display and prove their Serbianness. One of the most important aspects of Eastern Orthodox Christianity is fasting done during Lent. There are four Lent periods throughout the year, and Lenten fasting is required from Eastern Orthodox Christians. While not the only aspect of Lent, fasting is an important aspect of it, and the one that people in Serbia most commonly practice. During fasting, Orthodox Christians are not allowed to eat any meat, except fish and seafood on specific days. In addition to this, other animal products, such as milk and eggs, are also forbidden. The long periods of Lent and fasting mean that Orthodox Christians are forbidden to eat meat for over 200 days per year. Food suitable for fasting (*posno*) does not contain meat, milk, and eggs. Despite meat being emphasized as the most important food group in Serbia, there exists a great interest in fasting food. These things are not seen as the opposites, since both, in their own ways, fit the idea of traditional Serbian food culture.

There is a complex history between tradition, ethnicity, and religion in the region. Because atheism was strongly encouraged during socialism, many Serbs only (re)learned Eastern Orthodox Christianity after the breakup of Yugoslavia. Furthermore, Church and religion were always strongly attached to ethnicity in the Balkans; more than religion itself, Serbs view Serbian Orthodox Church as a marker of their ethnicity (Raković 2015). Being an Orthodox Christian is strongly attached to

Serbian national identity. There is a high degree of performative religiosity that might not match people's actual religious beliefs. Because of these complexities, it is not easy to tell how many people in Serbia truly observe Lent and practice everything it includes: frequent prayer, going to church, fasting, and spiritual cleaning. However, this is not of particular importance here; what is important is that fasting is something that many people in Serbia (claim to) practice. There is enough interest in food suitable during fasting (*posno*), and this food is readily available in stores throughout the year.[9] "*Posno*" is a common label on many products in Serbia, which guarantees (in theory, at least) that the product is free from animal products, milk and eggs. In recent years, more and more restaurants followed suit, so many menus contain meals with the "*posno*" label.

It is not surprising that Serbian vegans and vegetarians often seek food labelled "*posno*" to fulfill their own dietary needs. "*Posno*" is a way for vegans and vegetarians to find appropriate foods more easily and without much questioning. Some vegetarians admit in online comments that they seek products labelled "*posno*" or that they order such meals in restaurants. In a situation where veg(etari)an choice is extremely limited, *posno* comes in handy and as a helpful label for Serbian vegans and vegetarians to use.

Fasting dietary rules are close to vegan dietary rules so much that Eastern Orthodox Lenten fasting is sometimes called "periodic vegetarianism" (Puskar-Pasewicz 2010). Fasting is referred to as "periodic vegetarianism" in (Western) journals, with numerous studies examining the effect of Orthodox fasting on the organism (Morcos et al. 2013; Sarri et al. 2005, 2007). It is not unusual for Orthodox fasting to be used as an example of religious vegetarianism, along with Buddhist and Hindu vegetarian traditions. However, fasting is rarely, if ever, referred in such a way in Serbia. Religious Serbs and vegetarians alike generally agree that *posno* and vegetarianism are not the same thing. These two terms are considered to be so different that they are rarely connected; indeed, many Serbs will never say that Serbia has a "vegetarian tradition," let alone a religious one. However, this connection is mentioned from time to time, often by vegetarians seeking appropriate foods, or, at a rarer chance, by religious people themselves.

One of the best illustrators of this attitude is a post published on website Ispovesti.com. "*Ispovesti*" (confessions) is a website allowing people to post their anonymous confessions for others to comment on. While it is not possible to verify that the website's users are primarily from Serbia, there are many references to the Serbian Orthodox Church and Serbia as "we" and "ours" to allow for its inclusion.

An anonymous confession from November 2017[10] says:

> I am totally annoyed by these new Vegans. I just want to let them know that nutrition without any animal products existed before them, and it used to be called, and it's called today: FASTING (*posno*). I don't see that people who fast expect from their hosts to make food specially for them. In addition to this, I don't see that they, if there is not specially prepared food for them, take that as an attack on them or their religion, the way vegans see if hosts don't prepare "vegan food". Also, I don't see a need to label products, or in cafes and restaurants "vegan" when it's always been "*posno*" [...] You could stop copying the West for once; try to impose something on them, and not just let them impose things on you.

This confession generated a large number of replies. Both vegetarians and non-vegetarians participated. Typical arguments over the benefits and harms of vegetarianism were exchanged, along with common complaints. However, what makes this discussion particularly interesting is the equation of vegan food with fasting food (*posno*). This is not a commonly made connection, which caused numerous replies, on both sides:

> I disagree. Lent is a physical, emotional and spiritual restraint and cleaning; avoiding meat is not lent. The fact that you don't eat meat and fat food doesn't mean that you are fasting. I don't like veget[arians] either, but don't mix things. (Blacky 2017)

> There are so many things wrong with this confession. So, you think it's fine if someone doesn't eat animal products because of religious reasons, but if they don't eat because of moral reasons, you are bothered by it? None of it has anything to do with the West that you (all) attack so much, and even if it does, why wouldn't we take positive things from the West? (Anonymus 2017)

> Lent and veganism can't be the same in any way. [...] To observe lent, the person needs to believe in God and follow certain rules. A vegan is just a person who doesn't eat animal food and that's all. (An 2017)

> Fasting is one thing and veganism is another. [...] I am a vegan and I've been trying for years to teach people that just because something is *posno*, doesn't mean that I eat it, and they keep offering it to me. (cicos micos 2017)

> This makes no sense whatsoever. [...] Fasting is done for religious reasons. Veganism is just a way of nutrition. (niko 2017)

> I see absolutely no harm in labeling something as *posno* and something as vegan, because some things that are *posno* are not vegan, such as fish. (Veg (prvi dio) 2017)

> People forget that in Serbia you have a bunch of fasting (*posno*) food, and that they are the best for local people, because of the climate. I noticed with many vegetarians a tendency to emphasize some Eastern vegetarian meals and ignore ours and spend huge amounts of money on all of that. [...] when Serbian vegetarians [...] ignore *posno*, food prepared and tested for centuries, it says a lot. (Gosn.shofershajbna 2011)

Ideas about hospitality are used once again to emphasize the proper behavior in terms of guests' demands, food choice for the guests, and host obligations:

> It was clear in the past: someone observes lent, someone doesn't. But those who are fasting never got angry if the food is not *posno* because they know there is always someone who doesn't do fasting. And those people didn't run to everyone and said: "hey, I am fasting; I this, I that." You are a vegan, I am not, and why do you care that I eat meat? I don't criticize you for not eating it. (niko 2017)

> In my family, it's always been that if you invite someone over who is fasting, to prepare *posno* food for that person, buy *posno* cookies etc. Similarly, when someone invites me, a vegetarian, as a guest, they prepare something that I can eat. That is basic etiquette and not a trend. (Veg (prvi dio) 2017)

I see no point in inviting someone over for a lunch and then to prepare something that the person doesn't eat. "*Posno*" is not the same as vegan, because most people eat fish during fasting. Fasting is related to religion, always, and veganism is not. Labels on products are there so you don't have to read all the ingredients every time. And why would anyone impose anything to the "West"? I get annoyed by people who, like you, believe that other people's lives are their business. (Lena 2017)

"Us" Versus "Them": Vegetarianism as a Display of Foreign Values

It is clear that religious people and vegetarians do not see *posno* in the same light as vegan food, nor do they perceive each other as belonging to the same group of people. *Posno* and veg(etari)an might be similar in terms of foods and ingredients, but they rest on completely different contexts. *Posno* is part of Lenten fasting, and, as such, part of Serbian Orthodox tradition. Vegetarianism carries completely different connotations and the opposite context. *Posno* is part of the tradition, of Serbianness; vegetarianism is a synonym for the foreign and anti-Serbianness.

The analysis of online comments showed that Serbian people, in part, share common attitudes and concerns about vegetarianism with many others around the world (Twine 2014). A lot of the emphasis is put on discussions regarding health and ethical issues around meat consumption. However, many comments focus on specific Serbian issues regarding vegetarianism, such as hospitality, traditional food, Eastern Orthodox fasting, and Serbian values. Comments displayed a great degree of collectivism: everybody should eat the same, a group should not accommodate one person, a kindergarten should not prepare a special meal for one child, and so on. Vegetarians are often seen as people who display their individual whims at the expense of the group: they are seen as spoiled, picky eaters, doing things out of a whim and caprice and making unreasonable demands. These people are perceived as not "playing for the team" and for putting their individual preferences over the needs of the group. A common sentiment among Serbian commenters is that the group should not change to accommodate a single person. On the contrary: individuals should be ready to adjust their preferences (for example, food choices) in order to accommodate a group. Vegetarians are perceived as ignoring this social consensus. This is in line with collectivist sentiments in which individual choices can disturb the group dynamic, particularly in cultures that put a lot of emphasis on food, meal sharing, and hospitality (Yoo and Yoon 2015).

In this sense, vegetarians in Serbia break the established cultural roles between a host and a guest. As a result, they are met with hostility. Part of this hostility comes from the abovementioned collectivism that vegetarians ignore through their personal choices (Yoo and Yoon 2015). The other part comes from the violation of Serbian cultural norms: through their dietary choices and behaviors, vegetarians are perceived as attacking the established social order and "proper" ideas about food. Many hostile comments focus on criticizing vegetarians for not doing things the way "we" do them and for ignoring the "Serbian way." Vegetarians and vegetarianism are treated as synonyms for the foreign, for the opposite of "Serbianness" and for power imbalances attributed to Western hegemonies. It is not a coincidence that images about gay rights, European Union, and Western values are often evoked. Vegetarianism is seen in the same light, as something that is being forced from the West into Serbian culture. Those who adopt vegetarianism are seen as "not proper Serbs," as someone buying into harmful Western propaganda that destroys Serbian traditional values.

Serbia's attitude toward the West is ambivalent (Volčič 2005) and often antagonistic, particularly in the light of postsocialist, nationalist shift after the 1990s. During the years of Yugoslav socialism, the region enjoyed a relative peace and good relations with both the Eastern and the Western bloc. For the most part, socialist Yugoslavia enjoyed more freedom of movement and international respect from both sides than other socialist countries. It all changed after the breakup of Yugoslavia in the early 1990s and war(s) in Croatia, Bosnia, and Kosovo. Serbia faced economic embargo and hyperinflation during the early 1990s and Milošević's regime. The West is commonly seen as an antagonistic, anti-Serb force, particularly in the light of the embargo and 1999 NATO bombing. In the light of these sentiments, it is not surprising that anything perceived to be "from the West" is seen as harmful, dangerous, and the opposite of traditional Serbian values.

This is why it was so important for many commenters to make a distinction between "*posno*" and vegetarian food. The food in question might be similar (or virtually the same in many cases), but these two terms evoke completely different sets of values and contexts. In terms of tradition and Serbianness, *posno* and vegetarians are the opposing terms representing two conflicting principles. In this sense, vegetarianism is not seen as bad simply because of the food choices, but because it is seen as an attack on Serbian traditional values.

Notes

1. Both were reports about the same incident.
2. B92 (http://b92.net), N1 (http://n1info.com), 021 (http://021.rs).
3. The main exception are articles about parents who fought to have vegetarian meals for their children in the kindergarten.
4. Malenica, Branka. 2017. "Veganski vodič kroz Novi Sad," Moj Novi Sad, December 2, 2017. http://www.mojnovisad.com/vesti/veganski-vodic-kroz-novi-sad-foto-id19340.html

   "Vegani U Zemlji Mesa: Ovo su najbolji veganski restorani u Beogradu," Restorani Beograd, February 9, 2017. https://www.restoranibeograd.com/rs/aktuelnosti/vegani-u-zemlji-mesa-ovo-su-najbolji-veganski-restorani-u-beogradu/
5. Potatoes might be a choice, but not necessarily: many are fried in the same oil as the meat, or generally together with meat.
6. Adria News. 2018. "Deca vegetarijanci ipak će dobijati obroke u vrtiću," N1, September 19, 2018. http://rs.n1info.com/Vesti/a421222/Deca-vegetarijanci-ipak-ce-dobijati-obroke-u-vrticu.html

   Radio Televizija Srbije. 2018. "Vegetarijanci u vrtiću – egzibicija ili pravo na izbor?," RTS, September 23, 2018. http://www.rts.rs/page/stories/sr/story/125/drustvo/3266880/vegetarijanci-u-vrticu%2D%2Degzibicija-ili-pravo-na-izbor.html and
7. D.Z. 2013. "I Nakon Ulaska u EU: Srbi Će Moći Da Peku Rakije Koliko God Hoće!" Telegraf, October 14, 2013. https://www.telegraf.rs/vesti/846827-i-nakon-ulaska-u-eu-srbi-ce-moci-da-peku-rakije-koliko-god-hoce

LOTIS. 2018. "Evropska unija i srpska šljivovica," Kruševac Press, August 1, 2018. https://krusevacpress.com/evropska-unija-i-srpska-sljivovica/

8. Serbian farmers were appalled by The European Commission standard (now revoked) to prevent irregularly shaped fruits and vegetables from being sold in Europe.
9. Generally speaking, products labelled "*posno*" are available even outside of lent. For example, cookies and margarine are typically available in several variants, some of which are labelled *posno*. The same is true for many other products: they are readily available throughout the year.
10. 2304118. 2017. Ispovesti, November 14, 2017. https://ispovesti.com/m/ispovest/2304118

Bibliography

B92. 2009. Povratak povrća nepravilnog oblika. *B92*, July 1. https://www.b92.net/zivot/vesti.php?yyyy=2009&mm=07&dd=01&nav_id=368859. Accessed 13 Aug 2019.

———. 2016. Evo koliko je vegetarijanstvo zastupljeno u Evropi (MAPA). *B92*, October 14. https://www.b92.net/zivot/vesti.php?yyyy=2016&mm=10&dd=14&nav_id=1187896. Accessed 1 Aug 2019

Baltic, Milan Z., Jelena Janjic, Milka Popovic, Tatjana Baltic, Vesna Jankovic, Marija Starcevic, and Danijela Sarcevic. 2018. Meat in Traditional Serbian Cuisine. *Meat Technology* 59: 54–62. https://doi.org/10.18485/meattech.2018.59.1.7.

Bilić, Bojan. 2016. Whose Pride? LGBT 'Community' and the Organization of Pride Parades in Serbia. In *The EU Enlargement and Gay Politics*, Gender and Politics, ed. Koen Slootmaeckers, Heleen Touquet, and Peter Vermeersch, 203–220. London: Palgrave Macmillan.

Boyle, Joseph Edward. 2007. *Becoming Vegetarian: An Analysis of the Vegetarian Career Using an Integrated Model of Deviance*. Unpublished doctoral dissertation, Polytechnic Institute and State University.

Bulić, Dušan. 2007. Vegetarijanac u Srbiji. *B92*, October 15. https://www.b92.net/info/komentari.php?nav_id=267829. Accessed 8 Aug 2019.

Čajkanović, Veselin. 1973. *Mit i religija u Srba*. Beograd: Srpska književna zadruga.

Chelcea, Liviu, and Oana Druţa. 2016. Zombie Socialism and the Rise of Neoliberalism in Post-Socialist Central and Eastern Europe. *Eurasian Geography and Economics* 57 (4–5): 521–544.

Cofnas, Nathan. 2019. Is Vegetarianism Healthy for Children? *Critical Reviews in Food Science and Nutrition* 59 (13): 2052–2060. https://doi.org/10.1080/10408398.2018.1437024.

D.Z. 2013. I Nakon Ulaska u EU: Srbi Će Moći Da Peku Rakije Koliko God Hoće! *Telegraf*, October 14. https://www.telegraf.rs/vesti/846827-i-nakon-ulaska-u-eu-srbi-ce-moci-da-peku-rakije-koliko-god-hoce. Accessed 10 Aug 2019.

Đorđević, Dragoljub. 1999. Anti-Cult Movement: A Possible Expression of Religious Intolerance. *Facta Universitatis – Philosophy, Sociology, Psychology and History* 6: 257–270.

Eisewicht, Paul, and Heiko Kitschner. 2015. Giving in on the Field: Localizing Life-World Analytic Ethnography in Mediatized Fields. *Journal of Contemporary Ethnography* 44: 657–673. https://doi.org/10.1177/0891241615595437.

Fox, Nick, and Katie Ward. 2008. Health, Ethics and Environment: A Qualitative Study of Vegetarian Motivations. *Appetite* 50 (2–3): 422–429. https://doi.org/10.1016/j.appet.2007.09.007.

Gagić, Snježana, Ana Jovičić, Dragan Tešanović, and Bojana Kalenjuk. 2014. Motives for Food Choice Among Serbian Consumers. *Economics of Agriculture* 1/2014: 41–51.

Ghodsee, Kristen R., and Julia Mead. 2018. What Has Socialism Ever Done for Women? *Catalyst: A Journal of Theory & Strategy* 2 (Summer, 2): 100–133.

Greenberg, Jessica. 2011. On the Road to Normal: Negotiating Agency and State Sovereignty in Postsocialist Serbia. *American Anthropologist* 113: 88–100.

Hallett, Ronald E., and Kristen Barber. 2014. Ethnographic Research in a Cyber Era. *Journal of Contemporary Ethnography* 43: 306–330. https://doi.org/10.1177/0891241613497749.

Hine, Christine. 2000. *Virtual Ethnography*. London: Sage.

Ispovesti. 2017. Komentari. *Ispovesti*. https://ispovesti.com/m/ispovest/2304118

Jabs, Jennifer, Jeffery Sobal, and Carol M. Devine. 2000. Managing Vegetarianism, Identities, Norms and Interactions. *Ecology of Food and Nutrition* 39: 375–394. https://doi.org/10.1080/03670244.2000.9991625.

Jansen, Stef. 2000. *Antinationalism: Post-Yugoslav Resistance and Narratives of Self and Society*. Unpublished PhD dissertation, Hull.

Klub putnika. 2011. Vegetarijanstvo – za i protiv. Klub putnika. http://www.klubputnika.org/forum/topic/4454-vegetarijanstvo-za-i-protiv/

Kluever Romo, Lynsey, and Erin Donovan-Kicken. 2012. "Actually, I Don't Eat Meat": A Multiple-Goals Perspective of Communication About Vegetarianism. *Communication Studies* 63 (4): 405–420. https://doi.org/10.1080/10510974.2011.623752.

Kozinets, Robert. V. 2010. *Netnography: Doing Ethnographic Online*. Thousand Oaks: Sage.

Kuburić, Zorica. 2008. Images of the Religious Other in Serbia. In *Images of the Religious Other: Discourse and Distance in the Western Balkans*, ed. Christian Moe, 167–198. Novi Sad: CEIR.

Kuhar, Roman. 2013. Introduction to the Issue: "In the Name of Hate: Homophobia as a Value". *Southeastern Europe. L'Europe du Sud-Est* 37 (1): 1–16. https://doi.org/10.1163/18763332-03701001.

Leahy, Eimear, Seán Lyons, and Richard S.J. Tol. 2010. National Determinants of Vegetarianism, ESRI Working Paper No. 341. Dublin: The Economic and Social Research Institute (ESRI).

LOTIS. 2018. Evropska unija i srpska šljivovica. *Krusevac Press*, August 1. https://krusevacpress.com/evropska-unija-i-srpska-sljivovica/. Accessed 15 Aug 2019.

Malenica, Branka. 2017. Veganski vodič kroz Novi Sad. *MojNoviSad*, December 2. http://www.mojnovisad.com/vesti/veganski-vodic-kroz-novi-sad-foto-id19340.html. Accessed 15 Aug 2019

Mikuš, Marek. 2011. "State Pride": Politics of LGBT Rights and Democratisation in "European Serbia". *East European Politics & Societies* 25 (4): 834–851. https://doi.org/10.1177/0888325411426886.

Morcos, Nadia Y.S., Dina M. Seoudi, Iman Kamel, and Marian Youssef. 2013. Effect of Coptic Orthodox Christian Church Fasting on Healthy and Diabetic Subjects. *International Journal of Nutrition, Pharmacology, Neurological Diseases* 3 (4): 375–382. https://doi.org/10.4103/2231-0738.119853.

N1. 2018. Deca vegetarijanci ipak će dobijati obroke u vrtiću. *N1*, September 19. http://rs.n1info.com/documents/1149383/comments/Vesti/Deca-vegetarijanci-ipak-ce-dobijati-obroke-u-vrticu.html. Accessed 15 Aug 2019.

Puskar-Pasewicz, Margaret. 2010. *Cultural Encyclopedia of Vegetarianism*. Santa Barbara: Greenwood.

Rajaram, Sujatha, and Joan Sabaté. 2000. Health Benefits of a Vegetarian Diet. *Nutrition* 16: 531–533.

Raković, Slaviša. 2015. Secular and Orthodox Subjectivities in Serbia: A Trajectory of a Conflict over the Notion of the West. *Етнолошко-антрополошке свеске* 26 (н.с., 15): 7–28.

Restorani Beograd. 2017. Vegani U Zemlji Mesa: Ovo su najbolji veganski restorani u Beogradu. *Restorani Beograd*, February 2. https://www.restoranibeograd.com/rs/aktuelnosti/vegani-u-zemlji-mesa-ovo-su-najbolji-veganski-restorani-u-beogradu/. Accessed 15 Aug 2019.

Rozin, Paul. 2004. Meat. In *Encyclopedia of Food*, ed. Solomon Katz, 666–671. New York: Scribner.

Sarri, Katerina, M. Linardakis, Siobhan Higgins, and Anthony G. Kafatos. 2005. Is Religious Fasting Related to Iron Status in Greek Orthodox Christians? *The British Journal of Nutrition* 94: 198–203. https://doi.org/10.1079/BJN20051472.

Sarri, Katerina, Manolis Linardakis, Caroline Codrington, and Anthony Kafatos. 2007. Does the periodic vegetarianism of Greek Orthodox Christians benefit blood pressure? *Preventive Medicine* 44(4): 341–348. https://doi.org/10.1016/j.ypmed.2006.11.009.

Sloboda za zivotinje. 2018. Važno: saopštenje povodom diskriminacije dece vegetarijanaca i vegana u predškolskim ustanovama. *Facebook,* September 18. https://es-la.facebook.com/SlobodaZaZivotinje/photos/va%C5%BEno-saop%C5%A1tenje-povodom-diskriminacije-dece-vegetarijanaca-i-vegana-u-pred%C5%A1kols/10155772971025936/. Accessed 13 Aug 2019.

Sovilj, Miodrag. 2011. Vegetarijanci i dalje ne nailaze na razumevanje. *021,* August 10. https://www.021.rs/story/Novi-Sad/Vesti/29017/Vegetarijanci-i-dalje-ne-nailaze-na-razumevanje.html. Accessed 13 Aug 2019.

Stojanović, Danijela, and Slobodan Čavić. 2018. Critical Analysis of the Gastronomy Offer of the Serbian National Cuisine Restaurants. *Tourism in Function of Development of the Republic of Serbia, Tourism in the Era of Digital Transformation,* Thematic Proceedings II. University of Kragujevac.

Twine, Richard. 2014. Vegan Killjoys at the Table—Contesting Happiness and Negotiating Relationships with Food Practices. *Societies* 4: 623–639.

Valić-Nedeljković, Dubravka, R. Ruard Ganzevoort, and Srđan Sremac. 2017. The Patriarch and the Pride: Discourse Analysis of the Online Public Response to the Serbian Orthodox Church Condemnation of the 2012 Gay Pride Parade. https://doi.org/10.1007/978-3-319-43406-3 5.

Volčič, Zala. 2005. The Notion of 'the West' in the Serbian National Imaginary. *European Journal of Cultural Studies* 8 (2): 155–175. https://doi.org/10.1177/1367549405051842.

Yoo, Taebum, and In-Jin Yoon. 2015. Becoming a Vegetarian in Korea: The Sociocultural Implications of Vegetarian Diets in Korean Society. *Korea Journal* 55 (Winter, 4): 111–113.

5

Ancient Text, Modern Context: Patanjali's *Yoga Sutras* and the Twenty-First Century Veg(etari)an

Sharon Lauricella

Lisa: You don't eat cheese, Apu?

Apu: No, I don't eat any food that comes from an animal.

Lisa: Oh, then you must think I'm a monster!

Apu: Yes, indeed, I do think that. But I learned long ago, Lisa, to tolerate others rather than forcing my beliefs on them. You know you can influence people without badgering them always. It's like Paul's song, "Live and Let Live."

Paul McCartney: Actually, it was "Live and Let Die."

Apu: Whatever, whatever. It had good rhythm.

The Simpsons. "Lisa the Vegetarian." Season 7, Episode 5. October 15, 1995.[1]

S. Lauricella (✉)
Ontario Tech University, Oshawa, ON, Canada
e-mail: sharon.lauricella@ontariotechu.ca

C. Hanganu-Bresch, K. Kondrlik (eds.), *Veg(etari)an Arguments in Culture, History, and Practice*, The Palgrave Macmillan Animal Ethics Series,
https://doi.org/10.1007/978-3-030-53280-2_5

Like Lisa Simpson and her neighborhood friend Apu, many vegetarians are asked to address both their conviction and conversion to becoming and staying vegetarian. Some people become vegetarian or vegan after they are affected by an event (such as Lisa's visit to a petting zoo in *The Simpsons*), health reasons (such as a desire to reduce cholesterol or increase fiber), media (including documentaries such as *What the Health* and *Cowspiracy*), or embracing a life philosophy (such as yoga). No matter the reason, vegetarians are often asked how and/or why they embraced a cruelty-free diet or lifestyle.

Well-recognized individuals have expounded on their convictions regarding vegetarianism; Jonathan Safran Foer's *Eating Animals* (2009) addresses how vegetarianism is a powerful response to the significant problems with contemporary factory farming. John Robbins (ironically, heir to the Baskin-Robbins ice cream empire) has written powerfully about the economic and environmental consequences of consuming farmed animals and makes claims that his work caused a significant decrease in meat consumption (Robbins 1987). However, some of the most notable vegetarians have outlined that becoming and staying vegetarian is about morals and personal conduct rather than one's health, environmentalism, or economic benefits/ drawbacks. Paul McCartney explains that he and his family became vegetarian "because we like animals, it's an ethical thing, not really about health" (Slater 2007). Notably, like Lisa Simpson of *The Simpsons*, Paul and Linda McCartney became vegetarian after witnessing a group of lambs frolicking joyfully outside their window, and made the ethical decision to avoid meat in the 1960s. One of the world's most famous vegetarians, Mohandas Gandhi, considered debates about the health benefits or drawbacks of vegetarianism as irrelevant, and believed that if health is why one adopts a meat-free diet, that they would "fall back" to meateating (Gandhi 1949). Rather, Gandhi argued that one must believe with firmness and conviction that a plant-based diet is one which is the right and moral thing to do, and one that builds one *spirit*, not necessarily one's *body*.

Yoga and Vegetarianism

One of the most pro-vegetarian philosophies in the West can be found in the yoga community. An increase in the popularity of yoga in North America, in particular, runs concurrent with an increase in self-declared vegans and vegetarians. As the number of yoga practitioners in the United States increased from 20 million to 36 million between 2012 and 2016 (Yoga Alliance 2016), the proportion of both vegetarians and, in particular, vegans, has also increased significantly. Veganism has been identified as a "top trend" (*The Independent* 2018) and "plant based" is now considered a key feature of contemporary food and dining according to international food consultancy firms (Baum + Whiteman 2018). *Forbes* even dubbed 2019 the "Year of the Vegan" (Parker 2019).

A national study of yoga and vegetarianism found that about 10% of yoga practitioners follow a vegetarian diet, meaning that they consume no meat, fish, or poultry (Ross et al. 2013). Indeed, researchers have found that the likelihood of yogis becoming vegetarian increases significantly the more they practice, whether they do so in studio, at home, and whether they increase such practice weekly or if their practice grows over time (Ross et al. 2012). Yoga has been shown to be beneficial in recovering from breast cancer (Van Puymbroeck et al. 2013), improved mental health in schools (Khalsa et al. 2012), and musculoskeletal and cardiopulmonary functions (for review see Raub 2004) among myriad other physical or medical benefits.

Many yogis consider vegetarianism both beneficial to their health and integral to the practice and lifestyle of yoga. Celebrated yoga instructor B. K. S. Iyengar argued that vegetarianism is "essential" to the practice of yoga, partly because vegetables "are healthy, they are exposed to air, earth, water, fire, and wind" and that the vegetarian diet causes the least harm to animals and the planet (Manuel 2017). Rosen (2011) argues that yoga and vegetarianism are "necessary bedfellows) (p. xiv) given the long tradition of nonharming in the yoga community. Similarly, MacGregor (2011) outlines how ashtanga yoga, a particularly vigorous style of practice, demands that one adheres to vegetarianism as the best diet to develop one's spirituality and peaceful way of being. Thus, according to some contemporary practitioners and leaders in the yoga community, yoga's tenets of nonviolence and its aim toward spiritual development suggest that vegetarianism is the most pure and helpful diet to fully embody what it means to be a yogi.

It is important to note that yoga is not a religion: yoga is a philosophy. Yoga seeks to free the practitioner of rumination and unhelpful thoughts, and to bring the practitioner stillness and a sense of self in the world. Most of what Westerners understand as "yoga" is otherwise known as hatha yoga, a series of physical movements with the underlying purpose of bringing the practitioner a sense of emotional freedom. However, yoga's roots can be traced to religious texts some 4000 years ago. Yoga grew from the *Vedas*, or ancient Indian texts that were composed about 2000 years BC. From the Vedas emerged three significant religions: Jainism, Buddhism, and Hinduism. Although some physical yoga postures (such as the sun salutation or surya namaskara) reference what have become contemporary Hindu gods, yoga is not tied to Hinduism as a religion. Given that some of the ancient texts, including Patanjali's *Yoga Sutras* of about 200–400BC, reference terms that became Hindu thousands of years later, some Westerners consider yoga anti-Christian or anathema to Islam. To that end, it is essential to understand the historical and philosophical nature of yoga, and that the physical aspect of this philosophy is intended to train the mind. Many Westerners now consider yoga a purely physical practice, though others have gone so far as to adopt names of postures to suit Christian values (Kremer 2013) so as to avoid any potential criticism of practicing anything that could be considered anathema to Christian doctrine.

The belief in and adherence to an animal-free diet can be identified as following the core principles of living as outlined in the *Yoga Sutras* of Patanjali. An ancient text which serves as an instruction to yoga practitioners regarding their personal conduct, the *Yoga Sutras* are laid out in a series of "chapters" known as the yamas and niyamas. These prescribed behaviors and practices serve as guidance for yoga practitioners seeking a fulfilling life. The Sutras address an eight-limbed "path" or framework through which yoga practitioners are encouraged to conduct their lives; embodying this framework should lead one to a fulfilling life of freedom and truth. These limbs or paths include ethical and moral principles (known as the yamas and nimayas), physical movement (asana), breathing exercises (pranayama), direction toward silence (pratyahara), inner awareness (dharana), sustaining awareness (dhyana), and the return of one's body and mind to inner silence (samadhi) (Hartranft 2003).

Most yoga practitioners in the West focus on asana (a physical movement in a yoga class, often in expensive, moisture-wicking clothing). However, the central tenet to truly embodying yoga as a philosophy (not a physical practice or even a religion) can be found in Patanjali's instructions in the yamas and niyamas, and are what is considered in this chapter. The yamas and niyamas are of primary importance to the conviction of vegetarianism, and will be outlined each in turn here. The series of yamas and niyamas consists of ten moral and ethical guidelines. If embodied in the practitioner's life in practical ways, Patanjali suggests that one will be closer to peace with oneself, family, and in the community. The yamas and niyamas are much more complex and nuanced than a simplistic list of "dos and don'ts." In essence, the yamas and niyamas are a group of sentiments that allow yoga practitioners to explore their own human nature as congruent with peace, generosity, and honesty. This chapter includes both description and examples of how each of the yamas and niyamas as outlined by Patanjali in the *Sutras* are relevant and applicable to the contemporary practice of a vegetarian diet. How much of the historical aspects of yoga, and in particular, the Sutras, are present in contemporary yoga classes and practices will vary among yoga traditions, specific studios, and even particular instructors. Much of a practitioner's experience with the Sutras will come via one's own independent study, or by means of a teacher training program.

The Yamas: Wise Characteristics

Ahimsa

Ahimsa is the highest duty. Even if we cannot practice it in full, we must try to understand its spirit and refrain as far as is humanly possible from violence. – M. Gandhi

The most obvious translation of the word "ahimsa" is "nonharming" or "nonviolence." The prefix a- indicates a negative, and the root word "himsa" means violence or injury, so together the inference is the opposite or absence of violence, injury, or harm. However, a true interpretation of ahimsa goes well beyond "lack of violence" or "noninjury", and assigning a negative interpretation to another word does not do the concept justice. For example, in English, there is no single word for "nonviolence." The word implies lack of harm, or absence of violence. If interpreted without depth or analysis, the word can sometimes infer apathy, passivity, or putting one's head in the sand. However, the word abhaya means "non-fear," though there are other words to apply, such as courage or bravery. Ahimsa means not simply a lack of violence, but it can be interpreted as "love in action" (Metta Center for Nonviolence n.d.). Other powerful interpretations of ahimsa include kindness, compassion, and/or connection with other sentient beings. Ahimsa also infers nonviolence to the self, and this self-kindness is perhaps the most difficult manifestation of the concept.

The yama of ahimsa is arguably the most obviously applied moral principle in adopting vegetarianism. If one were to consider kindness or nonviolence to animals, then the slaughter of animals simply does not fit with this precept. Ahimsa suggests that the killing of animals for humans' selfish consumption is not moral or ethical. This "love in action" does not have room for killing, particularly of innocent creatures, such as the lamb in Lisa's visit to the petting zoo in *The Simpsons*, and the lambs on Paul and Linda McCartney's farm. Ahimsa can even be taken further than simply not eating animals. Ahimsa can be interpreted to infer that innocent animals need protection, so not only must vegetarians not consume animals, but they also must behave with conviction to actively protect them from harm. Awareness and accountability for one's actions, and the consequences therein, is an evolved form of ahimsa, and that to which Patanjali directs practitioners.

Satya

There are only two mistakes one can make on the road to truth; not going all the way, and not starting. – Buddha

The second yama, satya, can be translated as "truthfulness" or "commitment to the truth." This yama, when applied to the conviction of vegetarianism, means that practitioners must recognize not only the truth in that an animal was killed for human consumption but also a full understanding of how one makes informed choices about food.

For example, one might not personally kill an animal in order to eat it, and thus rationalize that eating an animal that someone else killed is acceptable. After all, it's in the supermarket, it's already dead, and someone's got to eat it, so the notion of ahimsa could still be met. In the modern West, and in particular in urban centers, killing for one's own consumption is uncommon and meats are generally purchased in markets, already skinned, de-boned, and prepared in an orderly fashion on clean shelves. However, satya requires the practitioner to see that meats are the flesh of dead animals, whether one did the killing work themselves or whether another person (or machine) conducted the death by proxy.

The term "carnism," coined by Joy (2003), suggests that those who consume meat do so by means of psychic numbing, thereby blocking empathy toward animals, and referring to animals as food objects. Terms such as "bacon" or "burger" do not identify the animal of origin, but the finished food to be consumed. Joy further conducted extensive research on the psychological wellbeing of employees in slaughterhouses and learned that exposure to the violence of animal death caused significant and ever-increasing psychological distress (Joy 2010). Accepting the satya or truth of animal slaughter, together with real acceptance of ahimsa can lead to a life of nonviolence and ultimately, as the Sutras endorse, freedom and moral wholeness.

Satya is important when considering the truthfulness associated with food advertising in contemporary culture, for the suppression of truth is seen in modern advertising relative to animal products. For example, modern milk advertising often features a frothy glass of pure white milk, athletes consuming a hefty glass of ice-cold milk, or mothers dutifully providing milk to their young children and teens. There is a very clear disconnect between the cow who is milked for hours per day, having been injected with hormones to continue to produce milk and the vibrant human who is consuming the presumably healthy product. Further, few modern consumers pause to think that cow's milk is constructed to grow a calf into a 350-pound animal, not to grow a 7-pound human infant into an adult. Further, the health benefits of milk and its potential benefits to athletic performance are largely debated (Warner 2005). The "whitewashing" (Keon 2010) of milk as a healthful product by governments and advertisers alike distorts the truth of this animal product.

The practice of misleading consumers and obscuring the truth about factory milking is another example of a lack of satya, or truth, in consuming animal products. The deception in advertising takes on another level of mistruth when people fail to acknowledge the manipulation inherent in food advertising, and make excuses to buy and consume products such as dairy; milk is a government-subsidized industry and was only introduced to consumers as "necessary" in the early twentieth century when refrigeration and pasteurization permitted its safe consumption. This manipulation lends further relevance to Patanjali's instruction that satya is required in order to live a peaceful, clear life.

Asteya

When we feel connected to the vastness of life and are confident of life's abundance, we are naturally generous and able to practice the third yama, non-stealing (asteya). – Donna Farhi

Asteya is translated from Sanskrit as not stealing and not acquiring things from others that do not belong to us. The concept of asteya is directly relevant to vegetarianism in the most simple of ways: killing and eating an animal is stealing its life. So too is using other animal products such as leather, wool, and fur. Thus, eating and using an animal is the most egregious violation of the morality of vegetarianism and veganism.

Gandhi offered meaningful convictions about "stealing" food in the way of overconsumption. Though it is unknown whether he was familiar with Patanjali's *Sutras*, Gandhi's reflections about moderation are a helpful example in illustrating the concept of asteya. He believed that one must not take something that they do not need, even if it is given to them freely; this kind of overindulgence runs anathema to the concept of asteya. Gandhi wrote that it is "theft for me to take any fruit that I do not need, or to take it in a larger quantity than necessary" (Johnson 2006,

119). According to Patanjali's instructions about asteya, vegetarians must understand that consumption of food should reflect one's need for nourishment. Taking more than one needs simply for taste or out of fear of not having enough at a later time is considered greedy.

In contemporary consumer culture, fast food restaurants can be considered as a violation of asteya. For example, the Mighty Angus Original burger at McDonald's contains 790 calories, 43g of fat, and 1310mg of sodium. Similarly, the Wendy's Dave's Triple Cheeseburger boasts 1060 calories, 67 grams of fat, and 2020mg of sodium. Consumption of one product with such high (and potentially destructive) numbers, is a simple example of asteya. The practice of overeating is common in North American culture, including at family dinners, sporting events, or parties, and has even ushered in medications such as Alka-Seltzer and Prilosec, which are marketed as relief from overeating. Asteya dictates reflection on the moral concept of what one needs, consideration of others, and the emotional concept of having enough and sharing with others.

Asteya is also integral to the concept of vegetarianism in that newborn animals spend little or no time with their mothers. Degrazia (2011) describes inhumane conditions in factory farms, including the separation of weeks-old pigs from their mothers, even just hours after birth. In 2013, dairy cows, separated from their calves, made such sounds of distress that residents nearby to a farm in Massachusetts phoned the police to report "strange noises coming from the farm" (Just 2013). Clearly, taking animals away from their infants causes distress for both mother and newborn animal. Kool and Agrawal (2012) argue that empathy binds us to others, and is integral to the understanding and promotion of nonkilling. This concept can be taken further to argue that when we recognize the bond between mother and child, even in the animal world, we are thus bound not to steal infant animals from their mothers under any conditions, and certainly not for the purposes of meat consumption.

Brahmacharya

You demonstrate love by giving it unconditionally to yourself. – Paul Ferrini

Of all the guidelines in the Sutras, brahmacharya is the least understood and perhaps most open to interpretation, particularly to Western practitioners. In general, brahmacharya is translated as "good sex" or "right sex." It is also interpreted as chastity or continence. This guideline instructs the practitioner to understand the sheer potential and power of sexual energy in order that one can direct it with wisdom and mindfulness. The *Sutras* teach that in directing sexual energy with careful thought and feeling, one can experience communion with another and with God. Without such wise direction in the practice of brahmacharya, the practitioner can encounter exploitation, humiliation, manipulation, and feelings of separateness.

The guideline of brahmacharya holds important meaning in the contemporary practice of vegetarianism. Gandhi, for example, suggested that the practice of brahmacharya demanded that one's food be "limited, simple, spiceless, and if possible, undercooked" (1949, 17) in order to preserve one's sexual chastity. It is not a leap to argue that few mainstream Westerners would likely adopt a vegetarian diet to preserve their own sexual continence. However, the concept of respect for sex and sexual energy can be applied directly to the animals that many modern individuals consume. Animals in contemporary factory farms do not experience normal sexual behavior with their own species. Instead, females—cows, pigs, goats, sheep, and the like—are artificially inseminated and produced offspring over and over again until they are no longer fertile. The animals are then slaughtered and sometimes thereafter are eaten. Male animals serve as sperm donors, and their sperm is collected by farmworkers and then taken to artificially inseminate the females. Such practices, when considered in context of brahmacharya, are degrading and exploitative; animals are treated as sexual vessels whose reproductive lives are artificial and controlled by humans. Thus, the modern yoga practitioner would adhere to the vegetarian diet by practicing brahmacharya, whereby they do not endorse and condone conditions upon animals which are sexually exploitative.

Aparigraha

Be content with what you have; rejoice in the way things are. When you realize there is nothing lacking, the whole world belongs to you. – Lao Tzu

Patanjali addressed aparigraha, or greed, as a challenge to be addressed particularly when one's own happiness comes at a cost or damage to others. Alternative ways of interpreting aparigraha are "nonhoarding" or "nongrasping." In other words, in order to be truly free, one must let go of all that he or she has or wants; when that release happens, the needs and wants will, by natural course, decrease.

By means of practicing aparigraha, one can see how all beings, including animals, are connected. The application of aparigraha to vegetarianism is that practitioners have the power to release the desire for saving or hoarding food. Taking more food for oneself at a particular meal is an example of violating aparigraha: not only is this destructive to the one eating because he or she has overconsumed, but this hoarding and greed deprive others the opportunity to eat what is available.

At other times, aparigraha demonstrates the greed associated with an animal losing its life. Greed in insisting on eating meat when other food options are available is a simple example. Why must one eat farmed cow, lamb, pigs, or chicken when there are so many other food choices available? The approximate recommended daily intake of protein is 50 grams for sedentary people, and while many nutritionists and dieticians recommend more, the Harvard Health Blog at Harvard University School of Public Health recommends that "more protein" does not mean "more meat" (Pendick 2019). Sources such as nuts, beans, and dairy (if one is not vegan) are sound sources of protein.

The highest purpose of aparigraha is to acknowledge the connection and consciousness in all beings. Seeing and respecting the consciousness of animals and refusing to kill (or have killed) a "fellow creature" for consumption is the manifestation of greed. In contemporary context, overconsumption, buying in bulk, and the obsession with consuming large portions of any food are examples of the violation of aparigraha.

The Niyamas: Codes for Living Soulfully

Saucha

The body is your temple. Keep it pure and clean for the soul to reside in. – B. K. S. Iyengar

Purity and cleanliness are the essence of saucha. Patanjali taught that when one embodies the practice of saucha, one maintains cleanliness of the mind as well as the environment. In physical terms, cleanliness of the body and surroundings is one way of interpreting the precept of saucha. In spiritual terms, the notion of cleanliness extends to one's internal being and lifestyle. For example, consumption of food that is unprocessed, free of pesticides, and free of hormones allows one's body to function more efficiently and positively contributes to the maintenance and preservation of one's pure physical state.

Patanjali's instruction to embrace saucha is of particular importance in the modern diet, which is laden with sodium, additives, preservatives, and processed foods. Aisles in supermarkets are dedicated to sugary, processed "treats," many which are vegan, including chips, cookies, or candies. The largest meat recall in the US history, in 2008, included 143 million pounds of beef (roughly enough for two hamburgers for all residents in the United States, including children), and was ordered because inspectors were not available to examine the cattle for chronic illness (Pifer 2008). The largest meat recall in Canada occurred in 2013, when more than 200 people got sick from eating E. Coli-tainted beef linked to XL Fine Foods's plant in Brooks, Alberta (Canadian Press 2013).

Documentaries such as *Food, Inc.* (2009), *What the Health* (2017), and *Forks Over Knives* (2011) reveal shocking details about food, including the extent to which some factory farms do not provide sanitary conditions for animals or contaminants in food. *Food, Inc.*, for example, captures cows standing for long periods of time in stalls with inches of feces, and chickens in dark, cramped conditions, also on feces-laden floors. Further to the concern about the saucha, or cleanliness, of food is that animals in factory farms (particularly in the United States) are often injected with growth hormones and/or antibiotics. This is particularly true of milk-producing animals, for the secondary absorption of these hormones and antibiotics is potentially problematic, and further research on potential health risks is necessary (Jeong et al. 2010). Unsanitary conditions, together with the impurity of product, violate the moral principle of saucha. Vegetarians maintain the moral precept of saucha by avoiding the consumption of animals farmed and treated in unsanitary, medically influenced farming conditions.

Santosha

Be content with what you have, rejoice in the way things are. When you realize there is nothing lacking, the whole world belongs to you. – Lao Tzu

Considered by some to be the most important Niyama, santosha is a combination of "sam" (completely) and "tosha" (contentment). Patanjali's instructions in the *Sutras* encourage practitioners to be grateful, accept, and be satisfied with their present circumstances and experiences. This sense of contentment does not imply complacency, as it does not imply a lack of growth or careless acceptance of the current state of affairs. By contrast, santosha implies that one can make peace with where one is and the current situation. If one is not entirely pleased with what "is," then a sense of patience and hope ought to be the focus. When one practices santosha, the ability to remain whole and balanced even in the bleakest of circumstances is possible. Santosha implies that faith and hope are possible, and even the most challenging of situations can improve so that one has the possibility to enjoy life even when it feels impossible to do so. Showing up for life with equanimity is the foundation of santosha—living as wholly and fully as possible given one's conditions allows for even more growth and positive experiences in the future.

An illustration of santosha can be found again with Gandhi. When he lived in London, he was dismayed with the lack of good restaurants that could satisfy his desire for a dish beyond boiled vegetables. Gandhi recalled in *Diet and Diet Reform* (1949) that he went to the Holborn Restaurant with a friend, and when he asked about the ingredients in the soup, his friend told him that he was mannerless and must leave the restaurant immediately. He went without dinner that evening. He kept faith that he would find somewhere that could offer nutritious and tasty food in the city. Some days later, he approached the manager of the Holborn to offer a vegetarian plate, and his vegetarian friends "hailed the new experiment with delight" (Gandhi 1949, 12). Gandhi manifested santosha because he was content to go without food, however challenging it was to do so, though he had faith and hope that he could ask for what he wanted and needed. The acceptance of his current circumstances, combined with his faith that they could change, resulted in pleasing not only himself but also others.

The principle of santosha is applicable to the modern Western vegetarian in similar ways. Only about 5% of Americans report that they are vegetarian, and 3% report that they are vegan, for a total of 8% of Americans consuming a plant-based diet (Reinhart 2018). Vegetarians may feel alienated or "othered" because it may be difficult to find suitable food in restaurants, or people may ridicule them for being difficult or foolish (a quick Google search for "vegetarian memes" will reveal the type of ridicule directed toward vegetarians by some meateaters). Rather than experiencing frustration or anger, the precept of santosha guides practitioners to accept what is with hope and faith that it may change. For example, family and friends (and memes, of course!) may not be supportive of one's vegetarianism and may call them difficult or problematic. The practice of santosha would prescribe that the individual accept the resistance of friends, family, and the general public while holding hope that others can come to accept the conviction demonstrated by adhering to a vegetarian diet. In this case, santosha, or contentment, is apparent in one's realization of the current state of their reality, and the practitioner does not rail against reality in a futile attempt to change it. The virtues of patience, resolve, and hope are inherent in the precept of santosha so that practitioners can dwell in a peaceful condition no matter the challenging circumstances.

Tapas

Set your life on fire. Seek those who fan your flames. – Rumi

The third Niyama in the Sutras is tapas, which can be translated as "heat," "fire," and "discipline." The Sanskrit word "tap" means burn, and forms the root of this precept. Tapas is not translatable in simply one word, but perhaps best as "firey discipline," or the fierce, intense commitment that is necessary to "burn away" negative thoughts, challenges, or doubt along the way to achieving one's personal wellness and greatness. Farhi (2000) suggests that tapas, or "burning enthusiasm," is shown when we generate an attitude of commitment or devotion to a cause or issue. The vigor of our convictions creates a current that carries us forward—this commitment and current is the manifestation of tapas. Farhi makes an analogy to a fire: a fire can take significant effort to start up, though once it is roaring, it is easy to sustain.

The commitment to vegetarianism is a clear demonstration of tapas—the enthusiasm and steadfast adherence to a plant-based lifestyle shows "fiery discipline" for a diet that is kind to oneself, animals, and to the planet. Patanjali made clear in the *Sutras* that tapas is not energy that is directed toward suffering or deprivation; rather, tapas allows one to focus energy on a meaningful, clear, enjoyable life. The practice of tapas is demonstrated by the vegetarian when one recognizes the physical and moral benefits of honoring one's own and the planet's life. In the West, given that the great majority of the population are meateaters, vegetarians will certainly find challenges in not only finding food if one is not at home or at the market, but also encounter resistance from others. One must have the conviction to seek further, ask for their needs to be met, or communicate with patience and truth.

If a vegetarian were to cease a plant-based lifestyle, the fire, as Farhi (2000) describes, will dwindle, will be more difficult to get going again. Tapas, to the contemporary vegetarian, means staying with the moral conviction to vegetarianism despite the myriad challenges which one may encounter. When the moral conviction is held, the fire keeps burning, and the path becomes clearer.

Swadhyaya

Your vision will become clear only when you look into your heart. Who looks outside, dreams. Who looks inside, awakens. – Carl Jung

Swadhyaya, or "one's own reading" or "self-study" means being aware of oneself or engaging in any activity that facilitates awareness of one's own being and/or consciousness. For some practitioners, this self-awareness may be found through art, music, cooking, nature, spending time with animals, or in physical activity. Whichever means through which one becomes more aware of him or herself is not important; any specific activity can bring the opportunity to know oneself, and to one oneself better. Swadhyaya is the practice of identifying an activity, committing to it, and becoming aware enough of the self to observe its evolution. The most important aspect of this precept in the *Sutras* is self-awareness in a way that identifies and celebrates the interconnection of one with others and in a way that dissolves a sense of separation in order to make way for recognizing interconnectedness. Swadhyaya is a moral practice in that it helps to create an understanding of the interrelation of all living beings, and also the interplay between one and the greater universe.

Patanjali taught swadhyaya with the objective that once one knows oneself, then one cannot help but be aware of others. In the case of eating animals, swadhyaya encourages an awareness that recognizes and celebrates life and therefore does not permit the consumption of living creatures. Swadhyaya's aim toward awareness of the self and awareness of all life would lead many vegetarians to agree with Will Kellogg, American food manufacturer and founder of Kellogg's cereals, when he said, "How can you eat anything with eyes?" When one looks into the eyes of another, it is possible to see and acknowledge the self in us that is also in others.

The awareness of self and then the more powerful, deep awareness of others is evident both in arts and with animals. For example, Artist Marina Abramovic, in her 2010 retrospective, "The Artist is Present" at the Metropolitan Museum of Art, shared a period of silence and eye contact with each individual who voluntarily sat across from her at a simple table. The majority of participants were strangers, though Abramovic was surprised when she opened her eyes and her former partner, Ulay, sat across from her. Upon eye contact, the emotional and spiritual connection between the two was undeniable. The same kind of emotion is possible —if one chooses to allow it— with a friend, child, pet, animal in a zoo, or even a wild animal. If a meateater were presented with the necessity of killing an animal for consumption with one's own hands, it is possible or even likely that swadhyaya would prevent such slaughter. Similarly, Marsella argues that when we recognize the interconnectedness of humans, animals, and the natural world, we then in turn consider how we kill animals for food, the destruction of the oceans, air, land, and nonhuman life around us, and such carnage "can no longer be tolerated or justified" (2012, 362).

Ishvara pranidhana

> *You surrender to a lot of things which are not worthy of you. I wish you would surrender to your radiance, your integrity, your beautiful human grace.* – Yogi Bhajan

Literally translated, Ishvara pranidhana means "surrender to God." This final moral precept in the *Sutras* can be problematic, particularly to Westerners, because some traditions interpret Ishvara as "Lord," thus inferring worship of a specific deity, while others suggest that Ishvara pranidhana refers to a divine entity with which each practitioner relates to in his or her own way. In order to avoid the nuances of religion and dogma in the interpretation of this moral precept, many traditions understand Ishvara pranidhana as the practice of giving up the need to control, and rather, having faith that events and situations will unfold as they are meant to in the divine order of life. In this interpretation, the practitioner can recognize that there is a force larger and more powerful than the self, and that one can let go of the self-centered way of believing that we know what we want and what is best. The practice of Ishvara pranidhana is not a passive practice. By contrast, relinquishing the need to control or orchestrate people, events, or feelings takes great strength. In committing to this niyama, practitioners must balance strength and surrender—one can feel the drive toward a particular cause or goal, though one must simultaneously have the courage to let go of the need to manipulate as much as possible and to surrender the details and course of events to the divine. Thus, the niyama of Ishvara pranidhana can be best described as having faith.

In the case of contemporary vegetarianism, one is often asked how or why they came to vegetarianism, and the moral reasons for doing so. Vegetarians in the modern West can practice Ishvara pranidhana by knowing and trusting their own conviction and having faith that their actions and words will have a positive impact. Relinquishing or surrendering the outcome of one's actions to the divine order of things takes great strength. Vegetarians can embody the niyama of Ishvara pranidhana by experiencing this strength of conviction. The precept goes further in that like Lisa Simpson's friend Apu, one can release the need to preach to others or be, in Apu's words, "badgering them always." One can trust that their actions, even in their own immediate communities, can have a positive effect on the unfolding of the ethical treatment of animals.

Conclusion

Patanjali's yoga *Sutras* were communicated thousands of years ago, yet the application of the yamas and niyamas can be seen as meaningful and applicable principles even in contemporary Western culture. The yamas and niyamas are but one of the eight limbs of yoga in the *Sutras*. Yet these ten moral precepts serve as a framework for practitioners in the goal to live a fulfilled, joyful, and positive life. These "wise characteristics" and "codes for living soulfully" include nonviolence, truth, nonstealing, continence, nonhoarding, purity, contentment, burning enthusiasm, self-study, and surrender to God. If the practitioner takes the outlook that vegetarianism is not a choice made primarily for *physical* health, but one made as a *moral* decision, the yamas and niyamas serve as a guide to making sound decisions. Further, they serve as principles to help the practitioner have faith and true conviction in their actions and beliefs.

Adhering to a vegetarian or vegan lifestyle comes when one commits to moral principles that encompass how one treats oneself, how one treats animals, and how one treats fellow human beings and the earth. Recognizing the ethical responsibility of acknowledging animals as living creatures, caring about how animals are treated, and a healthy attitude toward the consumption of food, one cannot help but realize the essence of the vegetarian lifestyle. When concepts such as knowing the self and trusting the divine order of events are added to the practitioner's toolbox, one is prepared to lead a life of conviction with happiness, wisdom, and freedom. The *Sutras* offer a framework to help embody and practice these wise characteristics and codes for living with peace and freedom. Although many yoga practitioners, particularly in the West, consider yoga a physical practice conducted inexpensive brand-name clothing, these principles are available for the contemporary yogi's study and reflection.

Note

1. Indeed, if the credits to this episode of *The Simpsons* are played backwards, one can hear Paul reading Linda's recipe for lentil soup (Siegel 2015).

Bibliography

Andersen, Kip, Keegan Kuhn, and Fernando Arce. 2017. *What the Health*. New York: A.U.M. Films & Media.

Baum + Whiteman. 2018. Consultants Predict 11 Hottest Food and Beverage Trends in Restaurant and Hotel Dining for 2018. https://www.baumwhiteman.com/trend-reports

Canadian Press. 2013. Judge Approves Class-Action Lawsuit Against XL Foods. *Global News Online*, October 9. https://globalnews.ca/news/892790/judge-approves-class-action-lawsuit-against-xl-foods/

Degrazia, D. 2011. The Ethics of Confining Animals: From Farms to Zoos to Human Homes. In *The Oxford Handbook of Animal Ethics*, ed. T.L. Beauchamp and R.G. Frey. New York: Oxford University Press.

Farhi, D. 2000. *Yoga Mind, Body, and Spirit: A Return to Wholeness*. New York: Holt.

Fulkerson, Lee, John Corry, Joey Aucoin, Neal D. Barnard, and Gene Baur. 2011. *Forks Over Knives*. New York: Virgil Films.

Gandhi, M.K. 1949. *Diet and Diet Reform*. Ahmedabad: Navajivan Publishing House. Retrieved from http://www.mkgandhi.org/ebks/diet_and_diet_reform.pdf

Hartranft, C. 2003. *The Yoga-Sutra of Patanjali*. Boston: Shambhala Publications.

Jeong, S. H., D. Kang, M.-W. Lim, C. Soo Kang, and H. Jung Sung. 2010. Risk Assessment of Growth Hormones and Antimicrobial Residues in Meat. *Toxicological Research* 26 (4): 301–313. https://www.ncbi.nlm.nih.gov/pmc/articles/PMC3834504/.

Johnson, R.L. 2006. *Gandhi's Experiments with Truth: Essential Writings by and About Mahatma Gandhi*. Oxford: Lexington Books.

Joy, M. 2003. *Psychic Numbing and Meat Consumption: The Psychology of Carnism*. PhD dissertation (OCLC729946807), Saybrook Graduate School and Research Center, San Francisco, CA.

———. 2010. *Why We Love Dogs, Eat Pigs, and Wear Cows: An Introduction to Carnism*. San Francisco: Conari Press.

Kenner, Robert, Richard Pearce, Eric Schlosser, Melissa Robledo, William Pohlad, Jeff Skoll, Robin Schorr, et al. 2009. *Food, Inc.* Los Angeles: Magnolia Home Entertainment.

Keon, J. 2010. *Whitewash: The Disturbing Truth About Cow's Milk and Your Health*. Gabriola Island: New Society Publishers.

Khalsa, S.B.S., L. Hickey-Schultz, D. Cohen, N. Steiner, and S. Cope. 2012. Evaluation of the Mental Health Benefits of Yoga in a Secondary School: A Preliminary Randomized Controlled Trial. *Journal of Behavioral Health Services & Research* 39: 80–90.

Kool, V. and Agrawal, R. 2012. From empathy to altruism: Is there an evolutionary basis for nonkilling? In *Nonkilling psychology*, ed, D. J. Christie and J. E. Pim, 145–173. Honolulu: Center for Global Nonkilling.

Kremer, M. 2013. Does Doing Yoga Make You a Hindu? *BBC News*. Retrieved from https://www.bbc.com/news/magazine-25006926

MacGregor, K. 2011. Ashtanga Yoga Body: Feel Your Way to Enlightened Eating. In *Food for the Soul: Vegetarianism and Yoga Traditions*, ed. S. Rosen, 49–62. Santa Barbara: ABC-CLO.

Manuel, M. 2017. A Vegetarian Lunch with B. K. S. Iyengar: Remembering the Master on International Yoga Day. *Huffington Post.* Retrieved from https://www.huffingtonpost.in/mark-manuel/a-vegetarian-lunch-with-b-k-s-iyengar_a_22491098/

Marina Abramovic e Ulay – MoMa. 2010. https://www.youtube.com/watch?v=OS0Tg0IjCp4

Marsella, A. J. 2012. Lifeism and nonkilling: I am what I am. In *Nonkilling psychology*, ed, D. J. Christie and J. E. Pim, 362. Honolulu: Center for Global Nonkilling.

Metta Center for Nonviolence. n.d. Ahimsa. http://mettacenter.org/definitions/gloss-concepts/ahimsa/

Parker, J. 2019. The Year of the Vegan: Where Millennials Lead, Businesses and Governments Will Follow. *Forbes*. Retrieved from https://worldin2019.economist.com/theyearofthevegan

Pendick, D. 2019. How Much Protein Do You Need Every Day? *Harvard Health Blog*, June 25. https://www.health.harvard.edu/blog/how-much-protein-do-you-need-every-day-201506188096

Pifer, J. 2008. USDA Orders Recall of 143 Million Pounds of Beef. *CNN*, February 18. http://www.cnn.com/2008/HEALTH/02/17/beef.recall/

Raub, J. 2004. Psychophysiological Effects of Hatha Yoga on Musculoskeletal and Cardiopulmonary Function: A Literature Review. *Journal of Alternative and Complementary Medicine* 8 (6). https://doi.org/10.1089/10755302605111810.

Reinhart, R.J. 2018. Snapshot: Few Americans Vegetarian or Vegan. *Gallup*, August 1. https://news.gallup.com/poll/238328/snapshot-few-americans-vegetarian-vegan.aspx?g_source=link_NEWSV9&g_medium=NEWSFEED&g_campaign=item_&g_content=Snapshot%3a%2520Few%2520Americans%2520Vegetarian%2520or%2520Vegan

Robbins, J. 1987. *Diet for a New America: How Your Food Choices Affect Your Health, Happiness, and the Future of Life on Earth*. Tiburon: New World Library.

Rosen, S.J. 2011. Introduction. In *Food for the Soul: Vegetarianism and Yoga Traditions*, ed. S. Rosen, vii–xvi. Santa Barbara: ABC-CLO.

Ross, A., E. Friedmann, M. Bevans, and S. Thomas. 2012. Frequency of Yoga Practice Predicts Health: Results of a National Survey of Yoga Practitioners. *Yoga as Therapeutic Intervention*. https://doi.org/10.1155/2012/983258.

———. 2013. National Survey of Yoga Practitioners: Mental and Physical Health Benefits. *Complementary Therapies in Medicine* 21 (4): 313–323. https://doi.org/10.1016/j.ctim.2013.04.001.

Siegel, A. 2015. Celebrating 'Lisa the Vegetarian,' the *Simpsons* Episode that Changed the Image of Vegetarians on TV. *Slate*, October 12. https://slate.com/culture/2015/10/the-simpsons-lisa-the-vegetarian-episode-changed-the-image-of-vegetarians-on-tv.html

Slater, N. 2007. When the McCartneys Came for Lunch. *The Observer*, April 29. https://www.theguardian.com/lifeandstyle/2007/apr/29/foodanddrink.features4

The Simpsons. Season 7, Episode 5, "Lisa the Vegetarian," directed by Mark Kirkland, written by David X. Cohen, created by Matt Groening, aired October 15, 1995 on Twentieth Century Fox Entertainment.

Van Puymbroeck, M., B.N. Burk, K.J. Shinew, M.C. Kuhlenschmidt, and A.A. Schmid. 2013. Perceived Health Benefits from Yoga Among Breast Cancer Survivors. *American Journal of Health Promotion* 27 (5): 308–315.

Warner, M. 2005. Chug Milk, Shed Pounds? Not so Fast. *New York Times*, June 21. www.nytimes.cm/2005/06/21/business/media/21adco.html?pagewanted=print&r_r=0

Yoga Alliance. 2016. 2016 Yoga in America Study Conducted by *Yoga Journal* and Yoga Alliance Reveals Growth and Benefits of the Practice. https://www.prnewswire.com/news-releases/2016-yoga-in-america-study-conducted-by-yoga-journal-and-yoga-alliance-reveals-growth-and-benefits-of-the-practice-300203418.html

Part II

Veg(etari)anism as Embodied Practice

6

The Accidental Vegetarian: Object-Oriented Ontology at the Intersection of Alpha-Gal Mammalian Meat Allergy

Elizabeth Baddour

A cool February walk in the Tennessee woods intended as a respite from the rigors of my doctoral program became instead the impetus for an unintended, dramatic change in lifestyle. The forest walk did, in fact, prove restorative; the close sighting of a herd of unwary bedded-down deer was a feast for my eyes. But unbeknown to me at that moment, two lone star ticks (*Amblyomma americanum*) carrying the alpha-gal molecule hitched a ride on my body and made a buffet of my blood. Feeding unnoticed for hours, the ticks' saliva subsequently altered my immune system, causing my body to become violently intolerant of mammal meat for what my physician predicts may be the remainder of my life.

The link between the tick bites and the onset of mammalian meat allergy took many months of testing before doctors diagnosed my odd conglomeration of symptoms as commensurate with the "red meat allergy," officially named Alpha-Gal Syndrome (AGS). Discovered in 2009, AGS symptoms range from swelling hands and feet, severe

E. Baddour (✉)
The University of Memphis, Memphis, TN, USA

C. Hanganu-Bresch, K. Kondrlik (eds.), *Veg(etari)an Arguments in Culture, History, and Practice*, The Palgrave Macmillan Animal Ethics Series,
https://doi.org/10.1007/978-3-030-53280-2_6

gastrointestinal issues, and frightening episodes of anaphylactic shock that are delayed in onset, seemingly defying explanation. Because my symptoms were inconsistent and the condition was then extremely rare, a blood test became my physician's last diagnostic resort. A couple of years into the AGS experience, a televised documentary made me aware of the institutional cruelty of intensive animal factory farming around the same time I lost my appetite for fish and fowl—the only animal groups I could safely ingest without repercussion. Armed with a new perspective on the animals our culture commonly consumes, I became an accidental vegetarian through the confluence of circumstance—ironically via the agency of an animal. After years of iterations of fish and chicken in lieu of red meat, my disgust at the mouthfeel of flesh on my palate combined with an aversion to animal cruelty drove me unexpectedly to a plant-based diet.

Through the lens of Object-Oriented Ontology (OOO) and phenomenology, this essay explores vegetarianism as a rhetorical intervention in challenging a peculiar illness that renders red meat—a staple of the American diet—toxic to those with alpha-galactose sensitivity. I examine the strangeness that AGS visits upon the body (with its wide range of bizarre and inconsistent symptoms) as a means of understanding the interrelatedness of objects in the vast web of connectivity that informs meaning and makes meat consumption bodily and ethically problematic. This chapter briefly provides an overview of the history and complexity of AGS in the context of my personal experience before diverging into two paths. The first path explores AGS through the lens of philosophy, alienation, and the unhomelike being that the body experiences in the state of unwellness. The second path considers illness narratives as a heuristic for contextualizing the alienation of unwellness/otherness of AGS and the usefulness of narratives in ideating change through the specific mediation of vegetarianism as a form of somatic resistance to illness and animal commodification.

Countless studies have detailed the various motivations for adopting a vegetarian regimen; most notably, incentives run the gamut from concerns regarding personal health and animal welfare (Hoek et al. 2004, 266), emotional responses to flesh consumption (Kenyon and Barker 1998, 189), environmental impacts concurrent with intensive animal farming (Twine 2010, 137–140; Taylor 2012, 1–12) along with a myriad of personal, religious, and political reasons. However, vegetarianism as a solution to a health situation in which meat is the allergen may be an understudied aspect of the practice. In a nod to Actor-Network Theory, I frame my experience as an exercise in literally "Learning to Feed Off Controversies," as Bruno Latour's (2005) chapter title in *Reassembling the Social* connotes. In this chapter, Latour writes, "it's always the paradoxical presence of something at once invisible yet tangible, taken for granted yet surprising, mundane but of baffling subtlety that triggers a passionate attempt to tame the wild beast of the social" (31). My experience with AGS forced an accidental inversion of normalcy into my existence which, in turn, propelled me into the community of the animal—because, through an allergy to their flesh, animals became visible.

Conflating the violent AGS symptoms that accompany the involuntary ingestion of red meat engenders empathy for violence enacted upon the millions of animals subjected to the butcher's abattoir. For example, Annie Potts (2017) writes that Tyson Foods, one of many of the meat industry's key companies, reports "over 42 million chickens, 170,000 cattle and 350,000 pigs are killed each week" (2). Multiply the aforementioned weekly numbers by years and decades to realize the astounding ecological and environmental toll exacted upon this partial list of fellow living beings.

Typically, the choice to become vegetarian is just that—a choice—rooted in deeply held moral conviction(s) or aspirations for improved health. My research fills a gap in scholarship for its recognition of vegetarianism as a stabilizing remedy to an unusual and unpredictable illness that can be controlled only through absence—the absence of mammalian meat—in order to mitigate the loss of the fragile state of health in a situation where the alternatives are extremely limited. My altered mindset toward flesh food is directly attributable to the non-human elements intertwined in my existence on a particular day in February. An invisible actant in the Tennessee wilderness ushered in an awakening to the invisible—a sensitivity to animals represented by the umbrella term "meat" brought about through my odd allergy to mammals. The presence of animals in my conscience along with their absence from my plate were the manifestations of a seismic shift in my conception of animal muscle fiber as something absolutely necessary—to something to painstakingly avoid.

Identity and Food

The decision to adopt a vegan/vegetarian diet involves considerations including identity, ethics, morality, and health at the junction of nourishment. Food, and our relationship to it, evokes passion and ambivalence rooted in social, cultural, and ideological factors because food and intimacy are intertwined in the experience of *being* from birth until death. However, AGS sufferers not only face the possibility of a life without steak, burgers, or hot dogs but are at risk when coming in contact with seemingly innocuous animal-based products such as lanolin or gelatin. Through the lens of Object-Oriented Ontology (OOO) and phenomenology, I explore the embodied rhetoric of illness, the uncanny nature of the experience of unwellness, and the unhomelike character of AGS largely through my lived experience with the red meat allergy. Chronicling the waypoints in my improbable journey from omnivore to vegetarian, I conclude with the prospect that AGS poses for enlarging a deeper understanding of the meaningfulness of absence while living with a chronic illness in the present.

Alpha-Gal Description

Alpha-gal syndrome is described in medical literature as "a novel and severe food allergy" linked to the bite of a lone star tick whose saliva induces an immune system response in humans that creates intolerance to mammalian meat (Flaherty et al. 2017, 1). The condition is thought to be initiated when a tick introduces its infected saliva into its host during the process of biting and feeding. The transmission includes a sugar molecule injected into the human body that delivers Galactose-alpha-1,3-galactose oligosaccharide (alpha-gal), a carbohydrate found in non-primate mammals (Commins et al. 2011, 1287). While not all ticks carry the alpha-gal carbohydrate, the lone star tick, so named for the white Texas star on its back, is the main vehicle of alpha-gal transmission to humans in the United States. Currently, 6 continents and 17 countries report cases of mammalian meat allergy: Australia; the United States of America; Europe—France, Spain, Germany, Belgium, Switzerland, Sweden, United Kingdom, Italy, and Norway; Asia—Korea, Japan; Central America—Panama; South America—Brazil; and Africa (Kwak et al. 2018). In addition to the American lone star tick (*Amblyomma americanum*), the castor bean tick in Europe (*Ixodes ricinus*), and Australia's hard tick (*Ixodes endopalpiger australiensis*), the Australian paralysis tick (*Ixodes holocycus*) is also implicated in the emergence of red meat allergy (Craig 2019).

Ticks that cause AGS are believed to transmit alpha-gal molecules into humans from animals such as deer who are unwitting hosts to ticks. When an alpha-gal carrying tick bites a person, the offending foreign sugar is introduced into her bloodstream, and her human body mounts a vociferous, albeit delayed allergic response. An immune response to alpha-gal sugar results in a cascading allergic reaction in some, but not all humans whose bodies respond to exposure to galactose-alpha 1,3-galactose. The alpha-gal sugar overloads the body with immunoglobulin E (IgE) antibodies on contact with carbohydrates found in mammalian meat or products made from mammals inducing a reaction (Commins and Platts-Mills 2013, 354).

Compounding the difficulty of alpha-gal diagnosis is the hallmark delayed reaction which comes anywhere from three to ten hours after consuming red meat. By contrast, other more common allergies such as latex or peanut allergies produce reactions that typically follow soon after exposure to the allergen. In addition to the delayed reactions, alpha-gal researchers note that diagnosis of alpha-gal syndrome is further complicated by "negative skin prick tests to meat" that obscure the corollary to mammalian meat consumption (Wolver et al. 2012, 322). The difficulty in correlating the causation of symptoms to mammalian meat consumption is just one of the conundrums surrounding the bizarre and very unpredictable allergy. Its recent discovery in 2009 underscores the strange nature of the condition; not only is the illness peculiar in its origins and manifestations, its relative newness requires constant vigilance when AGS sufferers deal with medical personnel or those in the foodservice industry whose inattentiveness may cost someone her life. Oddly, the discovery of AGS is linked to clinical trials of cetuximab, a life-saving cancer drug whose ingredients caused severe reactions and even deaths among patients who unknowingly had a sensitivity to galactose-alpha 1,3-galactose or alpha-gal.

Alpha-Gal: Both Medicine and Menace

In 2004, the Bristol Meyers Squibb drug cetuximab became available for clinical trials in the treatment of metastatic colorectal cancer in the United States (Steinke et al. 2015, 589). Cetuximab contains the identical alpha-gal sugar found in mammalian meat. During clinical trials of cetuximab, allergy researcher Thomas Platts-Mills discovered that some cancer patients experienced a strong reaction to cetuximab therapy, and upon further investigation, Platts-Mills found that those cancer patients reacting to the drug shared two commonalities: they lived in the southeastern United States and they had high levels of alpha-gal antibodies in their blood (Commins et al. 2009, 426). Patients who received the intravenously-administered cancer drug and reacted did so in very serious, life-threatening ways, including anaphylactic shock. Around this time, a patient in Bentonville, Arkansas, received the first dose of cetuximab and collapsed and died (McKenna 2018). Researchers along with the drug company discovered an oddity; the reactions to cetuximab were not occurring in New York or California, but specifically throughout the southeast where nearly one-quarter of the patients had a severe reaction "upon first infusion of the medication" (Foronda 2019). The mysterious reactions appeared to be contained within a specific geographic region.

Meanwhile, Scott Commins and Thomas Platts-Mills listened to allergy patients increasingly express their concerns regarding possible allergies to beef, pork, and lamb. But because Commins and Platts-Mills's patients' reactions were inconsistent and delayed, Commins explains, "it sort of flew in the face of conventional food allergy at that point" (Foronda 2019). Relying upon decades of prior research, the researcher-physicians were puzzled at the possibility of a human reaction to mammal meats, because "humans typically make Immunoglobulin (IgE) antibodies against protein, [such as that contained in peanuts, shellfish, egg, milk, fish] and not carbohydrate determinates" (Wolver et al. 2012, 322). Around 2008, Platts-Mills published a paper regarding his findings involving mice cells and the cetuximab cancer drug. Platts-Mills discovered that when cetuximab was administered to the test mice, the mouse cells responded by coating the cancer drug with their alpha-gal sugar. At that point, Platts-Mills made the connection between the alpha-gal sugar produced by lower mammals and cetuximab and its role as a reactive agent in the cancer patients undergoing clinical trials (Foronda 2019).

Until Commins and Platts-Mills discovered the connection between ticks and alpha-gal in 2009, allergies to beef were a virtual anomaly, but the researchers were intrigued by a possible connection to the geography of the southeast and the presence of a sudden onset of a meat allergy among patients who had experienced no prior reactions to meat in prior decades. Then, a technician in Commins and Platts-Mills's laboratory went hiking in the Virginia mountains and was bitten by several ticks. Soon afterward, the technician developed a sudden sensitivity to beef (Foronda 2019). Through this serendipitous event, researchers connected the tick bites to the geographic incidence of reactions to alpha-gal and began unraveling the mystery. Commins states, "It does show you that serendipity, it favors the prepared mind" (Foronda 2019). Commins linked the causality of ticks to AGS through his familiarity with his patients' health histories and his awareness of the presence of ticks in the southeastern United States.

AGS affects each of its sufferers in diverse ways, and the condition for some, "not only affects what you can eat, [but] what you wear, how you relax, even [dictating] which medicines are safe" because alpha-gal induced reactions are not limited to the ingestion of meat products (McKenna 2018). Contact with even minute amounts of mammal products or the fumes from sizzling grilled meats can possibly trigger an assortment of delayed and potentially deadly responses ranging from life-threatening anaphylaxis to systemic hives, swelling of the throat, lips, and tongue; precipitous drops in blood pressure, abdominal pain, vomiting, difficulty breathing, and crippling digestive issues. In short, AGS has the potential to affect the corners of one's *being*.

Among the many frustrations for those with AGS is that simply giving up eating mammals is often not enough to mitigate the body's vigorous histamine response to mammal products, a situation that prompts many to adopt a vegan or vegetarian diet. Unlike the flu or other illnesses that present with common symptoms, alpha-gal's symptoms are not universal, acting on individuals in different ways (Wilson et al. 2019). Due to the unpredictability of AGS, the body may react wildly to micro-exposures of allergen triggers one day, then produce little or no reaction to an obvious one another day. Contributing to the syndrome's anxiety-provoking peculiarity is the mercurial nature of the reactions it spawns; the AGS body may respond with anaphylactic shock to the simple wafting odor of grilling meats on one occasion, and not at all to the consumption of a small portion of refried beans unknowingly processed with beef fat another day.

AGS patients suffer varying degrees of reactivity to mammalian products which can produce delayed symptoms with the consumption of milk products such as ghee butter or cheese, each containing small amounts of alpha-gal. Sweets and cake frosting pose additional hazards. Bone char is used in the sugar refining process, and gelatin—a protein derived from boiling the skins and connective tissue of cows and pigs—is used as a common thickening agent in an alarming array of foods and even some vaccines. French fried potatoes or vegetables cooked in lard induce unforeseen sickness and suffering. Lipsticks, soaps, or shaving creams whose list of ingredients includes beef tallow pose a latent danger. Porcine ingredients frequently used in pharmaceutical pill capsules or in the casing of turkey sausage offer yet other hidden sources of sickness. Fatty meats such as pork are also implicated in stronger and more delayed reactions because fats take longer for the body to digest. Complicating the diagnosis of AGS is the delayed onset of symptoms which can appear as late as eight to ten hours after mammalian products are consumed. Consequently, AGS symptoms are often attributed to other medical maladies because of the corollary to mammal meats is obscured by the relatively long interval between consumption and the appearance of symptoms.

In addition to avoiding the minefield of mammal products on grocery shelves, AGS patients are additionally vexed by widespread unawareness of the syndrome within the medical community to whom we turn for help. Equally as frightening are episodes in which our requests are ignored by those in the restaurant industry who simply do not acknowledge that an allergy to red meat is an actual medical condition warranting attention and respect. Perhaps most daunting is the prospect of living life precariously avoiding mammalian products, whose derivatives are found in everyday items as varied as refined sugar, cosmetics, marshmallows, gummy bears, and gelatin-coated capsules—because as of this writing, abstinence is the only means to control the mysterious condition that can present life-threatening danger in the safe spaces of the home and beyond. In all, mammal products are a source of tension to the AGS body for the potentially deadly health risks they pose.

The unusual and peculiar nature of AGS situates it uniquely within a rhetorical and philosophical framework: because AGS remains widely unfamiliar to healthcare providers, the US food industry and the public at large, AGS patients must assume increased agency in combatting a life-threatening allergy whose existence more often than not elicits incredulity, surprise, or varying degrees of empathy from omnivores at the thought of forced estrangement from meat. However, alienation from meat offers the AGS body the opportunity to step back from the "routinized norms of commodification and violence" of meat production to view alimentary adjustments as a personal way to contrast and resist the climate change, cruelty, and conformity attendant with our culture's obsession with meat (Twine 2014, 624). Val Plumwood (2000) argues the very term "meat" is problematic in its reductive categorization of animals as something commodified as edible (295). Plumwood writes, "no being should be treated reductionistically as meat, but we are all edible (food), and humans are food as much as other animals, contrary to deeply entrenched beliefs of human identity in the west" (295). Embodying vegetarianism—accidentally or otherwise—is a means of understanding animals as something other than "meat" by acknowledging their sentience as fellow beings.

Object-Oriented Ontology and Alpha-Gal Syndrome

Nourishment has an ontological component because eating fosters being; eating *is* being. From this perspective, AGS and Object-Oriented Ontology (OOO) intersect at the junction of knowledge. OOO and AGS share their respective roles as newcomers in entirely different disciplines. OOO is a relatively new philosophy rooted in speculative realism and dates to the late 1990s work of Graham Harman. AGS traces its recent discovery to 2009 and the serendipitous discovery by physicians Commins and Thomas Platts-Mills who related the causational factor of ticks to the red meat allergy. As different as OOO and AGS are, the philosophy and the illness converge at the crossroads of unexpected, life-altering change. In this case, the change involves the willful avoidance of mammalian meat in order to sustain well-being—well-being that is threatened through the catalytic bite of an alpha-gal bearing tick. A tick has made my own body entirely *other* to me, for the symptoms that suddenly and unexpectedly appear hours after unintentionally coming in contact with red meat enact a degree of violence within the AGS body—as if the body has gone to war against itself. The alpha-gal carbohydrate, as an object, has upended my *being* through the agentic power of a tick.

Although many variations of speculative realism exist, for purposes of this chapter, OOO is loosely defined as the interrogation of "objects" or "things" and their capacity as agents in the *being*, that is, the human experience. Influenced by twenty-first-century phenomenologist Martin Heidegger, OOO rejects the privileging of human existence above non-human objects. OOO's essence is its opposition to the anthropocene or the belief that humans are the central elements in an ecosystem. In considering AGS, the tick represents the agentic power of a "thing" over humankind through its opportunistic resiliency. For example, evolutionary biologists point to the 300 million year existence of ticks and their ability to grow 2880 times their larval size by consuming only three meals during the course of their two-year life cycle and their adaptability in vacillating between long intervals of dehydration and starvation to surfeit, an evolutionary feat far beyond the scale of human capacity to duplicate (Zimmer 2013).

Closely related to OOO is phenomenology, whose fundamental tenet is "knowing the world is inseparable from being in the world" (Høsieth and Keitsch 2015, 35). Frederik Svenaeus (2010) defines phenomenology as "a method by which an individual engages with, interprets and conceptualizes the world based on his [*sic*] own perspective and lived experience" (340). A phenomenological approach is useful in understanding the alpha-gal afflicted body because it is through the unusual and rare experience of a red meat allergy that AGS sufferers embody their condition while negotiating *being* in an environment that is all too often rife with triggers that carry life-threatening potential. A vegan/vegetarian diet is one of the very limited options that the AGS body has in safely navigating *being* through the terrain of mammalian hazards that form the contours of our existence in the push-to-place mammalian products to the periphery of our existence. Whereas typically, the decision to adopt an animal-free diet is a choice based on ethical and ecological considerations, vegetarianism is a vehicle through which empowerment over our environment is wrested from the condition visited upon us by a tick and returned to the AG body.

Central to the safety of the AGS body is awareness—an awareness that reflects the need for the embodied voices of those with AGS to engage in community in order to negotiate and renegotiate our place in ecology made *strange* by a tick. The alpha-gal body tells the story of how a tick, as an object—injects other objects—the alpha-gal carbohydrate—into our human body and through that act, challenges our feigned human privilege. A tick drastically and permanently makes food choices a matter of life and death. As we consider the stories of those with AGS and other mysterious illnesses, we are reminded that stories rooted in lived experience have the propensity to deepen the meaningfulness of our existence and our relationship to the world in which we find ourselves. And all stories need an audience, particularly stories told by the body as it finds itself at the crux of illness, change, and the need to adapt to a world irretrievably altered by a tiny non-human object that has asserted powerful agency over it.

It is at this juncture that phenomenology is helpful in understanding *how* our bodies are engaged participants in a shared world of objects (Aho and Aho 2008, 24). In contrast to the culturally pervasive Cartesian duality that privileges the division between the mind and the body, Heidegger offers a decidedly different perspective in the interrogation of the meaningfulness of the body to health and illness. In a truncated synopsis, I briefly apply several of Heidegger's concepts from *Being and Time* to AGS. By contrast to Descartes's theory of *being* which separates the mind and the body and asks, *what* is the body? Heidegger asks the question, *how* is the body? In *Being and Time,* Heidegger proposes that the lived body, complete with its range of afflictions, occupies a reciprocal relationship with the social and cultural environment of which it is a part, and extends "beyond its skin" into the tightly interwoven context that is the material and social world framing our ecology.

In the Skin of the "Strange Stranger"

Timothy Morton writes of the relevance of OOO to ecology by arguing "OOO is a form of realism that asserts that real things exist—these things are objects, not amorphous 'Matter,'" (Morton 2011, 165). Morton extends Jane Bennett's concept of the vital materiality of matter and the binary we humans tend to impose upon things as "passive stuff" as contrasted with ourselves, which we tend to view as vital and vibrant by comparison (Bennett 2010, viii). New materialist theorists including Bennett and Bruno Latour assert that as humans, we are active in complicated networks of being or assemblages. Bennett argues that all matters are "vibrant," and as such coexist with human actors in the web of our collective existence (13). Because OOO, as a complex emerging field of study, is impossible to treat in great depth here, I turn to a couple of OOO's key concepts to connect it to the ontology of AGS and the role of non-human actants in establishing agency over the anthropocene. Two key principles of OOO are particularly valuable in analyzing AGS in the philosophical context of speculative realism and phenomenology: the rejection of the centrality of humans to ecology and our inability to "know" an object absolutely. Each of these principles is embodied in the understanding and analysis of AGS, for the illness is epitomized in OOO's rejection of the anthropocene and the profound affirmation of the unknown.

First, OOO's rejection of anthropocentrism is central to this analysis. The principle that humans are the focus of ecology is a concept that OOO dismisses for speciesism, defined as the privileging of one's own species above others. Rather, OOO asserts that humans and non-humans exist on an equal ontological footing with each other in complicated webs of being. The relationship of ticks to poodles, pollen, windmills, or paper clips is no different to the kind of interactions these objects have with each other because they exist independently of human perception (Harman 2005, 1). In the realm of existence, objects—from parasites, pennies, *E. coli*, to hurricane winds—exhibit agentic influence over humans and non-humans alike because all objects deserve equal attention. In consideration of AGS, a disease-carrying tick exhibits its life-altering agency over the life of human, whose body cannot adapt to the changes its bite inflicts apart from substantial dietary and lifestyle alterations. A tick wields enormous influence over the alpha-gal body through the molecular contents of its saliva, which, in turn, wreaks systemic havoc within its susceptible victim—even as the tick goes about its business of eating, living, and being.

Secondly, OOO asserts that objects have an unknowable, irreducible dark side. Morton writes that "object is profoundly 'withdrawn'—we can never see the whole of it, and nothing else can either" (Morton 2011, 165). Morton offers the example of a coin, whose underside we never see *as* the underside; we always already see *this* side of a coin, which perpetually produces yet other undersides as it is flipped (165). In many ways, the AGS body represents one side of a coin; there is the seemingly healthy exterior—then there is the interior that may or may not be tormented by an overreactive immune system hotwired by a tick. The chaos attending the mysterious, delayed symptoms that are typical with alpha-gal reactions illustrate what Heidegger describes as an object's "irreducible dark side," or the ambiguity a thing possesses (Morton 2011, 165). Heidegger's "irreducible dark side" perfectly describes the sudden onset of the terror of anaphylactic reaction which can accompany a severe reaction to red meat, its byproducts, or cooking fumes. Or, there is the uncertainty whether the salad bar tongs at the restaurant buffet have made contact with neighboring ham cubes. An alpha-gal reaction reveals its irreducible dark side by making the AGS body strange to its inhabitant; the experience mimics what Morton terms "the strange stranger" (166). Morton describes the strange stranger experience as "the uncanny, radically unpredictability of life forms" made manifest by their complete mercuriality (166). Morton argues, "to think the strange stranger as an independent, solid, predictable object in advance of an encounter is to have domesticated it (or her or him) in advance" (166). Alpha-gal and its accompanying reactions will not be domesticated in the ubiquity of the elements that have the potential to trigger a reaction. AGS causes its host's body to feel what Hans-Georg Gadamer et al. (1996) describes as "a disorienting absence within oneself" due to illness (55).

The strange stranger whose body I occupy is *me*.

The body, as it goes about its ordinary work of taking me through my day, is inconspicuous as it involuntarily pumps blood, breathes, and does the business of making my existence possible. It is my host in the world, and, as a good host, my body withdraws, unnoticed—allowing me the opportunity to interact in a meaningful way in the web of being, that is, the context of my life. That is until my body demands attention through the arrival of unexpected AGS symptoms—the sudden feeling of doom that is a precursor to anaphylactic shock, for example, and the absolute fright the onset of anaphylactic symptoms imposes on the alpha-gal body—head to toe burning, hives, itching palms—the harbingers of nausea, diarrhea, and loss of consciousness that sometimes follows.

My body and her reactions are foreign and unpredictable.

Thomas Rickert (2013) argues, "it is not a given that we are simply at home, in ourselves, in our lives, in our world" (8). Bodily disruption in the form of pain or discomfort refocuses attunement to its condition in order to facilitate a response. Rickert notes that English speaking scholars translate Heidegger's term *Stimmung* as "attunement" or "mood" to "indicate one's disposition in the world, how one finds oneself embedded in a situation" (9). In other words, attunement is a referent to bodily awareness or the embodiment of the condition of sickness. Jean-Paul Sartre, another phenomenologist, takes up the subject of attunement via the example of a headache.[1] When plagued with a headache, the focus of my mind is tinged with pain—making reading, concentrating, or thinking, difficult. The discomfort of a headache imparts a "mooded" aspect to the activities in which I am engaged (Sartre 1957, 437). Heidegger conflates attunement and mood as two human qualities that lie dormant in our subconsciousness until aroused. Illness, Hans-Gadamer suggests, makes itself (and thereby its mood) known, penetrating our experience, and bringing us back to our "plagued embodiment" that is a result of AGS (Gadamer et al. 1996, 56).

Svenaeus (2000a) suggests Heidegger's perspective conveys an understanding that we inhabit a world of shared existence in which "the world I live in is first and foremost *my* world, but to this very 'mineness' also belongs otherness" (125). The consubstantial space between the "mineness" and "otherness" Heidegger describes illustrates the duality represented in the invisible illness of AGS. The AGS body maintains the appearance of "normalcy" as long as a careful regimen of mammalian meat avoidance is practiced. Heidegger's phenomenological approach is useful in understanding illnesses such as AGS in the context of the existential anxiety accompanying the complete or partial loss of health. Wellness is akin to a homelike being-in-the-world—a state of stasis in which the ankle, bladder, or big toe remains unnoticed—until the body calls attention to itself. By contrast, illness disturbs the equipoise of health, announcing itself through the intrusion of drastic symptoms and the suspension of the feeling of being at home in the world—invoking a state of otherness. The state of otherness is accentuated and reinforced around the topic of food, for inevitably, the source of my sustenance is also the source of my status as an outsider in the world of omnivores.

My relationship to food is ineluctably changed. A tick continues to daily perpetuate the power of its presence onto my humanity many years after walking into its web of being.

A Tick(et) into the Kingdom of the Sick

The experience of change through sickness, while unsettling, presents the opportunity for reflection and analysis as a means through which to process and resituate the unwell body in the context of its relationship to the world. The body, as it changes with age, or the insults of injury and/or sickness, reminds me of its impermanence; its temporal nature is exacerbated when the state of well-being is interrupted. Confrontation and adaptation to bodily change through the experience of illness is a journey addressed in many books, two of which are important in contextualizing the "otherness" of AGS.

In *Illness as Metaphor*, Susan Sontag (1978) writes of the embodiment of sickness: "Illness is the night-side of life, a more onerous citizenship. Everyone who is born holds dual citizenship, in the kingdom of the well and in the kingdom of the sick" (3). The bite of an alpha-gal bearing lone-star tick ushers in the metaphorical duality Sontag references because food becomes the mechanism through which we are forced to "identify ourselves as citizens of that other place" by inadvertently consuming or inhaling mammalian particulates that wreak havoc upon the alpha-gal body (3). Studies of rare illnesses indicate levels through which an ill body progresses during the course of sickness. For example, Arthur Frank (2013) writes in *The Wounded Storyteller* of three narratives the unwell body "tells" itself and its listeners to structure, interpret, and deal with bodily affliction (vix). Frank's narratives—restitution, chaos, and quest—are useful in the discovery of how a person with an illness negotiates life with a diminished body—even as that body gives voice to its own story. Each of the struggles encountered in the restitution, chaos, and quest narratives illustrate one of Graham Harman's (2018) key principles of OOO: "objects are not identical with their properties, but have a tense relationship with those properties, and this very tension is responsible for all of the change that occurs in the world" (9). The discordance illustrated in the restitution, chaos, and quest narratives underscore the tension required in order foment change—accidental or purposeful change—to contribute in a small way to environmental wellness through vegetarianism.

While *The Wounded Storyteller* explores illness from the perspective of those suffering from chronic conditions like multiple sclerosis and cancer, its arguments are useful in understanding and coming to terms with the rare, challenging, and potentially fatal condition of AGS through the stories of its sufferers, like me. The restitution, chaos, and quest narratives of *The Wounded Storyteller* provide a glimpse into AGS illness as embodiment, as alienation (homelessness), and the experience of what Heidegger describes as the uncanny strange stranger that is sickness.

Frank's *Wounded Storyteller* first describes the restitution narrative from the perspective of an "insider" in Sontag's metaphorical "Kingdom of the Sick." New relationships and hierarchies are established as help is sought to identify and rectify recurrent bodily malfunction. In the restitution narrative of illness, the physician is the protagonist and the patient is the object of the doctor's heroism. The restitution narrative involves the patient's subject position; it is the journey, fraught with innumerable obstacles on the uncertain, but much-desired recovery to health. However, AGS complicates the journey metaphor. Unlike a typical journey in which one moves in a linear progression from point A to point B, the AGS journey is iterative, uncertain, and quite likely, never ending. With AGS as with many other chronic illnesses, there is no sequential prediction of a patient's progress from diagnosis to recovery; due to the relative newness of the condition, the wildly unpredictable symptoms which accompany AGS and the dearth of research on the illness, AGS sufferers must cultivate an attitude of abiding patience tinged with hope. Recognition of suffering is foundational in the restitution narrative, which focuses on the shift from normalcy to affliction (Frank 1995, xiv). The juncture dividing what *was* from what *is* offers a lacuna for introspection— introspection that is easily obfuscated by the complacency cultivated in the ordinary.

To cultivate support and nurture hope, community groups established through social media offer outlets for us to express our stories, seek help, information, and advice from others regarding our collective condition. The vehicle of writing and speaking our AGS stories to each other is a form of encouragement, support, and education which is invaluable as we share the panic that delayed allergic reactions induce within the alpha-gal body—a terror that renders our bodies strange and unpredictable to us. The sharing of stories of disruption offers an important rhetorical tool to AGS sufferers. As we share experiences of how the AGS body communicates its profound distress caused by alpha-gal intrusion, we come to appreciate the rhetoricity of the body as a platform of persuasion. The AGS body resists mammalian meat as a toxin, and in turn, the AGS body comes to realize the possibilities of living without it. Embracing a life that elides meat consumption is a means to foster well-being in the constitution of our own AGS restitution narrative. In seeing meat for what it is, the flesh of animals, the AGS body coheres with the life forces at work in nature to take command not only of our own wellness narrative but in the wellness narrative of the sentient beings with whom we share the planet. It is a dynamic shift in thinking and acting that transcends the body and affects the world beyond my skin.

Consider, for example, the lives of chickens in intensive factory farming operations. Tactics such as genetic engineering and selective breeding produce chickens who, within the short span of six weeks grow so large that their legs are unable to support their own weight. Denied sunshine or the feel of grass under their feet, chickens are "kept in almost constant light so they remain stimulated to eat" (Potts 2017, 14). The invisible life of factory-farmed animals replaces sentience with the chaos of commodification, reducing the animal to a "thing" instead of a being. Richard Twine (2014) suggests that animal products and the violence that is inherent to their journey to our plates "do not constitute an overt presence at the dinner table" (626). The invisibility of animals is made manifest "through language and material fragmentation" that is socially and culturally reinforced through animal referents such as the table talk word "protein" (626). The conception of "normalcy" surrounding meat culture conceals animal suffering and desperation—cloaking their misery in the curtain of the commonplace. Intensive animal farming virtually eliminates any notion of normalcy for animals unfortunate enough to experience it.

Following the restitution narrative and the recalcitrant embrace of the end of "normalcy" is *The Wounded Storyteller's* chaos narrative. The chaos narrative describes the "claustrophobia of confronting others' inability to see" what a patient is feeling (Frank 2013, vix). Chronic conditions such as AGS expose the stigma of invisible illnesses particularly within a culture that is biased toward normalcy. James Aho and Kevin Aho (2009) argue that healthy bodies are so interwoven into the world that they tend to disappear (105). Healthy bodies articulate a silent visual rhetoric that does not necessarily communicate deviance from cultural standards. However, when the body malfunctions, it speaks by drawing attention to itself through pain, allergic reaction, or other such disruption. Welsh

Talia (2016) writes, "In health, the body schema is largely hidden and allows the individual to extend him or herself into the world and become busy with projects ... In illness, the breakdown of normal functioning brings the body into focus" (341). Alpha-gal patients find ourselves perpetually explaining our unusual illness to those whose actions (or inaction) might inadvertently create disruption within the alpha-gal body. Interaction and communication with those whom we entrust with our nourishment highlight the sociological nature of chronic conditions such as AGS.

Additionally, the chaos narrative personifies the maelstrom of alpha-gal's bizarre array of symptoms and the incredulity with which those symptoms are often met—from people whom we, as AGS patients, empower and entrust with our bodies and our health—from restaurant servers and cooks to emergency room doctors, and for some, even family members who simply refuse to believe that such a strange condition exists. For many AGS patients, the diagnosis offers validation that we suffer from a "real" illness—an odd comfort against the fear of neurosis the delayed symptoms can create. However, lack of awareness of AGS—even among the medical community—maximizes the agency AGS patients must exert in maintaining our status as citizens in the "kingdom of the well" through proactive measures and an abundance of caution as the AGS body negotiates its way through the world. An unseeable, unknowable object has altered my body and my way of being in the world through the process of passive transfer of a molecule from one body to another. The simple act of a tick feeding upon my blood has initiated an active response in keeping my own body healthy from the toxic reality of the content of its saliva.

The Wounded Storyteller's third narrative, the quest narrative, is the recognition of illness as a source of possibility in casting the disappointment that comes with a drastic change in lifestyle into the shadows of a well-crafted life. However, AGS differs from many illnesses in that its invisibility and its relative newness to the field of medicine renders the condition perplexingly anomalous—leaving its victims with a greater need to maintain advocacy and agency over our own health particularly until the condition becomes more widely recognized and accepted. *The Wounded Storyteller's* three narratives of restitution, chaos, and quest are apt descriptors of coming to terms with a medical condition that is not only life-threatening, but additionally involves a long period of adjustment—adjustment for the alpha-gal patient, and for those whose daily lives intersect her own through the intimate act of frequently shared meals. Celebrations with friends and family involving table-laden feasts may portend disaster in the form of tiny pieces of flavor-enhancing bacon or prosciutto ham, a delight to most carnivores. Seemingly safe baked-beans at the backyard cookout could mask the cook's secret ingredient of beef, one swallow of which could initiate a long night in the emergency room for the AGS body. Parties, gatherings, or ball games where the wafting smell of grilled meats takes flight on the wind is enough to bring about fits of nausea, or worse. The hidden object brings attention to itself through the involuntary act of an alpha-gal sensitive body breathing in mammal particulate.

Eating out is not only a risky endeavor for alpha-gal sufferers, but it carries the element of embarrassment and the propensity for being misunderstood. Annie Potts and Philip Armstrong (2018) describe the perception of a vegan/vegetarian who finds herself among flesh-eaters as an "'affect alien' who struggles against the dominant affective order that is part of hegemonic 'meat culture'" (400). For example, asking a server to grill my eggplant on foil (to avoid dangerous contact with beef particulate) may be misinterpreted as a trifling demand from a vegetarian customer who is perceived as "morally superior" to indulge in the celebrated fare of the restaurant's meat-eating status quo. Twine (2014) writes of the "vegan killjoy" who, "like the feminist, … is also assigned to a category of difficulty and sometimes will want to avoid being outed, not to be difficult, or to cause a fuss" (626). Vegetarians, who comprise a mere 5% of the US population, usually find ourselves in the minority, and this is particularly so in the meat-loving American South (Gallup 2020).

In sum, AGS requires its human host to seek the newness of exploring gustatory adventures in vegan/vegetarianism and find joy in the ramifications of a meat-free existence. In seeking the wide array of possibilities that vegetarianism offers as an ethical, sustainable, salubrious way of living, animals emerge from the shadows and take form as more than units of commodification for the benefit of other animals.

Vegan "Killjoys," Carnivores, and Coexistence

The centrality of food to the human experience is as undeniable as our need for water, oxygen, and shelter. Its constancy in our daily lives makes food a natural hub around which we share commodities while building and strengthening relationships around the pleasurable activity of eating. Food, as the point of convergence, has its own implicit rhetoricity when contemplating perceptions of normalcy. For instance: when advised by restaurant waitstaff that all prepared foods in their kitchen come into some degree of contact with pork offal, would it be peculiar to order a peanut butter sandwich when dining with friends at a famed barbeque restaurant? Or, what's a vegetarian to do deep in the American South where vegetables are commonly seasoned with pork? Twine (2014) writes of the conundrum sometimes facing the vegetarian who is a perceived threat to the prevailing "happiness order" by her very presence at the table (626). He states, "the vegan [or vegetarian] does not even have to purposefully engage, by arguing against someone's animal consumption, to be a killjoy" before the tired scripts of "I could never give up meat" emerge (626). This refrain often rings true when explaining the boundaries imposed by AGS, particularly to males who often cringe at the thought of a lifetime of Saturday nights without steak.

What is important about the above-referenced scenarios is the underlying rhetoricity of normalcy that pervades even mundane situations that subtly communicate the deviance residing in the alpha-gal body. The aforementioned examples of the social relationship of food and eating unintentionally enforce the social isolation and feelings of alienation that AGS patients face on a continual basis. With so many potential reactive triggers lurking in the shadows of everyday life, it is easy to understand why many who suffer from AGS also develop varying degrees of anxiety. Each day is an exercise in cautionary survival when the specter of anaphylaxis looms as a distinct possibility. Case in point: recently, I enjoyed dinner with my husband at an upscale restaurant in Memphis. Taking pains to inform the waitstaff of my allergy to meat, I implored our server to make certain that the chef excluded any form bacon from my order, although the menu made no reference to bacon in the description of my dish. The first bites of my entree had an odd texture that, with horror, I recognized was bacon. Despite efforts to educate others of the potentially deadly effects of AGS, insouciance or error can nonetheless contribute to a night in the hospital or worse—an occasion I was spared from this time, due to my quick action.

Sharing our stories with others who are similarly afflicted creates a mechanism by which we can process, come to terms with, and even find meaning or purpose in the embodiment that is our alpha-gal condition, a condition from which escape may be little more than an illusion. The cessation of normalcy that accompanies the diagnosis of a chronic condition such as AGS exposes what Heidegger describes as our "being in the world" and how sickness changes our relationships to others and to the world we inhabit. Svenaeus (2000b) offers an important way to understand and think about health and unwellness through Heidegger's analogy of illness to *Unheimlichkeit* –unhomelikeness and how the experience of illness is akin to "not being at home in my own world" (9). The embodied feeling of homelessness is exacerbated with the onset of a strange and rare illness about which knowledge—even among medical professionals—is at the moment, limited. Before the alpha-gal laced tick saliva entered my bloodstream, altered my way of being, and made my body chronically and unexpectedly strange to me, I had not yet experienced the *otherness* of illness—the feeling of discomfit at inhabiting a body which now seems somehow foreign, unpredictable, scary. Health, by contrast, is the status quo—it is a homelike sense in which relatively few surprises appear as the lived body goes on unobtrusively about its business of *being*.

Conclusion

The bodily experience of illness and its accompanying "otherness" offers a unique opportunity for attunement, reevaluation, and philosophical awareness that is easily overlooked in the optimal condition of health. A phenomenological perspective of the hiccups in our existence—such as AGS—create lacunary possibilities for allowing things, objects, and ideas to reveal themselves in unexpected ways, to consider the presence involved in absence. AGS opened up opportunities for a former carnivore to reevaluate her relationship to animals—to view them as more than livestock processed for their muscles, bones, and connective tissue—and consider the "inconveniences" of AGS in contrast to the suffering imposed upon fellow mammals raised for food—factory farm animals that were once inconspicuous because of their ubiquity. The absence of red meat has made visible the oppression and suffering of the slaughterhouse "other"—something that may have remained invisible until change imposed opportunity.

Each of us comes to vegetarianism intentionally or accidentally, as I did, or as a manifestation of an ideology to which we are deeply committed. Deane Curtin and Lisa Heldke (1992) write, "Vegetarianism can be seen as a political act of self-empowerment that resists the externalizing pressures of society. To choose one's food and define oneself by that choice in opposition to a dominant conceptual scheme is empowering" (132–133). In cultivating empathy for each other and the world that we inhabit, we can enhance the experience of our mutuality of *being* in the world through vegetarianism.

What I've learned is that alpha-gal did not shrink the corners of my *being*, it enlarged it through an appreciation of absence—and what the absence of meat may mean not only to my body but to the planet we call home.

Note

1. For further description, see Jean-Paul Sartre's *Being and Nothingness: An Essay on Phenomenological Ontology; Translated and with an Introduction by Hazel E. Barnes*. London: Methuen, 1957.

Bibliography

Aho, James A., and Kevin Aho. 2009. *Body Matters: A Phenomenology of Sickness, Disease, and Illness*. Lanham: Lexington Books.

Bennett, Jane. 2010. *Vibrant Matter: A Political Ecology of Things*. Durham: Duke University Press.

Chung, Christine H., Beloo Mirakhur, Emily Chan, Quynh-Thu Le, Jordan Berlin, Michael Morse, Barbara A. Murphy, et al. 2008. Cetuximab-Induced Anaphylaxis and IgE Specific for Galactose-α-1,3-Galactose. *New England Journal of Medicine* 358 (11): 1109–1117. https://doi.org/10.1056/nejmoa074943.

Commins, Scott P., and Thomas A.E. Platts-Mills. 2012. Delayed Anaphylaxis to Red Meat in Patients with IgE Specific for Galactose Alpha-1,3-Galactose (Alpha-Gal). *Current Allergy and Asthma Reports* 13 (1): 72–77. https://doi.org/10.1007/s11882-012-0315-y.

———. 2013. Tick Bites and Red Meat Allergy. *Current Opinion in Allergy and Clinical Immunology* 13 (4): 354–359. https://doi.org/10.1097/aci.0b013e3283624560. Ovid Technologies (Wolters Kluwer Health).

Commins, Scott P., Shama M. Satinover, Jacob Hosen, Jonathan Mozena, Larry Borish, Barrett D. Lewis, Judith A. Woodfolk, and Thomas A.E. Platts-Mills. 2009. Delayed Anaphylaxis, Angioedema, or Urticaria after Consumption of Red Meat in Patients with IgE Antibodies Specific for Galactose-α-1,3-Galactose. *Journal of Allergy and Clinical Immunology* 123 (2): 426–433. https://doi.org/10.1016/j.jaci.2008.10.052.

Commins, Scott P., Hayley R. James, Libby A. Kelly, Shawna L. Pochan, Lisa J. Workman, Matthew S. Perzanowski, Katherine M. Kocan, et al. 2011. The Relevance ofTick Bites to the Production ofIgE Antibodies to the Mammalian Oligosaccharide Galactose-α-1,3-Galactose. *Journal of Allergy and Clinical Immunology* 127 (5). https://doi.org/10.1016/j.jaci.2011.02.019.

Craig, Courtney. 2019. Alpha-Gal Meat Allergy: A New Tick-Borne Illness. Blog. *Dr. Courtney Craig.com*. https://www.drcourtneycraig.com/blog/alpha-gal-meat-allergy

Curtin, Deane W., and Lisa M. Heldke. 1992. *Cooking, Eating, Thinking: Transformative Philosophies of Food*. Bloomington: Indiana University Press.

Flaherty, Mary Grace, Samantha Jan Kaplan, and Maya R. Jerath. 2017. Diagnosis of Life-Threatening Alpha-Gal Food Allergy Appears to Be Patient Driven. *Journal of Primary Care & Community Health* 8 (4): 345–348. https://doi.org/10.1177/2150131917705714.

Foronda, Jay. 2019. Alpha-Gal Allergy – With Dr. Scott Commins | Department of Medicine. *Department of Medicine University of North Carolina*. https://www.med.unc.edu/medicine/news/chairs-corner/podcast/alpha-gal/

Frank, Arthur W. 2013. *The Wounded Storyteller: Body, Illness, and Ethics*. Chicago: University of Chicago Press.

Gadamer, Hans-Georg, Jason Gaiger, and Nicholas Walker. 1996. *The Enigma of Health: The Art of Healing in a Scientific Age*. Stanford: Stanford University Press.

Gallup, Inc. 2020. Snapshot: Few Americans Vegetarian or Vegan. *Gallup.com*. https://news.gallup.com/poll/238328/snapshot-few-americans-vegetarian-vegan.aspx

Harman, Graham. 2005. *Guerrilla Metaphysics: Phenomenology and the Carpentry of Things*. Chicago: Open Court.

———. 2018. *Object-Oriented Ontology: A New Theory of Everything*. London: Pelican Books.

Hilger, Christiane, Jörg Fischer, Florian Wölbing, and Tilo Biedermann. 2019. Role and Mechanism of Galactose-Alpha-1,3-Galactose in the Elicitation of Delayed Anaphylactic Reactions to Red Meat. *Current Allergy and Asthma Reports* 19 (1). https://doi.org/10.1007/s11882-019-0835-9.

Hoek, Annet C., P.A. Luning, A. Stafleu, and C. de Graaf. 2004. Food-Related Lifestyle and Health Attitudes of Dutch Vegetarians, Non-vegetarian Consumers of Meat Substitutes, and Meat Consumers. *Appetite* 42 (3): 265–272. https://doi.org/10.1016/j.appet.2003.12.003. Elsevier BV.

Høiseth, M., and M.M. Keitsch. 2015. Using Phenomenological Hermeneutics to Gain Understanding of Stakeholders in Healthcare Contexts. *International Journal of Design* [Online] 9: 33–45. http://www.ijdesign.org/index.php/IJDesign/article/view/1947/709

Kenyon, P.M., and M.E. Barker. 1998. Attitudes Towards Meat-Eating in Vegetarian and Non-vegetarian Teenage Girls in England—An Ethnographic Approach. *Appetite* 30 (2): 185–198. https://doi.org/10.1006/appe.1997.0129. Elsevier BV.

Kwak, Mackenzie, Colin Somerville, and Sheryl van Nunen. 2018. A Novel Australian Tick Ixodes (Endopalpiger) Australiensis Inducing Mammalian Meat Allergy After Tick Bite. *Asia Pacific Allergy* 8 (3). https://doi.org/10.5415/apallergy.2018.8.e31. Asia Pacific Association of Allergy, Asthma, and Clinical Immunology (KAMJE).

Latour, Bruno. 2005. *Reassembling the Social: An Introduction to Actor-Network Theory*. Oxford: Oxford University Press.

———. 2017. On Actor-Network Theory. A Few Clarifications, Plus More Than a Few Complications. *Philosophical Literary Journal Logos* 27 (1): 173–197. https://doi.org/10.22394/0869-5377-2017-1-173-197.

McKenna, Maryn. 2018. Why Are so Many People Getting a Meat Allergy? Blog. *Mosaicscience.com*. https://mosaicscience.com/story/mammalian-meat-allergy-alpha-gal-allergic-lone-star-tick-bite/

Morton, Timothy. 2011. Here Comes Everything: The Promise of Object-Oriented Ontology. *Qui Parle* 19 (2): 163–190. https://doi.org/10.5250/quiparle.19.2.0163.

Plumwood, V. 2000. Integrating Ethical Frameworks for Animals, Humans, and Nature A Critical Feminist Eco-Socialist Analysis. *Ethics and the Environment* 5 (2): 285–322. https://doi.org/10.1016/s1085-6633(00)00033-4. Elsevier BV.

Potts, Annie. 2017. *Meat Culture*. Leiden: Brill.

Potts, Annie, and Philip Armstrong. 2018. Vegan. In *Critical Terms for Animal Studies*, ed. Lori Gruen. Chicago: University of Chicago Press.

Rickert, Thomas. 2013. *Ambient Rhetoric: The Attunements of Rhetorical Being*. Pittsburgh: University of Pittsburgh Press.

Sontag, Susan. 1978. Illness as Metaphor. *The New York Review of Books*, January 26. https://www.nybooks.com/articles/1978/01/26/illness-as-metaphor/

Steinke, John W., Thomas Platts-Mills, and Scott P. Commins. 2015. The Alpha-Gal Story: Lessons Learned from Connecting the Dots. *Journal of Allergy and Clinical Immunology* 135 (3): 589–596. https://doi.org/10.1016/j.jaci.2015.01.005.

Svenaeus, Fredrik. 2000a. The Body Uncanny-Further Steps Towards a Phenomenology of Illness. *Medicine Health Care and Philosophy* 3 (May): 125–137. https://doi.org/10.1023/A:1009920011164.

———. 2000b. *Medicine, Health Care and Philosophy* 1: 3–16. https://doi.org/10.1023/a:1009943524301.

———. 2010. Illness as Unhomelike Being-in-the-World: Heidegger and the Phenomenology of Medicine. *Medicine, Health Care and Philosophy* 14 (3): 333–343. https://doi.org/10.1007/s11019-010-9301-0.

Talia, Welsh. 2016. Many Healths: Nietzsche and Phenomenologies of Illness. *Frontiers of Philosophy in China* 11 (3): 338–357. https://doi.org/10.3868/s030-005-016-0026-3.

Taylor, Nik. 2012. *Reversing Meat-Eating Culture to Combat Climate Change*. Haselmere: World Preservation Foundation.

Taylor, Steven J. 2016. Before It Had a Name: Exploring the Historical Roots of Disability Studies in Education. In Forward to *Vital Questions Facing Disability Studies in Education*, ed. Susan Gabel and Scot Danforth, vol. 2, xiii–xxiii. New York: Peter Lang.

Twine, Richard. 2010. *Animals as Biotechnology*. London: Earthscan.

———. 2014. Vegan Killjoys at the Table—Contesting Happiness and Negotiating Relationships with Food Practices. *Societies* 4 (4): 623–639. https://doi.org/10.3390/soc4040623.

Wilson, Jeffrey M., Alexander J. Schuyler, Lisa Workman, Monica Gupta, Hayley R. James, Jonathon Posthumus, Emily C. Mcgowan, Scott P. Commins, and Thomas A.E. Platts-Mills. 2019. Investigation into the α-Gal Syndrome: Characteristics of 261 Children and Adults Reporting Red Meat Allergy. *The Journal of Allergy and Clinical Immunology: In Practice* 134 (1): 108–115. https://doi.org/10.1016/j.jaip.2019.03.031.

Wolver, Susan E., Diane R. Sun, Scott P. Commins, and Lawrence B. Schwartz. 2012. A Peculiar Cause of Anaphylaxis: No More Steak? *Journal of General Internal Medicine* 28 (2): 322–325. https://doi.org/10.1007/s11606-012-2144-z.

Zimmer, Carl. "The Rise of the Tick." *Outside Magazine*, April 30 (2013). https://www.outsideonline.com/1915071/rise-tick

7

"You Are What You Eat": Oprah, Amarillo, and Food Politics

Callie F. Kostelich and Heidi Hakimi-Hood

Food, eaten and digested is not rhetorical. But in the meaning of food there is much rhetoric, the meaning being persuasive enough for the idea of food to be used, like the ideas of religion, as a rhetorical device of statesmen (173).
—Kenneth Burke, Rhetoric of Motives

While Kenneth Burke's *Rhetoric of Motives* predates the foodie revolution that swept the United States in the late twentieth- and early twenty-first centuries, Burke's concept of food as a rhetorical device is perhaps more poignant than ever given the proliferation of food culture since 1980 when the term "foodie" entered the lexicon.[1] Food critic Gael Greene

---

C. F. Kostelich (✉)
Texas Tech University, Lubbock, TX, USA
e-mail: Callie.Kostelich@ttu.edu

H. Hakimi-Hood
Midwestern State University, Wichita Falls, TX, USA
e-mail: heidi.hood@msutexas.edu

C. Hanganu-Bresch, K. Kondrlik (eds.), *Veg(etari)an Arguments in Culture, History, and Practice*, The Palgrave Macmillan Animal Ethics Series,
https://doi.org/10.1007/978-3-030-53280-2_7

(1980, 33) wrote the term "foodie" for the first time in *New York Magazine* in reference to Chef Dominique Nahmias, the "Crayfish Queen": "She offers crayfish with white feed or red ... three ways, tends stove in high heels, slips into the small Art Deco dining room of Restaurant d'Olympe ... to graze cheeks with her devotees, serious foodies, and, from ten on, tout Paris, the men as flashily beautiful as their beautiful women" (33). "Foodies," in Greene's reference, were a small collection of privileged people—rich men—who knew where to dine, what chefs to follow, how to experience food not at their kitchen tables or local diners but at the brightest, flashiest establishments with the best chefs and alongside impeccable company. Or, at the least, these foodies were sophisticated enough to know which individuals, such as Greene, to follow and which publications, such as *New York Magazine,* to read to learn of their next culinary conquest.[2] In this 1980s context, food critics and writers were those who directed the culinary scene, and foodies were people of distinction—those who had the economic capital to partake in culinary delights and the cultural capital to know how to do so.[3]

Fast forward four-decades and our language for referencing food and "foodies" has greatly changed, as have the people who hold the power of food persuasion. When food writer Paul Levy (2007) posed the rhetorical question, "I wonder what the word [foodie] means today?", he acknowledged the evolution of the term and the ever-changing food landscape. Granted, while the term "foodie" connects to the late twentieth century, the food landscape in the United States has never been stagnant, and many of our food-based traditions and practices are not "timeless or universal" but rather "highly variable and only recently constructed" (Belasco 2008, 19). For example, vegetarians were some of the earliest food activist-influencers in the United States. English Bible Christians arrived in the United States in 1817 to further, as Shprintzen notes, their "meat-abstinence" practices, and church leadership "equated meat consumption with violent, cruel tendencies appealing to the most uncontrolled whims of human aggression" (15). For immigrant Bible Christians, cruelty-free ideologies were a foundation in their belief system, so much so that they moved to a new country to practice these beliefs in peace. While these Bible Christians represent a very different type of "foodie" from the 1980s, they serve as a reminder that food is—and has been—central not only to daily life but to how people experience it and for what purposes, be it for social distinction in the fine dining restaurants Greene visited or religious life and freedom in the United States.

In the twenty-first century, food has become central to pop culture with round-the-clock food shows on television, on-demand culinary podcasts, food-based publications dominating grocery store check-out lines, food trucks roaming the streets, and diners and home cooks alike posting meal pictures on Instagram and tutorials on Facebook Live. To be a foodie in this context is less about having the right access to the right restaurant, as in Greene's description, and more about direct engagement with food as "another form of entertainment and self-expression" (Turow 2015, 6). How we engage with food and who has the privilege to call themselves a "foodie" is now in the hands of the general public, quite literally with food bloggers via mobile devices and social media streams. In this regard, Levy's focus on how we define "foodies" seems trivial in this day and age and reflects gatekeeping efforts to distinguish those who are *really* into food from those who are somehow posing. It is the implied questions within his statement that are most intriguing: Who or what entity has a say in defining how people engage with food? What are the implications of foodies and food influencers expanding beyond the bourgeoisie to the masses?

In this chapter, we turn to one particular example of taking food culture to the masses as we revisit *Texas Beef Group v. Oprah Winfrey* 20 years posttrial, a trial that was as much a referendum on the talk show host herself as it was a backlash to her on-air rejection of beef. Following Oprah's televised proclamation on April 16, 1996, that "[the threat of mad cow disease] has just stopped me cold from eating another burger," members of the cattle industry in Texas filed suit, blaming the talk show host for violating the newly minted "veggie libel laws," which prohibited food disparagement. In this legal battle, two opposing food influencers were caught in a liminal space: Oprah sought justice for free speech and meat consumers, and the cattle industry desired advocacy for beef producers. During the trial, Oprah became a central figure in Amarillo, Texas, and across the globe, a powerful stateswoman of food culture. She simultaneously navigated the courtroom, the town, and her talk show, while defending her right to free speech and commandeering her position as an American icon in a space that was less than welcoming to an African-American, meat-disparaging woman.[4]

For our study, while the trial in Texas garnered significant national attention regarding free-speech rights, what ensued outside the courtroom is of particular importance for analysis: how concepts of food are directly connected to race, gender, and power dynamics. We look closely at the ways in which Oprah responded to criticism and crafted a counter-narrative cemented in her resistance by tactically affiliating (de Certeau 1984) with the local community through subversive and embodied moves, such as embracing local vernacular and donning Western attire. While the lead plaintiff in the suit stated that "we cleaned up her act," a gendered and patriarchal response, mapping the intersectional rhetorics of the trial allows us to uncover complex social intersections at a crucial point in the foodie revolution (Batheja 2018). We contend that Oprah's success navigating these complex spaces is apparent in not only the trial's outcome but in how the public has reflected on the trial 20 years later. For the current foodie movement, including the burgeoning vegan awakening in the United States, Oprah's example provides a rhetorical framework for foodies and food influencers, whoever they may be, to consider as they continue to pave the way for a broader, more inclusive, and equitable food culture.

"The only mad cow in Amarillo is OPRAH"

Following the fallout from the Alar scare in 1989 that greatly impacted apple growers' revenue, 13 states, including Texas, passed "veggie libel laws," which prohibited food product disparagement (Hayenga 1998, 13).[5] When Oprah discussed Bovine Spongiform Encephalopathy (BSE), aka mad cow disease, on *The Oprah Winfrey Show* segment, "Dangerous Food" on April 16, 1996, she set forth a controversy that proved to be the first legal challenge to the new veggie libel laws, particularly the False Disparagement of Perishable Food Products Act in Texas. During the segment, guests discussed mad cow disease and the possibility of an outbreak in the United States, with Oprah concluding publicly that "it has just stopped me cold from eating another burger" (Winfrey 1996).[6] Within an hour of the broadcast, cattle futures plummeted on the Chicago Mercantile Exchange, dropping below the limit for suspending trading ("Appeals" 2000). In the weeks to follow, beef producers, particularly the feedlot industry, lost approximately 87.6 million dollars, and the economic disaster became known as "The Oprah Crash" (Hayenga 1998; Hollandsworth and Colloff 1998).[7]

That Oprah alone can be solely credited with the public reaction is highly unlikely, as she was far from the only influential figure to cover the mad cow outbreak. As quoted in Hollandsworth and Colloff (1998), Teel Bivins, Texas state senator and rancher, recounted the complexity of Oprah as a contributing factor to the market downturn in their *Texas Monthly* coverage: "Cattle prices had already been slipping somewhat … but Oprah's show created a panic. The cattle market basically went into a free-fall." To substantiate Bivins' perspective, Hollandsworth and Colloff add that "one Texas A&M economist said that in the three weeks following the Oprah show the cattle-feeding industry lost $87.6 million, although other observers blamed the loss on a devastating drought and an already volatile market shaken by Britain's mad cow scare."[8] This data indicates that the connection between Oprah's broadcast and the downturn in the beef market was not coincidental. Beef producers were furious that show producers had edited out two of the guests' clarifications that beef feeders in the United States had voluntarily implemented safety procedures, such as banning "ruminant-to-ruminant" feeding ("Appeals" 2000). While Oprah hosted Dr. Weber and cattle producer Connie Greig a week later for the "Dangerous Food Follow-Up" segment on April 23, 1996 to clarify the cattle feeders' perspectives and reiterate safety protocols, the damage was already done. In Texas, angry cattle producers and feeders filed suit against Oprah and her production company for violating Texas Civil Practice and Remedies Codes §96.001, §96.002, §96.003 of the False Disparagement of Perishable Food Products Act.[9]

By taking Oprah to trial in January 1998, the beef feeders, particularly Paul Engler who filed the suit on behalf of Cactus Feeders Inc., put the new veggie libel laws to the test, and the trial became a spectacle of free speech rights and a space for what *Texas Monthly* writers Skip Hollandsworth and Pamela Colloff (1998) called "a battle for the heart and soul of the state." The outcome of the trial was overwhelmingly in

Oprah's favor. Judge and jury vindicated Oprah by finding that there was insufficient evidence that she had knowingly and purposefully disseminated false information. Moreover, the court found that the veggie libel law did not apply to cattle, and the jury unanimously voted in favor of Oprah on the claim of business disparagement. In short, the trial was an uncontested legal victory for Oprah and for free-speech rights ("Appeals" 2000).

As native Texans, we remember when Oprah came to town. Heidi was living in Amarillo, where her late father worked for a local industrial beef-processing facility. She witnessed the spectacle firsthand and remembers seeing the *BBC* and multitudes of other news outlets arrive to cover the trial in a normally quiet downtown. She also sadly recalls her students' parents disagreeing about the First Amendment as it pertained to Oprah. Callie was living in central Texas, in a beef-producing family that directly felt the impact of the declining cattle market. As young women, we remember the Oprah trial dominating our families' television sets and talk at the dinner table. As scholars, we now occupy an interesting position: we grasp the gravity and real-life implications of what happens when opinions are spread—or interpreted—as fact on national platforms. There were real-life consequences to how viewers interpreted the "Dangerous Food" episode, particularly for beef producers and feeders who felt the public backlash and economic impact when one guest, Howard Lyman, gave his opinion on the mad cow crisis, and Oprah responded with an emotional reply that indicated her choice—at least at the moment—to refrain from eating the product. Our families literally felt the impact of the crashing beef market that followed.

At the same time, we grasp the complexity of the issue: laws, such as veggie libel laws, that serve to silence advocates who counter narratives propagated by big agribusinesses are dangerous to free-speech rights in the United States. Moreover, these veggie libel laws beg the questions: who gets to say what information is "right" and "accepted" and in what ways does the legal system punish those who may deviate from that information? While the trial itself is of great interest, particularly for the study of free speech in the United States, what is of most interests to us are the ways in which Oprah navigated the trial and her time in Amarillo, Texas—how she entered a "battle" in the heart of the Texas Panhandle, where cattle is king, and left victorious, not only setting forth legal precedent against veggie libel laws but winning over the hearts and minds of local folk in the process.[10]

When Oprah arrived in Amarillo, a town of 170,000 residents, she entered a place dominated by cattle and the feedlot industry (Batheja 2018). Rose Farley (1998), journalist for *Houston Press,* offered a colorful—and for locals, fairly accurate—rendering of the town:

> Flat as a griddle and straddling Interstate 40, Amarillo is the very middle of middle America. The average worker here pulls down $24,000 a year in wages, and the city's housing stock offers everything from three-story colonial homes on the west side to rundown shotgun shacks on the north side. The local economy was built on meat: Amarillo is home to Texas's largest cattle auction, and some days, when the wind blows just right off the feedlots, the entire city is filled with the ripe scent of cow poop.

Amarillo was, and still is, heavily dependent on the cattle industry, and residents and visitors alike literally breathe in the smells of local livelihoods.[11] The local vernacular claims that the scent of cow manure is, indeed, "the smell of money." It is to this space that Oprah traveled to defend herself, to, in her own words, "defend my name … I feel in my heart I've never done a malicious act against any human being" (Hollandsworth and Colloff 1998). For Oprah, the trial was an opportunity to fight for First Amendment rights, while also trying, quite literally, to vindicate herself from the controversy. Yet to do this meant that she went to the heart of the issue, to the place that literally embodied the cattle industry in sight, sound, and smell. She was in the land of the Big Texan Steak Ranch, a restaurant whose claim to fame is a free 72-ounce steak for anyone who can finish the meat and all the fixings (The Big Texan). She was in the land of pickup trucks and cowboy hats, images that are not stereotypes but representative of the very ways of being for many local residents. To go on trial in Amarillo meant that Oprah could not escape residents' deeply embedded ideologies about what it meant to disparage a food so central, perhaps essential, to their livelihoods.

At the onset of the trial, public perception in Amarillo was largely against Oprah, with protestors reacting to the talk show host's arrival for trial with "The Only Mad Cow In Amarillo is OPRAH" bumper stickers, "Ban Oprah" buttons, and T-shirts with Oprah's face on the front beneath a large, red "no" symbol (Batheja 2018; Hollandsworth and Colloff 1998). Oprah's declaration that she would not eat "another burger" set forth an ideological debate on food choice and power, one that harkens back to Brillat-Savarin's signature: "Tell me what you eat, and I shall tell you what you are" (*The Physiology of Taste*, 1994). For supporters of the plaintiff, Oprah was a "mad cow." The phrase "mad cow" equates Oprah's decision to refrain from eating beef for some time to mental instability, insinuating that to not eat meat can only be attributed to "madness," not to a person's rational and reasonable choice. Moreover, in evoking Brillat-Savarin's framework, the protesters were quite visibly telling Oprah what they thought of her as they objectified her body by calling her a "cow." This gendered insult is an overt reference to Oprah's public weight management, which was a frequent topic on her television show and in the media. The protesters' reference stripped Winfrey of her personhood, reducing her to something that only served to benefit the protestors: she was a commodity that could be slaughtered for profit. To them, Oprah was neither a person with free-speech rights nor an individual with the ability to think critically and carefully about a subject. Instead, she was someone they discredited as mentally unstable, an animal fit for consumption.

If the reference to madness as a form of mental instability was not enough, the reference to Oprah as livestock—a cow—signifies the patriarchal culture that pervaded the area. To the plaintiff's supporters, Oprah's disavowal of beef, while significant, registered as more than a simple disavowal of a food (Shapiro 2018). Oprah undermined long-standing power structures by entering masculine space—the public, neoliberal sphere of the industrialized cattle industry—when she came to trial in Amarillo. Oprah's comment about beef and her hesitancy to eat meat during the mad cow crisis signified a rejection of an industry and a space that was largely dominated by a culture that valued and supported wealthy white men. In response to her stance, plaintiff supporters equated Oprah with bovines and used a "no" symbol over her face to reject not just her statement, but also her personhood. In essence, the plaintiffs' supporters were arguing against Oprah's basic humanity and dignity—completely rejecting her personhood and arguing against not just her opinion but her right to have one. And, given that Oprah is an African-American woman, this point becomes even more poignant. The rhetorical messaging was clear: not only can your opinions be put on trial, so can your very right to have them. By vocalizing her opinion of the commercial beef industry, Oprah trespassed on the masculine territory, and the plaintiffs' reactions to her comments on beef reflect how "food can obviously confine women to subservient roles, keeping them busy at home and 'quiet' in the public sphere" (Belasco 2008, 44).

In the context of this rhetoric that rejected the very dignity of Oprah's humanity, we turn to her response to the public outcry, particularly her thoughtful navigation of the Amarillo culture in ways that speak to her skillfulness as a rhetor and her power as a unifier. Instead of hiding out in a hotel, flying back and forth to be in town only for specific court dates, or lashing out at locals, Oprah chose to claim her power in this space by embracing Amarillo and its people, purposely choosing a path grounded in valuing their shared humanity—as opposed to participating in smear-campaign strategies. Oprah's counternarrative was cemented in her resistance, yet the ways in which she tactically affiliated with the local community ultimately contributed to her victory, both legally and in terms of public perception (de Certeau 1984).

The Power of a Good Haircut: Creating a Counternarrative

From the moment Oprah arrived in Amarillo, she greeted the press and the community with minimal but cordial remarks and gestures. When Oprah deplaned upon arrival, her first words about Amarillo were neither aggressive nor snarky. Instead, when asked what she thought of the place, she replied, "At least the sun is shining!" (Farley 1998). There was no reaction to the negative rhetoric from plaintiff supporters or negative talk about beef producers and their cause; instead, Oprah offered only smiles, waves, and remarks about the weather. In contrast to the anti-Oprah paraphernalia that circulated throughout Amarillo, Oprah's approach to the public was noteworthy: she was neither on the attack, nor was she on the defense. While she may have literally been the defendant, she did not set out to advocate for her innocence outside of court or disparage the plaintiffs, and her demeanor speaks volumes about how she chose to navigate the trial and her time in Amarillo with public, congenial restraint.

Importantly, Oprah did not spend her free time out of the courtroom in her hotel room, locked away from Amarillo and its citizens. Rather, she arrived in town ready for her trial *and* ready to work, for she brought her talk show to town with her and "framed parts of it as a homage to the city and state she suddenly found herself in" for the duration of the six-week trial (Batheja 2018). As difficult as it would be for anyone standing trial, court-ordered speech suppression was particularly challenging for Oprah since she was tasked with moving not just herself but her entire production from Chicago to Amarillo in order to continue filming for the duration of the trial.[12] Oprah had to refrain from discussing the very reason she was in Amarillo, a challenging task for a syndicated talk show host. During "Texas Celebrities," the first *Oprah in Texas* episode recorded in Amarillo (January 26, 1998), Oprah introduced her show by explaining her situation, referring to why her words about not eating beef had relocated her to feedlot country: "Well, I'm in Texas. I guess you heard. You also heard I'm not allowed to talk about why I'm here." In this brief statement, Oprah let her audience know that she was in Texas and that she could not discuss the trial by implying that the audience already knew. This was a power move, for Oprah's repeated "you heard" suggests that her fan base did not need her explanation; she was famous enough with a loyal enough base to let them make assumptions without her overt direction. Oprah was on trial, but she was still very much in control of her show and had significant sway in the realm of public perception.

Oprah's choice to not open the show with disparaging comments to Amarillo or its people made a clear statement: she was rising above the negativity and establishing herself as normal and neutral as possible. Further in the episode, guest and Texas native, Patrick Swayze, mentioned he had "roasted sides of beef." Oprah responded, "beef did you … Well now that's just fine by me" ("#11: Oprah on Taking the Show to Texas" 2012). Oprah's response was not to disparage Swayze or question his choices; nor was it to only invite guests that would avoid the topic altogether or that solely aligned with her comment about avoiding beef. Instead, Oprah highlighted how friendly, welcoming, and law-abiding she and beef-eating, leather-wearing Texans could be with one another. Oprah's choice to refrain from eating beef during the mad cow scare may have had vegetarian affiliations; however, her approach on the show demonstrated that a non-hamburger-eating individual and her carnivore companions could dialog with one another and that perhaps they shared a common enemy—the industry that was trying to silence free speech by bringing a Texas-friendly talk show host to trial.

Throughout her time in Amarillo, Oprah negotiated her response to the trial in a cordial manner by voicing her faith in her newly emerging relationship with the local studio audience made up of many Texans. She allowed conversations about food and food politics, while purposely avoiding any disparaging or isolating commentary, tactics that served both to meet court mandates and to make herself less polarizing and more relatable to local and national audiences. We use the word "tactic" purposefully in our analysis, drawing on Michel de Certeau's theory of tactics vs. strategies. According to de Certeau (1984), strategies are available to subjects of power because of their access to a space or location. A strategy "assumes a place that can be circumscribed as a *proper* (*propre*) and thus serve as the basis for generating relations with an exterior distinct from it" (xix). In this regard, the plaintiffs and supporters relied on strategies, such as the visual offensive rhetoric, to exert dominance and power, largely due to the trial's location in Amarillo and to the patriarchal culture of Texas and the beef industry. Oprah went to the mecca of cattle and entered the masculine, white space of the Texas Panhandle.

In contrast, de Certeau contends that other people, Oprah in this case, are not passive receivers of such strategic manipulation; on the contrary, "the place of a tactic belongs to the other" (xix). Granted, Oprah had significant power as a leading talk show host and an emerging media mogul. She had, according to the plaintiffs, the power to singlehandedly tank the cattle market with her "Dangerous Food" segment. Yet while Oprah had enormous cultural and economic capital—she was the leading talk show host of a nationally syndicated program with millions of viewers—she was the "other" in Amarillo, the defendant in a trial designated with evaluating—and devaluing—her position on the subject of food, particularly beef. Personally, Oprah viewed the trial and her reaction to it as representative of her moral and ethical code, her rights to free speech, and her integrity as a talk show host. During her in-trial examination, Oprah stated, "I am a black woman in America, having gotten here believing in a power greater than myself … I cannot be bought. I answer to the spirit of God that lives in us all" (Hollandsworth and Colloff 1998). Oprah openly acknowledged her position as the "other" in this statement, particularly as a Black woman. She could not—would not—be bought or made to alter her right to her opinion about beef consumption, even if it meant a long and costly legal proceeding. However, the very fact that Oprah had the financial means to fight the suit is indicative of the power and position that she, unlike many members of marginalized groups, had to advocate for herself. Yet, while her position as a woman of significant influence is noteworthy, while in Amarillo, she was on unfamiliar turf—she was the "other" by color, gender, and opinion—and she was legally restricted from directly discussing her position. The strategies of the plaintiff were not viable for her. And, in alignment with Oprah's persona, not applicable to her either. Instead, she purposefully, albeit subtly, used tactics in the court of public opinion; thus, de Certeau's concept of tactical movement is a framework for analyzing the ways in which Oprah maneuvered during her time in Amarillo that while not revolutionary, were essential to her success in navigating the trial and its contentious food politics.

De Certeau contends that "a tactic depends on time—it is always on the watch for opportunities that must be seized 'on the wing' … It must constantly manipulate events in order to turn them into 'opportunities'" (xix). When Oprah arrived in Amarillo, the chamber of commerce president, Gary Mohlberg, issued a memo to chamber staff that directly told them not to attend Oprah's tapings in Amarillo and to not give her "any red-carpet rollouts, key to the city, flowers" (qtd. in Hollandsworth and Colloff 1998). Mohlberg's strategy was to reject Oprah solely based on her opinion of beef; this was a strategy of rejection, a powerful way to shun the talk show host. While Mohlberg requested that his staff not attend Oprah's show or give her any welcome, Oprah countered this unkindness publicly as she frequently "rolled down the window of the car, and smiled benevolently at everyone" (Hollandsworth and Colloff

1998). Oprah brought her own Texas-friendly welcome wagon to remind people of what it meant to be a "Texan." Eventually, Mohlberg responded in kind by sending her yellow roses and rescinding his unwelcoming message (Hollandsworth and Colloff 1998). Mohlberg's public shunning provided just the opportunity for Oprah to take the high road, to engage the public, to show herself as welcoming and warm, even when chamber staff were encouraged to be as cold as ice.

Moreover, de Certeau notes that "many everyday practices (talking, reading, moving about, shopping, cooking, etc.) are tactical in character. And so are, more generally many 'ways of operating': victories of the 'weak' over the 'strong'" (xix). For Oprah, responding to invitations to engage with locals provided ripe opportunities for "everyday" tactical strategies. She forged bonds with local women, no matter how informal, and this unsettled the social and economic hierarchy in Amarillo. The cattlemen of Texas may have filed suit against Oprah, but she and women of certain Panhandle political clout extended hands of friendship to one another, the tactics of goodwill and amiability. One particular example is Oprah's response to a book club invitation, which arrived shortly following Mohlberg's memo. Skip Hollandsworth and Pamela Colloff (1998) reported:

> Nancy Seliger, … wife of [then] Amarillo mayor, Kel Seliger, sent Oprah a note to invite her to a book club … Every woman in Amarillo knew within days what happened next: Winfrey immediately picked up the phone, called Mrs. Seliger, graciously thanked her, and chatted with her for several minutes about—among other things—where to get her hair done in Amarillo.

To call on Nancy Seliger was to seize an opportunity to bypass the offensive rhetoric and public shunning through the everyday practice of talking. It was an easy strategy for the plaintiffs' supporters to name-call and don disparaging signage when they could think of Oprah an abstract figure, reduce her to an opinion, and devalue her right to even have it. By returning a call and chatting about something so commonplace as where to get her hair cut, Oprah made herself known as approachable, relatable, and real, and her everyday practices resulted in humanizing tactics that contributed to her evolving public perception.

Oprah did not restrict the tactic of friendliness solely to women who reached out to her in town. Rather, she used her platform on the talk show to provide a visual and verbal contrast to offensiveness. To illuminate how she had become more Texas-friendly than the cattlemen who sought to punish her for not choosing to eat their cash crop, Oprah invited her audience to become familiar with her on traditional Texas terms. If the spirit of the word *Texas* means *friends*, a hispanophone derivative of the indigenous Caddo word *Taysha*, then Oprah would simply harken back to Texas' roots and make friends with the locals. The gag order may have prevented her from talking about the trial, but it didn't restrict her from being professional, courteous, and compassionate. She may have voiced a decision not to eat a particular food, but she never indicated she would be unfriendly to those who lived in beef-producing areas. The cattlemen, plaintiffs, and de facto representatives of the Amarillo chamber of commerce immediately voiced an inhospitable approach toward Oprah upon her arrival. In response, Oprah offered "extravagant salutes to the very place she was accused of ruining" (Hollandsworth and Colloff 1998).

Moreover, Oprah chose to keep her work local and broadcast her show from the Amarillo Little Theatre, a nonprofit community organization. As noted, "Texas Celebrities" included well-known native Texans, Clint Black and Patrick Swayze. To demonstrate her friendship, Oprah danced a Texas two-step with Swayze, and she remarked to the audience that "two-stepping is big around here" ("#11: Oprah on Taking the Show to Texas" 2012). She often uttered "How are y'all," "Did you now," and "I reckon" while filming her Amarillo episodes, making herself at home with the local vernacular ("#11: Oprah on Taking the Show to Texas" 2012). She warmly connected with her new community by welcoming guests and the audience to her show with an inviting, tailored-for-Texas voice. When Céline Dion appeared on the show on February 18, 1998, Oprah noted that Dion would be "singing the love theme from *Titanic* for a billion people, but today, she's singing it just for you, Texas … She's … she's going to do that for us… here in Texas" ("Céline Dion-Oprah Winfrey Show 1998" 2017). In addition to celebrating Dion's performance of the Oscar-nominated song, Oprah invited Dion to sing "Love Can Move Mountains," a song that evoked spiritual messages such as one might hope to find in an Amarillo church.

In food studies scholarship, individuals typically use their food voices to communicate "world views on community, economics, gender, nutrition, ethnic identity, and tradition" (Hauck-Lawson 2004, 24). *Texas Beef v. Oprah Winfrey* illuminated competing food voices, and these distinct voices would remain strong throughout the trial since each legal team sought to prove to jurors—individuals whose livelihoods to some degree intersected with the local cattle economy—that Texas foodway had or had not been damaged.[13] Oprah responded by demonstrating that "For some people, not eating is an expression of voice" (Belasco 2008, 53). While the cattlemen took her comments as an attack on their industry, Oprah's influential food voice joined in a conversation that emphasized how "not eating can stigmatize individuals and upset social relations—as lone vegetarians discover when dining amongst carnivores" (Belasco 2008, 52). In the cattlemen's eyes, Oprah's food voice was a viable threat to industrialized meat production and, arguably, the patriarchy.

However, Oprah maneuvered by taking her voice local, rebranding the show as *Oprah in Texas*. This move placed her in the homes of regular Amarilloans, places the cattlemen had likely not engaged with on a daily basis. Jeff Blackburn, an Amarillo attorney, noted how the trial distanced the plaintiffs from the local people: "To these cattlemen, Oprah—a successful black woman from Chicago—seems like a foreigner. But the real comeuppance is that Engler is a lot more foreign to people here than Oprah. People in Amarillo watch *Oprah* every day" (qtd. in Hollandsworth and Colloff 1998). Her friendliness spoke volumes and was indeed persuasive. Charles Rittenbury, an Amarillo trial lawyer, stated: "I don't know if our local cowboys are going to come out on top of this damn deal. We've already got wives of respectable ranchers sneaking around town, trying to get tickets to Oprah's show. I'm telling you, Oprah's about to cause a lot of hell to break loose out here" (qtd. in Hollandsworth and Colloff 1998). The hell Oprah broke loose was to stand above intimidation rhetorics, to stand for her right to her opinion, and to stand against censorship. But she also stood for something more—she stood against devaluing and dehumanizing strategies and rose above the fray to ultimately win over the hearts and minds of both jury and locals alike.

Oprah's tactical practices of embracing, not defying, offending, or shunning the Texas culture were crucial to her ability to connect with locals and to build rapport. By adapting to the environment through the local vernacular, contributing to local nonprofits, and interacting with locals, Oprah changed the narrative of what people thought about her and how they conceived of the politics of the trial. She flipped Texan hospitality—the land of "yes ma'ams" and "no thank yous" and "y'all take good care now"—and proved that the outsider was in fact more hospitable than the locals. She used these "extravagant salutes" to humanize herself and to impact the ways in which audiences—local and national—viewed her as a public figure (Hollandsworth and Colloff 1998). As Jennifer Richardson (2007) has suggested: "The Winfrey trial evidenced some slippage between the forensic and the popular, between the rhetoric of the courtroom and the culture that contains it" (177). Through Oprah's tactically rhetorical, outside-the-courtroom moves during the trial, she persuaded individuals to see that she was more familiar with "Texas style" than the plaintiffs. In simplest terms, she was friendly and honest. She modeled "the Texas way" better than the Texans. When the verdict was read on February 26, 1998, Oprah left the courtroom and Amarillo legally victorious and personally triumphant.

Conclusion

In a posttrial interview outside the courtroom on February 26, 1998, Oprah declared: "I will continue to use my voice … I believed from the beginning that (the lawsuit) was an attempt to muzzle my voice, and I come from a people who have struggled and died in order to have a voice in this country. And I refuse to be muzzled" (qtd. in "Oprah: Free Speech Rocks" 1998). If, as Richardson (2007) argues, "the dominant ruling class, governmental agencies, and corporate financiers regulate not only what Americans put in their mouths but also what they can say about it," then Oprah complicated this notion by positioning herself as a dominant, Texas-friendly force who, like other individuals, had decided not to eat a burger (182). In her own way, Oprah arguably demonstrated that she possessed certain "rugged, heroic qualities that we used to look for in the great Texas men [and women] of the past" (Hollandsworth and Colloff 1998). It was Oprah's purposeful maneuvering, even in the midst of hateful, contentious rhetoric, that provides a ripe opportunity to reflect on anti-veg(etari)an discourse that existed 20 years ago and continues in the present day.

There is clearly a divide in how the plaintiffs and the public reflect on the Oprah trial. Despite the passage of time, the plaintiffs remain steadfast in their perspectives. In "Whatever Happened to the Man Who Sued Oprah?" (2002), cattleman Bill O'Brien reflected, "I think she was dishonest in this program, dishonest in they [*sic*] way she edited it, and presented it, and by what she did to the beef industry. And my feelings won't ever change about that. She snowed this community." O'Brien clearly believes that Oprah is guilty and that her ways of behavior in Amarillo were an act, not an honest reflection of her true persona. Moreover, the lead plaintiff, Bill Engler, told the *Amarillo Globe-News* in 2011 that "I think we did some good … I think we cleaned up her act" (qtd. in Batheja 2018). These comments suggest *Oprah* the show and Oprah the person are simply entertainment values, fictitious even. The plaintiff's comments are steeped in patriarchal strategies of the oppressor, those who are fixated on blame and vigilant in their claims to have taught Oprah a lesson by implying that Oprah's thoughts and opinions were bad fiction, instead of reality. However, at the end of the day, it was and is Oprah who rightfully claimed victory for free-speech rights and against veggie libel laws. It was Oprah who won over the public and was victorious in a case for not eating beef in beef country. As Hollandsworth and Colloff (1998) conclude, "at the end of every one of her shows, Texans stood and roared their approval, prouder than ever of who they thought they were."

We are now two decades into the twenty-first century, and food culture has evolved from the 1990s when Oprah was accused of disparaging beef. The increased attention to "foodies" and the democratized engagement with food has led to—or at least has the potential to lead to—increased access to food and awareness of food systems, points not associated with the original term from the 1980s. Granted, there is embedded privilege when referencing food, as we live in a world where food access and equality is directly connected to issues of social inequality. Yet food has become central to our daily lives, not only as a form of sustenance but as a means for pleasure, entertainment, and community building. How we think of food has greatly changed, for regardless of whether or not people like the term "foodie" or even agree on its definition, there is a rising consensus that to engage with food is to take on larger social issues, just as Oprah infamously did 20 years ago.

Marvin Hayenga (1998) reflected that the Oprah trial set forth a trajectory where "the ability to speak freely about concerns or issues regarding the safety of our food supply is very important, and many would be reluctant to see that freedom abridged because of such concern" (20). In the decades that followed, food writers contended that lay citizens, many of whom identify as "foodies," now have the potential to engage in issues of "environment, politics and health wrapped into one" (Turow 2015, 10). While Oprah may have gone to trial for her comments during the "mad cow" epidemic, she serves as a foremother for early twenty-first-century foodies to think about what they eat in a larger context than just what they put in their mouths, or as Josephine Livingston (2019) states, "being a foodie now, in 2019, requires thinking with more than your tongue." The role of foodies and food influencers, those who can shape opinions on food and food-related issues, has expanded from celebrities like Oprah to average American citizens who think about food in complex and diverse ways that connect to larger social and cultural implications. According to Eve Turow's (2015) research, "this trend isn't just changing the face of fine dining and upscale grocery stores; it is revolutionizing American culture" (7).

Just as how we think of food and who gets to critique food are concepts that have expanded in the past 20 years, the types of foods that have become mainstream are increasingly diverse, including a rise in vegetarian and vegan options. One notable example in Georgia is Aisha "Pinky" Cole's Slutty Vegan food empire, which offers vegan burgers to a cult following of vegans and non-vegans alike who line up to be "sluttified" after eating her food. Cole's patrons include such celebrities as Tyler Perry, Snoop Dog, and Senator Cory Booker (Ogunsola 2019) as well as "communities that would have otherwise never had these options," such as

Atlanta's historic West End community (Slutty Vegan 2019). However, food libel laws remain in the legal system, and big agribusiness stakeholders continue to lobby for food-related restrictions.[14] While Cole's vegan burger empire spans multiple restaurant locations and food trucks to meet patrons' demands, in nearby Mississippi the state legislature passed Senate Bill 2922 in January 2019, which bans plant-based meat providers from referring to their products as "burgers" or "dogs" (Amy 2019; Wimer 2019). This law has met with opposition by free-market groups, and as of September 6, 2019, lawmakers are considering revising the law so that the state might avoid expensive legal battles surrounding the First-Amendment rights, a flashback to the Oprah trial (Wagster Pettus 2019). This distinction between the demand for a food product and the access to it, as well as how foods are labeled, is indicative of the tensions that continue to surround food production, freedom of speech, and food access in the United States. However, Oprah's tactical navigation during her trial in the late 1990s provides a precedent for creating a productive discourse around food and food justice. As we continue to face food libel laws that restrict how we talk about and access vegan and vegetarian foods—often foods deemed threatening by agri-based industries and lobbyists—it will be imperative that we harness the food culture of the twenty-first century and engage in tactical opportunities and intentional dialogues that continue to support free-speech rights, food equity, and food diversity.

Notes

1. The term "foodie" originated in the early 1980s in both the United States and Great Britain. Food writer Gael Greene used the term in a 1980 restaurant review for *New York Magazine* (Greene 1980, 33). In 1981, Ann Barr issued an editorial call in *Harper's & Queen Magazine* for readers to respond to the changing food word. Subsequently, in an August 1982 anonymous article, edited by food writer Paul Levy, readers, in Levy's words, "derided [me] in the anonymous article (edited, as it happens, by me) as the ghastly, his-stomach-is-bigger-than-his-eyes, original, appetite-unsuppressed, lip-smacking 'king foodie'" (Levy 2007).

Barr and Levy popularized the term two years later with their publication *The Official Foodie Handbook: Be Modern-Worship Food* (Barr and Levy 1984). While the term originated in both contexts in reference to opulent diners with, in Levy's example, less than flattering food obsessions, the term gained traction following Barr and Levy's handbook and continues to be a popular descriptor, though with conflicting connotations.

2. While Burke referenced "statesmen" (1969, 173), we prefer the term "statespeople" for gender inclusivity. The term "statesmen" historically harkens to those with political power (e.g., Plato's *Statesmen*); however, we extend the term to include individuals, such as Oprah, who have significant cultural power that often leads to political power, even if the person, such as Oprah, is not a politician. While Oprah is not a statesperson in the traditionally-political sense, she has the capital to persuade the public and to influence not only what people think but how they act, including, for the purposes of this article, what people eat and why.
3. In the *London Review of Books,* Angela Carter describes the essence of foodie culture in the 1980s: "One of the ironies resulting from the North/South dichotomy of our planet is the appearance of this odd little book [*The Official Foodie Handbook*], a vade mecum to a widespread and unashamed cult of conspicuous gluttony in the advanced industrialised countries, at just the time when Ethiopia is struck by a widely publicised famine, and the rest of Africa is suffering a less widely publicised one… At a conservative estimate, eight hundred million people in the world live in constant fear of starvation. Under the circumstances, it might indeed make good 20th-century sense to worship food, but punters of 'foodism' (as Ann Barr and Paul Levy jokily dub this phenomenon) are evidently not about to drop to their knees because they are starving" (Carter 1985, 22). In this sense, foodie culture in the 1980s was steeped in opulence and culinary conquests with little to no regard for issues of food justice and distribution, sustainable agricultural practices, or other food-related concerns.
4. Across platforms, Ms. Winfrey's brand simply refers to her as "Oprah," for her influence is so great that she has reached the pinnacle of pop culture recognition: to be known solely on a first-name basis (Lowe 1998; Harris and Watson 2009).
5. The "veggie libel laws" came into law following the conflict between Washington state apple growers and the CBS broadcast, *60 Minutes*, in

1989. Apple growers filed suit following a segment that discussed the connection between Alar chemical, a product used on apples, and its potential to cause cancer when ingested. The suit was eventually dismissed, with the judge ruling that food could not be defamed. Thus, in reaction, "veggie libel laws" became popular to prevent false statements about food (Epstein 1998, 16). The first 13 states to enact veggie libel laws were Alabama, Arizona, Colorado, Florida, Georgia, Idaho, Louisiana, Mississippi, North Dakota, Ohio, Oklahoma, South Dakota, and Texas (Jonna 2012).

6. The guests on the "Dangerous Food" segment included Dr. William Hueston, USDA expert on mad cow disease; Howard Lyman, animal rights activist and lobbyist for the American Humane Society; and Dr. Gary Weber, National Cattlemen's Beef Association spokesperson (Hayenga 1998, 14).
7. See Loroz and Braig's "Consumer Attachments to Human Brands: The 'Oprah Effect'" (2015) for an analysis of brand personality appeals and marketplace responses.
8. For additional factors that contributed to the downward turn in the cattle market, see Hayenga (1998, 19–20).
9. *Texas Beef Group* et al., *Paul Engler and Cactus Feeders,* et al. *v. Oprah Winfrey, Harpo Productions, Inc., Howard Lyman and King World Productions, Inc.*, Case No. 2-96-CV-208 and 233, District Court, Northern District of Texas, Amarillo Division.
10. For existing scholarship on the trial, see Margot Fell's (1999) "Agricultural Disparagement Statutes: Tainted Beef, Tainted Speech, and Tainted Law" and Jennifer Richardson's (2003) "Cowboys and Celebrities: Reading Rhetorics at the *Texas Beef v. Oprah Winfrey* Trial."
11. Hollandsworth and Colloff (1998) note that "within 150 miles of Amarillo, six million head of cattle, a third of the nation's cattle supply, are fattened in feedlots," a point that underscores the dominance of beef in this region.
12. For example, during a taping of *The Oprah Winfrey Show* in Amarillo, country singer Clint Black commented about a fellow guest and wife Lisa Hartman Black's purple leather pants, and she responded: "I think this is the mad cow. I mean I'd be pretty mad if I turned purple." Immediately, Oprah reminded her guests that they could not speak "a word" about mad cow ("#11: Oprah on Taking the Show to Texas During Her Trial" 2012). A YouTube clip of the "Texas Celebrities"

Episode 18 from Season 10 is used as reference since the authors were unsuccessful in locating the original episode from the *Oprah Winfrey Network*.

13. The jury was all white and included "a woman who had been involved in cattle feeding 25 years ago and a descendant of one of Amarillo's oldest ranching families" (Hollandsworth and Colloff 1998).
14. In 2018, Oprah's attorney, Charles Babcock, told Aman Batheja with *The Texas Tribune* that the False Disparagement of Perishable Food Products Act is "still on the books but to my knowledge, nobody has used it since that [Oprah's] case" (Batheja 2018).

Bibliography

#11: Oprah on Taking the Show to Texas During Her Trial. 2012. *TV Guide's Top 25*. YouTube video 2: 56. Posted September 11. https://www.youtube.com/watch?v=KA1Ecbq8Trw&t=53s

Amy, Jeff. 2019. Mississippi Lawmakers Have Real Fears About 'Fake Meat.' *The Associated Press*, March 2. https://www.apnews.com/7f1b401dee9741a7901982abd69e1d0e

Appeals Court Upholds Win in 'Mad Cow' Lawsuit. 2000. The News Media & The Law (Spring 2000): 16. https://www.rcfp.org/journals/the-news-media-and-the-law-spring-2000/appeals-court-upholds-win-m/

Barr, Ann, and Paul Levy. 1984. *The Official Foodie Handbook: Be Modern, Worship Food*. London: Ebury.

Batheja, Aman. 2018. The Time Oprah Winfrey Beefed with the Texas Cattle Industry. *The Texas Tribune*, January 10. https://www.texastribune.org/2018/01/10/time-oprah-winfrey-beefed-texas-cattle-industry/

Belasco, Warren. 2008. *Food the Key Concepts*. Oxford: Berg.

Brillat-Savarin, Jean-Anthelme. 1994. The Physiology of Taste. Trans. Anne Drayton. London: Penguin.

Burke, Kenneth. 1969. *A Rhetoric of Motives*. Berkeley: University of California Press.

Carter, Angela. 1985. Noovs' Hoovs in the Trough. *London Review of Books* 7 (1): 22–23. https://www.lrb.co.uk/v07/n01/angela-carter/noovs-hoovs-in-the-trough

Céline Dion – Oprah Winfrey Show 1998. 2017. YouTube video 23: 38. Uploaded March 30. https://www.youtube.com/watch?v=2nZ1g1-qgGA

De Certeau, Michel. 1984. The Practice of Everyday Life. Trans. Steve Rendall. Berkley: University of California Press.

Epstein, Aaron. 1998. Where's the Beef? For Oprah, It's in Court – Due to Angry Ranchers. *Chicago Tribune*, January 4. http://www.chicagotribune.com/news/ct-xpm-1998-01-04-9801040201-story.html

Farley, Rose. 1998. Oprah Does Amarillo. *Houston Press*, February 12. https://www.houstonpress.com/news/oprah-does-amarillo-6570354

Fell, Margot S. 1999. Agricultural Disparagement Statutes: Tainted Beef, Tainted Speech, and Tainted Law. *Fordham Intellectual Property Media & Entertainment Law Journal* 9: 981–1125.

Greene, Gael. 1980. What's Nouvelle? La Cuisine Bourgeoise. *New York Magazine*, June 2.

Harris, Jennifer, and Elwood Watson. 2009. *The Oprah Phenomenon*. Lexington: University Press of Kentucky.

Hauck-Lawson, Annie. 2004. Introduction. *Food, Culture and Society: An International Journal of Multidisciplinary Research* 7 (1): 24–25.

Hayenga, Marvin L. 1998. Texas Cattle Feeders v. Oprah Winfrey: The First Major Test of the 'Veggie Libel Law'. *Choices* 13 (2): 13–20. http://www.jstor.org/stable/43625049.

Hollandsworth, Skip, and Pamela Colloff. 1998. How the West Was Won Over. *Texas Monthly*, March 1. https://www.texasmonthly.com/food/how-the-west-was-won-over/

Jonna, Jamil. 2012. Veggie Libel Laws: Attempts at Silencing Animal Rights Advocates. *Civil Liberties Defense Center*, January 9. https://cldc.org/aeta-veggie-libel/

Levy, Paul. 2007. What Is a Foodie? *The Guardian*, June 14. https://www.theguardian.com/lifeandstyle/wordofmouth/2007/jun/14/whatisafoodie

Livingston, Josephine. 2019. What Was the Foodie? *The New Republic*, March 18. https://newrepublic.com/article/153335/foodie

Loroz, Peggy Sue, and Bridgette M. Braig. 2015. Consumer Attachments to Human Brands: The 'Oprah Effect'. *Psychology & Marketing* 32 (7): 751–763. https://doi.org/10.1002/mar.20815.

Lowe, Janet. 1998. *Oprah Winfrey Speaks: Insight from the World's Most Influential Voice*. New York: Wiley.

Ogunsola, Jennifer. 2019. Pinky Cole: Born That Way. Bitter Southerner. https://bittersoutherner.com/born-that-way-aisha-pinky-cole-slutty-vegan

Oprah: 'Free Speech Rocks'. 1998. *CNN*. February 26. http://www.cnn.com/US/9802/26/oprah.verdict/

Richardson, Jennifer J. 2003. Cowboys and Celebrities: Reading Rhetorics at the *Texas Beef v. Oprah Winfrey* Trial. PhD dissertation, Washington State University.

———. 2007. Phenomenon on Trial: Reading Rhetoric at *Texas Beef v. Oprah Winfrey*. In *The Oprah Phenomenon*, ed. Jennifer Harris and Elwood Watson, 165–188. Lexington: University Press of Kentucky. JSTOR.

Shapiro, Laura. 2018. *What She Ate: Six Remarkable Women & The Food that Tells Their Stories*. New York: Penguin.

Shprintzen, Adam D. 2013. *The Vegetarian Crusade: The Rise of an American Reform Movement, 1817–1921*. Chapel Hill: UNC Press.

Slutty Vegan. 2019. About Us. http://sluttyveganatl.com/about-us/

The Big Texan. 2019. The Big Texas Steak Ranch. Home Page. https://www.bigtexan.com. Accessed 1 June 2019.

Turow, Eve. 2015. *A Taste of Generation Yum: How the Millennial Generation's Love for Organic Fare, Celebrity Chefs and Microbrews Will Make or Break the Future of Food*. Independently Published.

Wagster Pettus, Emily. 2019. Mississippi Considering Rules to Let Companies Continue to Use Labels like 'Veggie Burger.' *USA Today*, September 6. https://www.usatoday.com/story/money/food/2019/09/06/mississippi-meat-lawsuit-state-considering-new-food-labeling-rules/2236908001/

Whatever Happened to the Man Who Sued Oprah? 2002. *KCBD*, May 20. Updated July 3 [Year Not Disclosed]. https://www.kcbd.com/story/789750/whatever-happened-to-the-man-who-sued-oprah/

Wimer, Andrew. 2019. New Lawsuit Challenges Mississippi Labeling Law that Makes Selling 'Veggie Burgers' a Crime. *Institute for Justice,* July 2. https://ij.org/press-release/new-lawsuit-challenges-mississippi-labeling-law-that-makes-selling-veggie-burgers-a-crime/

Winfrey, Oprah. 1996. Dangerous Food. Segment Transcript from "Episode Dated 16 April, 1996." *The Oprah Winfrey Show*. Chicago: Harpo Productions.

8

Queer Hunger: Human and Animal Bodies in Djuna Barnes' *Nightwood*

Molly Mann

Djuna Barnes' 1937 novel *Nightwood* is one of the first works of literature to foreground an explicitly lesbian relationship. Barnes' skill with language and her portrayal of the fraught dynamic between protagonists Robin Vote and Nora Flood make *Nightwood* a classic of both modernist and queer fiction. Because Barnes aligns queer desire with a particularly animalistic hunger, *Nightwood* also invites scholarship at the intersection between animal studies and queer studies fields of literary inquiry. Through this shifting alignment, Barnes challenges binaries that construct human experience, including those of (non)human status, gender, sexuality, and ethnicity/nationalism.

Robin, a hybrid woman/animal figure, ends the novel in a brutal wrestling match with Nora's dog, one that collapses physical boundaries between her and the animal. Throughout the novel, Barnes describes Robin using animal language, particularly about the hungry desire that moves her from one sexual partner to another without any evident

M. Mann (✉)
St. John's University, New York, NY, USA
e-mail: mannm2@stjohns.edu

C. Hanganu-Bresch, K. Kondrlik (eds.), *Veg(etari)an Arguments in Culture, History, and Practice*, The Palgrave Macmillan Animal Ethics Series,
https://doi.org/10.1007/978-3-030-53280-2_8

emotion or attachment. Robin's avian name brings her even closer to the animal than human. By revealing categories of (non)human as shifting and unstable, Barnes invites a critique of all other binary categories. Ultimately, the novel asks how humans live in relationship with one another and with other beings once these binaries are dissolved. All of its characters live in conflict with this question, including the Baron Felix Volkbein (who finds himself caught between national and religious identities), the gender-fluid Dr. Matthew O'Connor, and Robin and Nora, whose visions of queer relationship and proximity to heterosexual marriage differ.

For Robin, animal appetite represents a private, internal self that is at odds with public expectation for how human beings interact with each other. Each in their own way, the other characters in *Nightwood* struggle with negotiating these public and private selves. By challenging the boundary between animal/human, public/private, Barnes renegotiates these constructs to accommodate identities previously excluded from public space, including Jewish identity, gender fluidity, and queer sexuality.

Nightwood, and especially the novel's final scene, troubles the concept of normative—white, cisgender, heterosexual, male, Christian, meat-eating—identity against which queerness and other categories of difference are defined. Barnes' critique of social constructions of identity and relationship emerges from an interrogation of the (non)human binary. Therefore, reading the novel through a theoretical framework that includes feminist and queer scholarship as well as a post-structural understanding of language's relationship to difference illuminates the novels' presentation of this critique. In *Nightwood*, Barnes presents an alternative future that makes room for difference without forcing it into a constructed binary.

The interdisciplinary field of animal studies examines the relationship between human beings and their animal fellows, interrogates the distancing categories that construct those relationships, and asks what responsibilities we have as humans to acknowledge the agency and sentience of other beings. In *The Animal That Therefore I Am*, Jacques Derrida considers the shame humans feel at having their nakedness witnessed in comparison to animals who are always naked and demonstrate no shame over that condition. Standing naked before his cat, Derrida articulates his sense of shame at having the cat observe his vulnerable human condition, referring to this distancing between animal and human sensibilities as a "rupture" (Derrida 2008, 36). To Derrida, the animal's gaze arouses shame because it confronts us with our own "*bêtises*," which in French most often refers to foolish actions or speech but which David Wills translates into English as "bestiality" (Derrida 2008, 53). Humans feel shame at being naked in front of animals because we are thereby reminded of the thin constructs that separate us—not always effectively—from the violence and precarity of animal existence. Clothes, which form part of that construction, separate us from animals and once they are removed, humans are forced to confront all of the other ways that dominion over animals may be challenged. Through his inquiries into the work of Kant and Levinas, Derrida also shows how this construct is translated into the marginalization not just of animals, but of groups of humans. Comparing Jews to animals—and thereby denying them the agency and sensibility attributed specifically to humans—is a common device in the service of anti-Semitism.

Our relationship to animals—and especially meat-eating—parallels other social constructions of identity, power, and dominance. In *The Sexual Politics of Meat*, Carol Adams challenges the rupture between women and animals, arguing that violence against animals is structurally related to violence against women. Therefore, the only way to opt out of the systems that oppress women is to eat a vegan diet. To eat meat, according to Adams, is to exercise power that operates along a continuum of exploitation of other bodies, whether those bodies belong to women, marginalized racial groups, or animals (Adams 1990, 4). This is why, historically, meat-eating has been a predominantly male practice in Western culture (Adams 1990, 4). It is both the product of a system of power, privilege, and othering and reifies that system.[1] Adams observes what she identifies as the role of language in creating distance between live animals and the meat products we derive from them; we don't eat "cow" or "pig" but rather "beef" and "pork" (Adams 1990, 21). Although I would argue that language is always already based on difference, Adams correctly highlights the participation of language systems in the structures of power and privilege upon which eating meat is based. Calling it "beef" or "pork" separates meat from animals, and further establishes an oppressive dynamic between human and animals. Adams gives another example of the idiom to treat someone "like a piece of meat"—especially in instances of sexual violence, which is Adams's primary focus (Adams 1990, 21). To be like a piece of meat is to become the product of distancing, to become separated from one's humanity just as meat inhabits a different imaginative space from a live animal. Derrida refers to this distancing as a "rupture" between humans and animals. The result of this rupture, according to Adams, is the animal as an "absent referent" or object (Adams 1990, 24). Through the same linguistic structures in which animals are made absent referents, so too are women and people belonging to marginalized racial groups.

To close the rupture between humans and animals is to allow for hybridity. In *A Thousand Plateaus*, Gilles Deleuze and Félix Guattari refer to the consumption of one species' body by another as a "becoming" (Deleuze and Guattari 1987, 10). In their example of a wasp and orchid, the becoming is a symbiotic alchemy in which the wasp contributes to the reproductive process of the orchid and the orchid is incorporated into the body of the wasp. This dynamic of mutual becoming, catalyzed by consumption, contributes to what Jon Roffe and Hannah Stark call the non/human turn in philosophical and literary studies, an examination of the interactions between concepts of humanity and the nonhuman (Roffe and Stark 2015, 2). To understand humans in mutual becoming with animals and the natural world is to critique the anthropocentrism that has historically placed humans in a dominant role in these relationships. In so doing, the constructs of subject and object (or absent referent) come into question and the human is relocated within a broader context of nonhuman life.

Homi Bhabha applies Deleuze and Guattari's becoming to cultural, political, and racial hybridization. Bhabha's "hybridity" emphasizes the interdependence of colonizer/colonized relationships in mutually constructing cultural identity. This mutual construct occurs within what Bhabha calls a "Third Space of enunciation" (Bhabha 1994, 37). The existence of this Third Space makes any claim to or belief in cultural purity indefensible, since culture is always already hybrid and mutually constitutive. Bhabha acknowledges the importance of difference in cultural creation without negating the role of hierarchical systems of dominance and oppression that shape colonial and post-colonial society. Uma Narayan establishes Bhabha's concept of hybridity in relationship to food and eating. In her work on "food colonialism" or "culinary imperialism," Narayan describes these relationships as "incorporating the Other into the self, but on the self's terms" (Narayan 1995, 67). The Other and self—dominant and colonized foodways, respectively—mutually constitute each other in a Third Space of enunciation, and yet the relationship between them is articulated according to a power dynamic that favors the dominant culture.

In relationship to eating meat, then, human beings find ourselves in mutual becoming with animals, and yet our narratives about that relationship are established on the human self's terms, not that of the animal Other or absent referent. This othering occurs along a continuum of compartmentalization that also includes gender, race, ethnicity, and sexual orientation. In *Nightwood,* Barnes explores the incorporation of self and others through hunger, particularly a carnivorous kind of hunger. It is what moves the novel forward, so much that the work itself is hungry for what is beyond its own pages. Barnes's rich visual and sensory language compels the reader forward, even as her unflinching presentation of the proximity between sexual desire and primal violence has repelled and befuddled readers since the work was published. A work of modernism, *Nightwood*'s hungers critique the ability of the novel form—enmeshed as it is with the heterosexual marriage plot—to accommodate queer desire and queer relationships. The character Robin Vote, in particular, moves like a hungry maw through the novel, devouring everything in her path. Her relationship to embodiment and appetite—one that Barnes describes as being closer to animal than human—drives the narrative in *Nightwood*, constantly confronting the reader with the novel's limitations in representing unstable categories of self/other and public/private, especially in the context of gender and sexuality. Hunger, which Barnes presents as a particularly predatory and insatiable sensation, offers a way of relating to the world that is inwardly focused, without regard for social and institutional structures, and that makes room for desires not otherwise accommodated within these structures.

Hunger—a primal, animal urge as Barnes presents it in her novel—recurs throughout *Nightwood* as a motivating force. The narrator's first introduction to Robin, who consistently appears as more animal than human, as "beast turning human," likens her to "human hunger pressing its breast to its prey…we feel that we could eat her, she who is eaten death returning, for only then do we put our face close to the blood on the lips of our forefathers" (Barnes 1937, 41). Robin doesn't just *have* hunger, she *is* hunger, and in this passage is meant to arouse the reader's own desire to consume her (through the text). Robin unsettles the reader by representing human *bêtises*; she is both "beast turning human" and human turning beast. She reveals not only how easily one might transgress that boundary, but that the boundary itself is a false construction. Robin evokes the "blood on the lips of our forefathers," blood they have spilled through the careful policing of this boundary between the human self and animal other, one that also enforces domination over other Others, including women, non-gender conforming people, racialized groups, and sexual minorities. She captivates first her husband, Baron Felix Volkbein, then Nora Flood, and subsequently Jenny Petherbridge, all of whom are fascinated by her as an elusive figure they can never fully know or possess.

So, too, does Barnes leave her readers hungry for details of Robin, whose inner monologue is never revealed to us, especially at the novel's inscrutable ending, in which she wrestles Nora's dog. Both the reader and the novel's other characters understand Robin as driven primarily by her hungers for food and sex. She moves from one sexual encounter to the next, to the dismay of Felix, Nora, and Jenny, and Nora describes, upon first meeting Robin: "She sat up in bed and ate eggs and called me, 'Angel! Angel!' and ate my eggs too, and turned over, and went to sleep" (Barnes 1937, 53). Robin eats and sleeps according to her own internal desires, and not any external or social cues. She is thereby akin to Derrida's cat, who does not reveal shame of its own, but instead reflects Derrida the condition of shame as particularly human. Barnes, by imbuing the character of a woman, Robin, with these qualities, destabilizes the binary between animal and human.

The hungers in *Nightwood* include, but are not limited to, sexual desire. Elizabeth Blake describes Robin's hunger as a kind of "radical emptiness and instability," one that determines both Robin's magnetism and her movement throughout the novel (Blake 2015, 163). Hunger was a powerful force in Barnes's own life. She suffered from disordered eating that, according to Anne Dalton, resulted from her grandmother's sexual abuse. Dalton reveals letters in which Barnes's grandmother encoded descriptions of genitalia as food objects ("Bacon Cakes" among the most horrifying of these euphemisms) (Dalton 1993, 122).[2] Such metaphorical representation disrupts the boundary between who/what is eating and who/what is eaten. The letters also contain prescriptive advice about how much Barnes should eat, as well as evidence of using food as treats for sexual favors, all part of an overall subordination of Barnes's will (Dalton 1993, 121). Later, as an adult, Barnes demonstrated an understanding of the relationship between food and bodily disruption by subjecting herself to forced feeding as part of an article about English suffragists (Dalton 1993, 123). In *Nightwood*, hunger retains its violence and becomes a signifier for "visceral and counternormative" desire (Blake 2015, 154). Barnes captures hunger's evocation of physical sensations that are half-remembered, half-imagined, and fully desired, its keen sense of longing that is sometimes specific but more often vague and wandering (Blake 2015, 154). This violent and disruptive craving is what drives Robin—and the novel—forward throughout *Nightwood*.

Hunger, in Barnes's novel, includes a wide range of longings. For Felix, a European Jew who has concealed his Jewishness and pretends to be a Baron based on a false lineage, hunger carries a longing for the country and belonging as the "wandering Jew":

> No matter where and when you meet him you feel that he has come from some place – no matter from what place he has come – some country that he has devoured rather than resided in, some secret land that he has been nourished on but cannot inherit, for the Jew seems to be everywhere from nowhere. (Barnes 1937, 10)

Felix "devours" country but cannot belong to it. As a "wandering Jew," true aristocratic and nationalist identity is inaccessible to him. He searches for the nourishment that his false lineage and phony title can never truly provide. Barnes likens this nourishment to the milk of a wet nurse, rather than a biological mother:

> The Christian traffic in retribution has made the Jew's history a commodity; it is the medium through which he receives, at the necessary moment, the serum of his own past that he may offer it again as his blood. In this manner the Jew participates in the two conditions; and in the like manner Felix took the breast of this wet nurse whose milk was his being but which could never be his birthright. (Barnes 1937, 13)

In this passage, Felix's equation of national belonging with nursing as a baby reveals the construction of nationhood that requires the consumption of other bodies, what Narayan calls the incorporation of Other into the self. To belong to a people requires a not-belonging to any other people, a binary that allows for the incorporation of that Other. This is the binary construction that similarly sustains the human/animal power dynamic that allows humans to eat meat, as well as all other binary constructions of Other/self that follow from it. Belonging in the category of human—defined by its dominance over other beings—requires a not-belonging to the category of animal. Yet, as Barnes's analogy in this passage shows, such constructed binaries cannot withstand examination. Felix, in describing the difference between mother's milk and a wet nurse here, experiences the instability of categories pertaining to nationhood, ethnicity, and religion. As a European Jew—an increasingly marginal identity during the 1930s—he presents a second, more publicly acceptable identity. Yet, he can never shed his true heritage and therefore participates in "two conditions" of selfhood. In the same way that milk from a wet nurse feeds an infant adequate nutrition, but without any real maternal bonding experience, Felix projects a public self that allows him to survive in an increasingly hostile Europe, but at the expense of a disconnect between his public and private sense of identity. He is an example of Bhabha's hybridity, existing in a Third Space that keeps him from feeling fully nourished by either cultural identity and revealing the actual instability of the binary upon which those identities are constructed.

For Dr. Matthew O'Connor, the novel's queer prophet, to whom Nora comes for advice when Robin has left her for Jenny, hunger is also a way of knowing. As a medical man, he echoes Enlightenment ideals of sensory observation and embodied knowledge. Emotions, too, are inseparable from the physical nature of bodies, according to the doctor:

> I, as a medical man, know in what pocket a man keeps his heart and soul, and in what jostle of the liver, kidneys, and genitalia these pockets are pilfered. There is no pure sorrow. Why? It is bedfellow to lungs, lights, bones, guts and gall! There are only confusions. (Barnes 1937, 25)

"Confusions," to the doctor are the mistaking of physical impulses for some higher meaning. He denies any emotional or spiritual qualities, including sorrow and desire, and locates those feelings instead in the "liver, kidneys, and genitalia." Whereas Felix finds himself caught between the "two conditions" of Christian and Jew, Dr. O'Connor denies these conditions any meaning whatsoever, adopting a decidedly atheist stance toward the embodied human condition, which he argues is closer to an animal state than we care to admit. He does not recognize any spiritual or emotional conditions like "pure sorrow" but rather reduces such feelings to the "confusions" of biology. Cultural institutions like religion are, to Dr. O'Connor, only another kind of appetite for power: "Haven't I eaten a book too? Like the angels and prophets? And wasn't it a bitter book to eat? .... And didn't I eat a page and tear a page and stamp on others and flay some and toss some into the toilet for relief's sake" (Barnes 1937, 135). In his articulation of cultural institutions and the hunger for knowledge and social belonging that drives them, Dr. O'Connor equates these ostensibly human impulses with a feral hunger, one that causes him to rend pages and actually ingest the books themselves. He locates institutional hunger—like emotions—within the biological, animal body.

Dr. O'Connor, whom Nora finds wearing women's clothing when he is alone in his room at night and who shares with her that "no matter what I may be doing, in my heart is the wish for children and knitting," also divides his public and private selves into two distinct identities (Barnes 1937, 98). His gender presentation is female, and yet because he has a man's biological body, he can only express his true gender under the cover of night. For Dr. O'Connor, as for Felix, this divided self comes from social pressure to be what one is not, and the doctor advocates doing away with all of the structures that serve as obstacles to the expression of identity. The non/human binary is that from which all other constructions—attempts to "incorporate the Other into the self, but on the self's terms"—follow (Narayan 1995, 67). What Dr. O'Connor describes in the following passage is a moment toward an embodied self that exists outside of language and social construction:

> though your normal fellow will say all are alike in the dark, negro or white....Your *gourmet* knows, for instance, from what water his fish was snatched...he knows one truffle from another and whether it be Brittany root or if it came down from the North. (Barnes 1937, 100–101)

He argues for a return to the freedom of animal impulse, without self-consciousness and without a sense of surveillance over one's actions and ideas. A *gourmet*—someone who follows his appetites—is able to understand the world at a sensory level, like an animal. This yields him a deeper form of knowledge than one who uses his intellect to perceive the world at a surface level, understanding human beings only according to category. This is why, according to Dr. O'Connor, the night offers more possibilities for fully-lived experience, "the hour when he had evacuated custom and gone back into his dress" (Barnes 1937, 86). Whereas "every day is thought upon and calculated .... the night is not premeditated;" it "does something to a person's identity" by allowing that identity to articulate itself most fully (Barnes 1937, 87). Night dims one's vision and allows one's other senses—hearing, touch, smell, and taste—to come to the fore, similar to the way certain animals rely primarily on these senses. Consequently, it allows human beings to do away with the perceptions that are influenced by social convention and access what is primal and immediate.

Nora struggles in her relationship with Robin because she is less able to access her primal nature. Whereas Robin drifts according to her desires, Nora wants a monogamous relationship, modeled on heterosexual marriage. She feels "Robin's absence" as "a physical removal, insupportable and irreparable. As an amputated hand cannot be disowned because it is experiencing a futurity, of which the victim is its forebear, so Robin was an amputation that Nora could not renounce" (Barnes 1937, 64–65). Robin, in contrast, walks "in formless meditation...directing her steps toward that night life that was a known measure between Nora and the cafés" (Barnes 1937, 65). In this configuration, Nora stands in opposition to the cafés and "night life," the unpremeditated period in which animal senses and true identity come out. Nora, on the other hand, cannot help but experience "futurity," precisely the opposite of the animal's present reliance on sensory perception rather than rational forward-thinking. Like Derrida, she feels anxiety over her attachment to the affective states and institutions that are meant to distinguish animal and human experience. Whereas Lee Edelman positions queerness in opposition to futurity, or the possibility of biological reproduction, Nora's anxious thoughts are oriented toward the future (Edelmen 2004). Nora is tormented by the "unreasoning fear" that Robin "might lose the scent of home" (Barnes 1937, 61). Michael Davidson sees the relationship between Nora and Robin as one between an anxious mother and a willful child (Davidson 2012, 218). Nora, with her belief in monogamy and the powers of domesticity to protect futurity, forms a triangle with Robin and the cafés, which here become emblems of all that is opposed to futurity.

This is not to suggest that Barnes was herself in opposition to the idea of heterosexual, monogamous marriage. On the contrary, she would have had reason to question liberatory theories of sexuality, having grown up in an abusive environment due to her father's attempts to transgress sociosexual norms.[3] Yet, Barnes knew better than most how marital and familial ties can break down, and I argue that in *Nightwood*, Barnes explores the limits of family and relationship structures, as well as their inability to accommodate queer love and desire. It also questions how the novel form—shaped as it is by heterosexual marriage plots—can represent those arrangements. Just as Robin cannot conform to monogamous relationships—either with a man or woman—and Dr. O'Connor cannot conform to cisgender expectations of masculinity, *Nightwood* often questions whether any relationships can be predicated upon discrete and concrete categories of identity or feeling. As an alternative, Barnes offers her novel, which seems to explode out of its own boundaries and includes characters who are ever-changing in their relationships to themselves and to each other. Her interrogation of gender and sexuality categories, as well as categories of nationhood, is of a piece with her interrogation of the (non)human binary.

In *Nightwood*, Barnes consistently subverts the idea of definitive gender categories, upon which heterosexual relationships are predicated.

Robin, for example, has "the body of a boy" (Barnes 1937, 50). She also rejects motherhood entirely, pushing her son away from her and exclaiming, "I don't want him!," contrary to expectations of women and maternal "instincts" (Barnes 1937 53). Dr. O'Connor, too, evades gender categories. Felix observes him applying makeup, "dusting his darkly bristled chin with a puff, and drawing a line of rouge across his lips" (Barnes 1937, 39). He tells Nora that he wishes for "a high soprano… and deep corn curls to my bum, with a womb as big as the king's kettle, and a bosom as high as the bowsprit of a fishing schooner" (Barnes 1937, 97). Both Robin and the doctor make concessions to gendered expectations, Robin through marriage, and the doctor by confining his cross-dressing to the private space of his room at night. As Judith Butler argues, these performances of gender underscore the ways in which all gender is performative (Butler 1990). Meghan C. Fox also notes that Barnes uses hybridity to articulate the varied forms of embodied female experience that fall outside representational norms, offering alternative models of gender and sexuality that move beyond binary construction (Fox 2016). Both Robin and the doctor are aware of hiding a second, more intimate identity from the world. Upon encountering O'Connor on one occasion, "the doctor threw off his unobserved self, as one hides, hastily, a secret life" (Barnes 1937, 117). Consequently, the doctor argues throughout the novel that doing away with socially constructed categories of identity would allow this public and private self to merge and yield a deeper truth about the human condition.

Although several scholars have pointed to Dr. O'Connor as primarily driving *Nightwood*'s narrative, including Meryl Altman, who considers him the voice of sexual (and racial) theory in the novel, I see Nora's voice as frontal (Altman 1993, 166). Meeting Robin, for her, is a catalyst for confronting queer identity and reconciling it with her expectations for coupling. Barnes describes Nora as "outside and unidentified" until she meets Robin and is able to name her desires (Barnes 1937, 59). Nora is at first "dismayed to have come upon the doctor at the hour when he had evacuated custom and gone back into his dress," but quickly adjusts her expectations and draws him in as a confidante (Barnes 1937, 86). The narrative arc of *Nightwood* follows Nora as she finds her own sexual identity in Robin, and then learns to queer her expectation both of people and of relationships. She realizes that Robin's "life was a continual accident" and that it is precisely this chaos she herself craves although she feared it previously: "Everything we can't bear in this world, some day we find in one person, and love it all at once" (Barnes 1937, 144). Whereas she initially fears Robin leaving her and experiences her absence as loss, Nora comes to understand, in dialogue with Dr. O'Connor, that love is not the happy ending of a marriage plot, but rather a more chaotic kind of hunger, a driving force behind human behavior. Though earlier in the novel, Nora is described as "an early Christian," she later agrees with Dr. O'Connor that "love is the first lie," and begins to echo his position that "Man…conditioning himself to fear, made God… And I, who want power, chose a girl who resembles a boy" (Barnes 1937, 56, 147, 145). She comes to understand, under Dr. O'Connor's tutelage, that the urge to distinguish oneself from Others gives rise to divisions of religion, as well as binaries of gender and sexuality. The (non)human binary is the first of these constructions and provides a genesis for all others.

Just as Robin transcends human/animal boundaries in acting upon unfettered desire, modernist writers like Barnes sought to free language from the constraints of form. *Nightwood* is a novel that seeks to escape its own genre, to interrogate the boundaries of time, space, and the written page. Whereas the nineteenth-century novel assumed a linear progression of the narrative, often toward the generic "marriage plot," *Nightwood* not only questions the ability of this narrative to encompass all gender expressions, sexualities, and relationships, but also the ability of the novel as a form through which such a story can be told. Barnes writes at a time when many new media come into use, and like other modernist writers, including her mentor James Joyce, she interrogates distinctions between "high" and "low" art, as well as between textual and popular culture, which were increasingly blending into each other.[4] Descriptions in *Nightwood* very often remain at the surface level, replicating a pervasive media spectacle that offers more of a surveillance of the characters than the interior monologue found in other modernist novels like *Ulysses* and *Mrs. Dalloway*. *Nightwood* does not attempt the interiority of Joyce and Woolf but rather questions the ability of language to approach consciousness. Just as the characters elude categorization by constructions of gender, sexuality, and even species, the novel itself similarly resists formal limitations.

Robin and Dr. O'Connor also evade gender categories are in their relationship to violence. Robin expresses her sexuality through violence, another intersection of the human and bestial. The beginning of Robin's relationship with Jenny begins with Jenny striking Robin, "scratching and tearing in hysteria, striking, clutching and crying. Slowly the blood began to run down Robin's cheeks" (Barnes 1937, 83). Just a few lines later, the reader learns that "not long after this … Nora and Robin separated; a little later Jenny and Robin sailed for America." (Barnes 1937, 83) The gruesome and slightly absurd scene in which Jenny and Robin draw blood from each other immediately precedes news of their coupling. These two elements of their interaction are juxtaposed so directly, in fact, that the reader understands violence to be *part of* their relationship and not separate from it. Their struggle also foreshadows the novel's final scene, in which Robin squares off with Nora's dog:

> … down [Robin] went, until her head swung against his; on all fours now, dragging her knees…The dog, quivering in every muscle, sprang back, his tongue a stiff curving terror in his mouth; moved backward, back, as she came on, whimpering too, coming forward, her head turned completely sideways, grinning and whimpering… (Barnes 1937, 179–180)

The scene is what Adams refers to as an "interruption," or a place where the human power dynamic over animals becomes unsettling to the reader (Adams 1990, 126). It also serves as a Derridean rupture of the categories that construct that dynamic. In this passage, Robin becomes even more animal-like in her movements and behavior, engaging in a kind of primal beastliness that has confounded scholars since the novel's first publication. She crouches "on all fours," "grinning and whimpering," "dragging her forelocks in the dust" and begins to "bark, crawling after [Nora's dog]," all descriptions that seem to fit an animal rather than a human character. Blake argues that critics' discomfort with this passage reveals how the "movements of marginal bodies" can be deeply unsettling for even academic readers (Blake 2015, 167). She observes that the scene presents a moment of climax that is bodily but not erotic, one in which embodied hunger and violence are so extreme that they transcend language's ability to represent them (Blake 2015, 161). The parallels between this final scene and the one in which Robin and Jenny come together as a couple suggest that by wrestling with Nora's dog—the animal extension of herself—Robin is reuniting with Nora. However, the novel ends only with the dog's submission to Robin. Blake reads Robin's hunger as a kind of contagion that the dog catches from her, binding her to it in this confrontation with her own bestial nature (Blake 2015, 162). Robin Blyn understands the novel's strange conclusion, in which both Robin and the dog lie exhausted, as the end of Robin's acting like an animal, a performance she cannot sustain (Blyn 2008, 521). The violence of the passage and our discomfort with it remind us brutality is not unique to animals. We too, like Derrida and his cat, are naked and share a condition of mammalian experience. The binary construction that allows humans to pretend we are not animals allows all other dynamics of power and domination to exist.

By contrast, the doctor shies away from violence. He describes joining the army—an attempt to make his father proud of his masculinity—as a "terrible predicament: to be shot for man's meat, but to go down like a girl, crying in the night for her mother" (Barnes 1937, 81). Violence, and conducting oneself "like a soldier" is part of being a man during this era between the first and second World Wars, when Americans and Europeans were both confronting the brutalities of the First World War while also facing the possibility of a second (Barnes 1937, 81). The phrase "man's meat" also evokes gendered distinctions between appetites for meat that were present during this time. Sherrie Inness identifies the association between masculinity and meat-eating as a deeply held belief throughout Western culture, one that is as much about power as it is about taste (Inness 2001, 26). Meat-eating necessarily evokes the violence inherent in slaughtering or hunting and butchering animals into their edible form, and so to eat meat is also to engage the power of men to destroy life. To call himself "man's meat" in the context of dying in combat, the doctor assumes both positions of the killer and killed, undoing the binary between animal object and human self in parallel Robin's fight with the dog. He performs what Rasmus Rahbek Simonsen identifies as the need for veg(etari)an discourse to make visible the death and suffering it opposes (Simonsen 2012, 72). Veg(etari)anism queers dominant meat-eating society the way that Robin and Dr. O'Connor queer the heteronormative form in which they appear, a process predicated upon exposing the violence in that dominance, as Robin and Dr. O'Connor also do.

Within *Nightwood*'s narrative of queer relationship is a broader critique of the binaries that construct experience as either heteronormative or queer. Barnes presents Robin as a figure of human/animal becoming to show that this binary construction, from which all others follow, is inherently unstable. Thus, similar constructions of male/female, queer/straight, Jew/Christian, American/European dissolve once interrogated. Following Dr. O'Connor's argument about the origins of religion and man-made institutions, these exist to establish and maintain power dynamics and distinguish self from Other. To do away with such othering and cultural forms of oppression, as Dr. O'Connor advocates, is to examine this original (non)human binary, what Adams would offer as a social justice-oriented argument for vegetarianism. Over and over in *Nightwood*, Barnes presents binaries that subsequently collapse beneath the novel's narrative complexities, thereby showing readers how all categories of otherness similarly collapse once we interrogate the structures of power and privilege supporting them. In this final scene between Robin and Nora's dog, which intertwines the descriptions of their bodies' movement so that the reader can hardly tell one from the other, Barnes also undermines the categorical separation of human/animal, calling humans' role as meat-eaters into question alongside all other social constructions of identity. Barnes, in *Nightwood*, asks readers to renegotiate a relationship to difference, one that creates space for it outside binary constructions of power. By presenting to readers a novel that is hungry for possibilities outside of itself, Barnes invites an alternative futurity that demands new linguistic and generic forms.

Notes

1. I myself eat meat on occasion, and while I agree with Adams' argument about the role of meat-eating in reproducing structures of oppression, I also acknowledge that a vega(tari)an diet is not accessible for all.

2. See also: Blake; Barbara Green, "Spectacular Confessions: 'How It Feels to Be Forcibly Fed,'" *The Review of Contemporary Fiction* 13, no. 3 (1993): 70–88; Karen Kaivola, "The 'Beast Turning Human': Constructions of the 'Primitive' in *Nightwood*," *The Review of Contemporary Fiction* 13, no. 3 (1993): 172–185.
3. For more, see Dalton.
4. For more on American modernism and the virtualizing effects of language, see Katherine Biers, *Virtual Modernism: Writing and Technology in the Progressive Era* (Minneapolis: University of Minnesota Press, 2013).

Bibliography

Adams, Carol J. 1990. *The Sexual Politics of Meat: A Feminist-Vegetarian Critical Theory*. New York: Bloomsbury.

Altman, Meryl. 1993. A Book of Repulsive Jews?: Rereading *Nightwood*. *The Review of Contemporary Fiction* 13 (3). MLA International Bibliography.

Arendt, Hannah. 1958. *The Human Condition*. Chicago: University of Chicago Press.

Barnes, Djuna. 1937. *Nightwood*. New York: Harcourt, Brace, & Co.

Bhabha, Homi K. 1994. *The Location of Culture*. Abingdon: Routledge.

Biers, Katherine. 2013. *Virtual Modernism: Writing and Technology in the Progressive Era*. Minneapolis: University of Minnesota Press.

Blake, Elizabeth. 2015. Obscene Hungers: Eating and Enjoying *Nightwood* and *Ulysses*. *The Comparatist* 39: 153–170. MLA International Bibliography.

Blyn, Robin. 2008. *Nightwood's* Freak Dandies: Decadence in the 1930s. *Modernism/Modernity* 15 (3). MLA International Bibliography.

Butler, Judith. 1990. *Gender Trouble: Feminism and the Subversion of Identity*. New York: Routledge.

Dalton, Anne B. 1993. *This* Is Obscene': Female Voyeurism, Sexual Abuse, and Maternal Power in *The Dove*. *The Review of Contemporary Fiction* 13 (3): 117–139. MLA International Bibliography.

Davidson, Michael. 2012. Pregnant Men: Modernism, Disability, and Biofuturity in Djuna Barnes. In *Sex and Disability*, ed. Robert McRuer and Anna Mollow. Durham: Duke University Press.

Deleuze, Gilles, and Félix Guattari. 1987. A Thousand Plateaus: Capitalism and Schizophrenia. Trans. Brian Massumi. Minneapolis: University of Minnesota Press.

Derrida, Jacques. 2008. The Animal That Therefore I Am. Trans. David Wills. New York: Fordham University Press.

Edelmen, Lee. 2004. *No Future: Queer Theory and the Death Drive*. Durham: Duke University Press.

Fox, Meghan C. 2016. 'Vivid and Repulsive as the Truth': Hybridity and Sexual Difference in Djuna Barnes's *The Book of Repulsive Women*. *Space Between: Literature and Culture, 1914–1945* 12. MLA International Bibliography.

Green, Barbara. 1993. Spectacular Confessions: 'How It Feels to Be Forcibly Fed'. *The Review of Contemporary Fiction* 13 (3): 70–88. MLA International Bibliography.

Inness, Sherrie A. 2001. *Dinner Roles: American Women and Culinary Culture*. Iowa City: University of Iowa Press.

Kaivola, Karen. 1993. The 'Beast Turning Human': Constructions of the 'Primitive' in *Nightwood*. *The Review of Contemporary Fiction* 13 (3): 172–185. MLA International Bibliography.

Muñoz, José Esteban. 2009. *Cruising Utopia: The Then and There of Queer Futurity*. New York: New York University Press.

Narayan, Uma. 1995. Eating Cultures: Incorporation, Identity and Indian Food. *Social Identities: Journal for the Study of Race, Nation and Culture* 1 (1). MLA International Bibliography.

Roffe, Jon, and Hannah Stark. 2015. *Deleuze and the Non/Human*. New York: Palgrave Macmillan.

Simonsen, Rasmus Rahbek. 2012. A Queer Vegan Manifesto. *Journal for Critical Animal Studies* 10 (4): 51–80.

Part III

Eco Versus Ego: The Transformative Potential of Veg(etari)anism

9

Laying Down with the Lamb: Abolitionist Veganism, the Rhetoric of Human Exceptionalism, and the End of Creation

David P. Stubblefield and Dynestee Fields

Today the animal rights community finds itself at a critical impasse. On one hand, the Abolitionist Approach argues that all forms of animal exploitation should be abolished and that veganism should become a moral imperative. On the other hand, the animal welfare movement limits itself to the more realistic goal of reducing animal suffering, a goal that stops short of calling for veganism. As the vegan movement attempts to gain traction inside of the animal rights community and the general public, it has to make arguments that address the more entrenched position of the animal welfare movement and explain why simply curtailing

D. P. Stubblefield (✉)
University of South Carolina, Columbia, SC, USA
e-mail: dstubblefield@swu.edu

D. Fields
University of Tennessee, Knoxville, TN, USA
e-mail: DynesteeFields@mail.swu.edu

C. Hanganu-Bresch, K. Kondrlik (eds.), *Veg(etari)an Arguments in Culture, History, and Practice*, The Palgrave Macmillan Animal Ethics Series,
https://doi.org/10.1007/978-3-030-53280-2_9

animal suffering is not enough. A brief sketch of the differences between these approaches can bring to light the rhetorical situation in which arguments for veganism exist and provide an understanding of the obstacles that veganism must address in order to become a moral imperative.

From its inception, the animal welfare movement has taken its theoretical underpinnings from the utilitarian perspective of Peter Singer. Utilitarians follow John Stuart Mill (2002) in asserting that actions are good that bring the greatest amount of happiness to the greatest number (7). Happiness, for utilitarians like Mill, is not the happiness of the virtuous soul or of the blessed contemplating a beatific vision, but the happiness of the empirical being and its sensible experiences. By focusing strictly on the realm of sensation, utilitarians determine the morality of an action by calculating the sum total of pleasure and pain an action will cause, and—crucially for the animal welfare movement—by attributing equal weight to the sensible experience of each party affected by a particular action.

This is precisely the moral philosophy that Peter Singer drew on in his classic work, *Animal Liberation* (2015), in order to make powerful arguments for the better treatment of animals. From a utilitarian perspective, Singer insisted that it was not the possession of language, reason, or a soul that was the criterion for having a moral interest, but the capacity to experience pleasure and pain. Every sentient being, he argued, has an equal interest in seeking happiness (pleasure) and in not suffering (pain) and, therefore, must be represented in moral calculations. Failing to consider the suffering of non-human animals in any moral calculation amounted to an instance of a special kind of bias that Singer called "speciesism," a term that called into question the long-standing tradition of "human exceptionalism" within Western society that asserts that moral interests are exclusive to human beings.

By making the suffering of non-human animals a moral issue, Singer's work opened the door for society to reconsider how animals are treated and set into motion questions that have led many to adopt veganism. However, interestingly enough, neither Singer nor the ensuing animal welfare movement built on his thinking called into question the actual killing of animals, only the pain and suffering they experience in route to this fate. Thus, a tension emerges in this position, as the very line of thinking that set itself against speciesism ends up forbidding the killing of innocent humans but claiming that it is morally permissible to kill animals for the purpose of consumption.

When addressing the issue of animal slaughter, Singer supports this conclusion by appealing to the phenomenological difference between a human death and an animal death. Unlike humans, animals do not possess self-consciousness. They do not experience themselves in terms of a past, a present, and a projected future, but live in "an eternal present." Consequently, unlike humans, they do not experience death as an existential event, entailing tragedy, anxiety, and upheaval, but simply cease to exist at a given moment. This means that their death is not experienced and certainly is not experienced as a painful event. As such, it is not an instance of suffering that the animal welfare movement opposes, not a moral issue for humans, not something that enters into the utilitarian pleasure/pain calculation.

From the perspective of animal rights activists who advocate veganism, this conclusion is bound to be unsatisfying. For one thing, it seems to raise the specter of human exceptionalism. Humans are exceptional because they have self-consciousness and therefore experience death as painful; other beings, however, lacking self-consciousness, can be killed without moral culpability. Such a conclusion removes animal slaughter from the realm of moral deliberation, leaving the vegan movement with little choice but to differentiate itself from or critique the welfarist position.

To this end, Gary Francione has pioneered the Abolitionist Approach that seeks to abolish all forms of animal exploitation and to establish veganism as a moral imperative. Francione's work borrows both from Singer's utilitarianism and Kant's deontological approach. Like utilitarianism, it insists that sentience is the only criterion for warranting moral consideration. But unlike utilitarianism, it asserts that sentience grants membership in what Kant famously called "the kingdom of ends." Crucially, in this ideal moral community, no one should ever be used as a means to an end (Kant 1997, 41/4: 433). Each member is *intrinsically* valuable and, therefore, should always be treated as "ends in themselves."

That is, they should be treated with dignity and respect and their mere existence should be seen as having value in and of itself.

This deontological component to the Abolitionist position provides a way for Francione to push past the issue of animal suffering and to criticize all forms of animal exploitation. The issue is not whether animals suffer, but whether they are exploited. As Francione and Charlton explain, "At the core of the Abolitionist Approach is the idea that the primary moral issue involves the *use* of animals, and not the *treatment* of them" (Francione and Charlton 2015, 4). This shift from moral *interests* to intrinsic moral *value* emerges upon a shift from Singer's utilitarianism to Francione's sentience-based deontological approach opens the door for animal slaughter, not just animal suffering, to become a moral issue. The slaughter and consumption of animals by human beings clearly involves treating animals as a means and not as an end.

We have rehearsed this existing conversation in the animal rights community in order to flesh out some of the core issues facing the contemporary vegan movement. From our perspective, there are two salient points that will affect any discussion of veganism. First, the belief that most strongly impedes the moral claims of veganism is human exceptionalism. Once Singer declares that humans are exceptional in their possession of self-consciousness and their subsequent experience of death as a painful event, the path has been cleared to devalue animal death or to dismiss it altogether. One can be concerned about animal suffering, can oppose cruelty to animals, can call for the human stewardship of animals, and can even participate in various animal welfare campaigns, but as long as a belief in human exceptionalism persists, veganism's capacity to persuade audiences, even audiences sympathetic to animal rights like the animal welfare community, of the immorality of slaughtering animals will be severely limited or even non-existent. In other words, as long as human exceptionalism persists, arguments for veganism will gain little if any traction.

Second, human exceptionalism is deeply ingrained in our society, so much so that even so-called radicals like Singer who are interested in penetrating the veil of speciesism and address animal suffering, are unable to push past it and eventually succumb to it. One of the primary means by which human exceptionalism has become ingrained in our society is Judeo-Christian doctrine. According to a Gallup survey conducted in 2017, a staggering 73% of Americans identify themselves as being Christians (Newport 2017). While there are secular based arguments for human exceptionalism, such as those offered by Wesley J. Smith, and other religions that support human uniqueness, such as Islam, we argue that Judeo-Christian doctrine has had the strongest influence on Western culture's treatment of animals. In our culture, human exceptionalism often appears as an unimpeachable form of common sense. As a result, if veganism is going to take hold in Western culture, it must go deeply into the Christian origins of this belief and question its foundations.

In light of this context, this paper has a two-part structure. First, we attempt to shake human exceptionalism loose from its historical sedimentation by critically examining the foundational sources of contemporary Christianity's belief in this idea. Second, we draw on alternative sources and traditions inside of Christianity to construct an alternative anthropology where veganism is shown to be a moral imperative. By providing an immanent critique of Christianity from within Christianity, not only will we claim that Christianity is misreading itself, but that rather than being the historical glue for human exceptionalism, it is potentially its undoing. Accomplishing these tasks will clear the space for establishing a new relationship between humans and animals and for a vibrant discussion of veganism as a moral imperative.

The Imago Dei: The Appeal to Genesis 1

When Christians invoke the doctrine of human exceptionalism, they almost always appeal to Genesis 1:26–28 [King James Version], which states that man was "created in the image and likeness of God." As Carol J. Adams (2012) notes, appeals to humanity's dominion based on biblical authority often "cut off discussion" and express "a desire to justify contemporary practices" (2). For a long time, this belief has been commonplace in Christian discourse, with voices such as conservative radio host Dennis Prager (2017) and theologian Harold O.J. Brown (2009) acting as modern advocates. The threads radiating from this idea range from

Prager's straightforward interpretation that human life is infinitely valuable to Brown's complex theory of human uniqueness being key to the greater good of all creation. As Norm Phelps (2002) explains, Genesis 1:26 is "a claim that defenders of animal exploitation seize on to explain why we are entitled to use and abuse animals for our own purposes" (44).

The claim in Genesis 1:26–28 that humans are made in the image and likeness of God appears to assert that humanity is unique among all creatures. If this is true, then man has a greater value than any other being in the order of creation. However, there are significant problems with moving from this assertion to the position of human exceptionalism.

First, using Genesis 1:26–28 as an argument for human exceptionalism has several logical shortcomings. Just because something has a particular property does not mean something else does not have it. For instance, if a particular being was created with antlers, this does not mean that another being was not also created with antlers. In fact, such a claim does not say anything about whether any other being has antlers or anything about how other beings were created nor allows us to conclude anything about this antlered being's place in creation vis-à-vis other creatures.

Second, nowhere in the Bible is there an actual assertion of human exceptionalism (Phelps 2002; Cunningham 2009). Therefore, any the biblical argument for this position must rely on what theologians call an argument *e silentio,* or an "argument from silence." Arguing on matters where the Bible is silent is not the most reassuring practice for Christians. David Clough (2009) has recently argued that the emergence of Darwin exposed many of the hitherto unseen Aristotelian assumptions of Genesis, highlighting not so much the need for the Bible to be scientifically accurate, but the ease with which cultural biases often enter into scriptural interpretation unseen. For Clough, while "reason has been displaced from the Aristotelian world view in modern accounts as demarcating the line between humans and all other living creatures" (149), theologians have often operated on the basis of earlier interpretations based on this world view, unaware that that this kind of thinking only came about on the basis of the science available to a particular culture. In any case, it is very tempting for the *e silentio* arguments to rely on suspect evidence based on particular cultural assumptions and as Clough's work shows this has often been the case. Thus, the authority of *e silentio* arguments is questionable since cultural assumptions of a particular time period often fill the silence and then become cemented in the theological tradition.

Third, beyond the problematic attempts to justify this position theologically, there are serious ethical and political concerns about this argumentative strategy. Cunningham (2009) strongly cautions us in regard to making these kinds of arguments about scripture, noting that historically such arguments have been ethically hazardous practices (102). In particular, *e silentio* arguments for human exceptionalism have provided the justification for the worst kinds of political oppression. The narrative is all too familiar. Humanity's likeness to God is determined on the basis of a particular characteristic. Certain human beings are then said to possess this quality while others are said to lack it so that some humans are more fully human and others are less than human. For instance, the marginalization of African-Americans, women, and the disabled all proceeded on the view that these groups lacked rationality. In these cases, *e silentio* arguments surrounding Genesis 1:26–28 often provided a direct or indirect sanction of bigotry and political oppression.

In a 2019 article, "Addressing White Supremacy on Campus: Anti-Racist Pedagogy and Theological Education," Steffano Montano demonstrates how this practice has continued into the present day. In the absence of any definitive statement on the meaning of the image and likeness of God, the Anglo-Saxon myth of cultural exceptionalism develops as what Montano calls "a counter–Imago Dei" (246). Racist assumptions, not simply scientific ones, fill the silence on what constitutes being made in the image of God, as a "theo-ideology of Anglo-Saxon exceptionalism posits that only whites, apart from other races, are capable of entering into a sense of unity with and full election by God" (246). This problematic history should make us skeptical about the ethical and political consequences of the practice of putting forward *e silentio* arguments for human exceptionalism. With these considerations in mind, Genesis 1:26–28 as a means of establishing human exceptionalism appears flawed on multiple levels: the move from premise to conclusion is logically unjustifiable, an absence of biblical evidence for this idea leads to *e silentio* arguments, and the historical practice of this strategy raises very real ethical concerns.

While these flaws do not prove that animals have moral interests or a moral value or that we should refrain from eating them, they raise significant questions about human exceptionalism. Where exactly is the theological evidence for this doctrine? Are our current beliefs about animal-human relations based on anything more than particular cultural and historical assumptions, many of which are considered outdated? Moreover, as Phelps (2002) asks, even if humans are made in God's image, how does this result in "a license to terrorize, torture and kill?" and if it does, then "what does that say about the God, whose image we are supposedly reflecting?" (46) But beyond these questions of scriptural interpretation, there are also questions as to whether trying to establish that humanity is made in God's image is doomed from the start.

One major issue surrounding interpretations of Genesis 1:26–28 centers around humanity's lack of epistemic access to the image of God. That is, it is far from clear that we are able to know what the image of God is. After centuries of interpretation, we have not reached anything that would even approach a consensus on the matter. Inside the Christian tradition there is the substantive approach, the patristic interpretation of the substantive approach, the medieval interpretation of the substantive approach, the relational approach, the function approach, and many, many more that attempt to describe what the image of God is. However, there is nothing like a consensus as to what this image is. Martin Luther was skeptical of appealing to any natural capacity of humanity, fearing that even Satan possessed these natural traits and maybe even in greater degree than humanity. Thus, he argues that the idea refers to a state before the Fall where humanity "only knew God and believed that God was good" and were, thus, able to live "without the fear of death and any other danger and was content with God's favor" (qtd in Lazareth 2001, 62). John Calvin, for his part, takes a more holistic approach that combines both mind and body arguing that in prelapsarian man "there was no part of him in which some scintillation of it did not shine forth" (qtd in Cunningham 2009, 108–109). Karl Barthes' (1958) interpretation was heavily influenced by the existentialist philosophy of Martin Buber and his concept of the I-thou relationship, leading him to argue that the *imago Dei* is our capacity for relationship with God and for community with each other (196). The point here is not simply that language and reason rarely figure into these Christian interpretations of the *imago dei*, but that the nearly endless interpretation of this ideas suggest that Christian theology has been unable to arrive at any sort of agreed upon knowledge of what the *imago dei* is.

But even if individuals could somehow have epistemic access to the image and likeness of God, it is far from clear whether they could discern what would count as an empirical instance of this image. The strategy of appealing to an image of God in man appears doomed on many levels. Cunningham (2009) demonstrates the point brilliantly when he asks us to imagine a well-known painting, then proceeds to ask what would count as an image of the painting (110). A very accurate reproduction from another artist? A slightly less accurate reproduction by another artist that did not have the same texture? An image printed from a home computer with below average resolution? A child's watercolor drawing of the painting? These questions become very difficult, as the exact line of where something ceases to be an image is not clear. For Cunningham, this is because "the word 'image,' as it is used in our language, is a designation of degree" (110). As such, the word "image" does not permit the kind of distinctions that claims for human exceptionalism have often wanted from it:

> The grammar of the word is based on degree, not on a simple distinction between attribution and non-attribution. The word 'image' is not a Boolean operator (like true or false); we use it more like we use words for a certain colour (say, yellow). Such a word does not specify only one colour of a certain hue and saturation, but rather an entire range of colours (in which the transition for into other colors is difficult to discern). (111)

Hence, it is not the case that something simply is or is not an image of something else, but rather that something is more or less of image.

But at this point in his analysis, Cunningham recognizes that this analysis still leaves open the possibility of a sliding, hierarchical scale of degrees of likeness. However, he argues that such a sliding scale is not possible "because of another feature of the word 'image': namely, that something can be an image of something else in a variety of different ways" (Cunningham 2009, 111). To this end, he asks us to consider a situation where various media such as a painting, a sculpture, a work of literature, and other media attempt to represent the image of a human being. An argument can be made on behalf of each form of representation that it provides the best representation. However, in truth, each medium does some things well and other things not so well. They are simply different ways to represent something, but none of them can claim to be inherently superior to the others. By analogy, Cunningham (2009) concludes that it is possible that "the birds are like God in their ease of movement; the bees are like God in their simultaneous unity and multiplicity; the rocks in their steadfastness; and cats, as T.S. Eliot reminds us, in their mystery (possesses as each of is of a 'Deep and inscrutable singular Name') (113). In this way, in the absence of a hierarchical understanding of the image of God, it is possible to conceptualize a non-hierarchical conception of difference where all of creation bears the mark of the Creator in different ways.

Dominion

However, even if logical, textual, ethical, hermeneutic, and epistemic problems plague the interpretation of Genesis 1:26–28 as a support for the idea that humans are exceptional based on their likeness to God, another common argument for human exceptionalism focuses on a unique function performed by humanity: dominion. Once again, the crucial passage is from Genesis, and because of its importance, it is worth quoting at length:

> So God created man in his own image, in the image of God created he him; male and female created he them. And God blessed them, and God said unto them, Be fruitful, and multiply, and replenish the earth, and subdue it: and have dominion over the fish of the sea, and over the fowl of the air, and over every living thing that moveth upon the earth. (Gen. 1.27–28 [KJV])

This is the functionalist interpretation of the *imago dei* takes: humans have a special function. Their function is to exercise dominion, and this involves *subduing* the earth, an exercise that sets humanity in *opposition* to nature as it rises above it.

While this passage contains what appears to be a straightforward designation of a task and, therefore, appears to require little interpretation, significant contextual issues surround its interpretation. First, the most immediate context is that this claim to dominion takes place inside of all of creation being declared good (Adams 2012, 5), a claim that suggests all creation is intrinsically valuable and cannot simply be disposed of at humanity's whim. Second, as multiple commentators have pointed out, the idea of dominion, like any other idea in the Bible, must be interpreted in the larger context of scripture and, in particular, in terms of the overriding spirit of Christian ethics that emerges in the New Testament (Phelps 2002, 44; Scully 2003, 97). Third, interpretations of dominion take place inside of different cultural and historical contexts, so that the forces that construct these contexts must be examined to understand any particular understanding of dominion.

If we consider the larger context of the Bible and the limits it sets on humanity's will, power, and conduct, then even if man has dominion over creation, we would expect that this power would be limited by the spirit of Christian morality. For instance, a Christian teacher may have a certain type of dominion in the classroom, but if she uses this power destructively and in ways that oppose the spirit of Christian ethics, then presumably this form of dominion would not be Christian. Hence, dominion cannot mean absolute license, but would denote an activity that is necessarily limited by morality. If humanity's exercise of dominion over nature is meant to share in or imitate God's dominion and if God is a benevolent God, we can surmise that any lawful form of dominion would also have to be limited by that same benevolence.

Indeed, humility resonates much more with the message of the New Testament than prideful assertions of human power and will. After all, the very meaning of Christian *kenosis* is the renunciation of the divine nature, at least in part, by Christ in the Incarnation, a meaning that emerges from the Greek word *kenōsis* that literally means 'an emptying."

Thus, the ethical subject of Christianity involves an emptying of its own will and power, and it is precisely this self-emptying that enables an overflowing of love and charity to others in excess of any legalistic framework. As Matthew Scully (2003) explains, "The whole logic of Christianity is one… of the higher serving the lower, the strong protecting the weak, the last being first, and all out of boundless love and generosity, "rights" having nothing to do with it" (97). Likewise, when one considers the New Testament call to humility, to self-emptying love, and to charity, Scully's claim that "kindness to animals may be a small yet necessary part of decent and holy life, essential if only as a check against human arrogance and our tendency to worship ourselves, our own works and appetites and desires instead of our Creator and his works" (99) seems to fit better with the general tenor of the New Testament, than the idea that humans are lords of the earth who can satisfy our desire for pleasure at the expense of other living things. In one way or another, unqualified appeals to dominion and an unbridled extension of human will seem hard to square with the ethical subject of Christianity and its underlying kenotic structure.

In light of this incompatibility between the self-emptying structure of Christian ethics and notions of dominion and power, the question becomes how dominion, or the unlimited extension of human will over nature, ever become associated with Christianity. Many have sought to answer this question. For instance, in his well-known essay 1967 essay, "The Historical Roots of the Ecological Crisis," Lynn White Jr. clearly lays the blame for the human domination of nature on Christian metaphysics:

> By destroying pagan animism, Christianity made it possible to exploit nature in a mood of indifference to the feelings of natural objects. It is often said that for animism the Church substituted the cult of saints. True; but the cult of saints is functionally quite different from animism. The saint is not *in* natural objects; he may have special shrines, but his citizenship is in heaven…The spirits *in* natural objects, which formerly had protected nature from man, evaporated. Man's effective monopoly on spirit in this world was confirmed, and the old inhibitions to the exploitation of nature crumbled. (1967, 1205)

For White, Christian metaphysics usurps animism and, like a gigantic vacuum cleaner, sucks all meaning and being out of nature and puts them into man. This intervention, then, opens the door for the unbridled domination of nature (and animals).

However, the problem is that virtually any Christian theologian who accepts the incarnation would have a difficult time recognizing Christianity inside of this radically dualistic picture. After all, the entire religion is based on the idea that word became flesh and, in doing so, overcame precisely the radically dualistic picture that White describes. Hence, rather than being a straightforward example of Christian metaphysics, it is likely that the force that White is recognizing is Cartesianism and the mechanistic science that grew from it. Mechanistic science arose with the advent of modernity and imposes a sharp divide between mind and nature, where mind is autonomous and free and the rest of nature is inanimate and wholly subject to machine-like laws. And while this force may have influenced Christianity at a certain point and later became absorbed by the larger culture, the idea that it is a Christian idea is highly suspect. Terry Eagleton (2010) points out, in the Christian tradition, "Evil involves a split between body and spirit—between an abstract will to dominate and destroy, and the meaningless piece of flesh that this will inhabits" (23). The evil that Eagleton describes involves an isolated will disconnected and disembodied, ceasing to feel or share the pain of the bodies it encounters. Furthermore, he explains it is "the rationalist who treats the world, including his own and others' bodies, as mere valueless stuff to be moulded by his imperious will" (23), a will which in becoming infinite is "a secularized version of God" (21). This hardly seems like a position that Christianity would advocate for, but much more like something Christianity would combat. Consequently, Cartesian rationalism and the mechanistic science that emerged from it, with its radically dualistic understanding of the distinction between mind and matter, may be a better explanation of what removed spirit from objects and not Christianity, which is founded on an incarnation that overcame this dualism.

There is significant etymological evidence to support this view. For one thing, prior to the rise of mechanistic science and Cartesian philosophy, there were no "animals" in the modern sense of the term. Stephen Moore

(2014) makes exactly this point, asserting that "Prior to the epochal changes in European culture synecdochically signified by the name 'Descartes,' there were no 'animals' in the modern sense. There were 'creatures,' 'beasts,' and 'living things,' a bionomic arrangement reflected in, and reinforced by, the early vernacular Bibles" (9). In the same way, Laurie Shannon notes that the term is curiously absent in the Great Bible (1539), the Geneva Bible (1560), or the King James Version (1611), three key books for the development and transmission of the standard lexicon. Likewise, in her article, "The Eight Animals In Shakespeare; or Before the Human," Shannon claims the word animal only appears eight times in Shakespeare's entire work and that "the OED confirms that the word 'animal' hardly appears in English before the end of the sixteenth century" (2009, 474). These absences suggest that the modern understanding of the animal that we have inherited did not exist in the premodern period and that "Descartes was the prime inventor of the animal" as a being "whose behavior is purely mechanical and as such altogether unlike human behavior" (Moore 2014, 6).

Indeed, when the term "animal" was used prior to modernity and the Cartesianism that defined it, it was not used as something *opposed* to man. In other words, according to Laura Hobgood-Oster (2008), up until the modern period, the line between the human and the animal was more supple, as evidenced by instances of "baptized lion" and "a chorus of birds praising God" (6). Susan Crane (2013) cites further evidence of a premodern continuity between humans and animals in John Trevisa's translation of *Bartholmaeus Ang* where he describes the human as a type of animal.

> All that is compounded of flesh and spirit of life, and so of body and soul, is called animal, a beast whether it be of the air like birds, or the water like fish who swim, or of the earth such as beasts that go on the ground and in field, like men and wild and tame beasts. (1–2)

Crane quotes Trevisa himself who notes that in Seville Etymology describes man "as a beast that resembles God" (2). All of these examples suggest that White's claim that traditional Christian metaphysics is radically dualistic is highly suspect. Such a metaphysics opens the door for the domination of nature by the will of humanity. However, this metaphysics seems to be largely a result of Cartesian modernity, and the incarnational model of Christianity problematizes exactly this kind of dualistic thinking. In doing so, it opens the door for the different understanding of the relationship between man and nature.

A Theology of Creation

One place where we can see this different, non-dualistic understanding of this relationship is in Genesis 2. While Genesis 1 has exercised an astounding amount of power over the western political and moral imagination, Genesis 2, which offers a second creation story, has garnered significantly less attention. However, Theodore Hibbert (1996) asserts that Genesis 2 represents "a truly distinctive tradition about the origin of the world" (16), one that not only does not support human exceptionalism, but also forecloses any possibility of this position. The key verses in this alternative account are Genesis 2:7, "And the Lord God formed man of the dust of the ground, and breathed into his nostrils the breath of life; and man became a living soul," and Genesis 2:15, "And the Lord God took the man, and put him into the garden of Eden to dress it and to keep it" [KJV]. In this second creation story, Hibbert finds an alternative Christian tradition and explains:

> In this tradition, the human being is positioned very differently within the world of nature. Here the archetypal human is made not in the image of God but out of topsoil, out of the arable land that was cultivated by Israelite farmers. As a result of this kind of creation, humans hold no distinctive position among living beings, since plants and animals also were produced from this same arable soil. (1996, 19)

Since all life has the same origin, Genesis 2 establishes a kinship among all creatures. Moreover, in this version of creation, humanity is not given dominion over creation, but is instead called to *serve* creation. To this end, Hibbert focuses on the King James translation of Genesis 2:9:

> Moreover, the role assigned humans within creation in this story is not to rule (*radah*) and to subdue (*kavash*) but rather to "serve" (*avad*; Gen. 2:15; 3:23). The Hebrew term *avad* is properly translated "till" in these verses [NRSV], since it clearly refers to the cultivation of arable land. But *avad* is in fact the ordinary Hebrew verb "serve," used of slaves serving masters and of humans serving God. (1996, 24)

Genesis 2, then, offers a different anthropology (man as a servant of nature) and a different relationship to nature, one where humanity is not opposed to nature or charged with the task of subduing it, but one where the relationship is reversed and humanity is the servant of nature.

How are we to understand the juxtaposition of two divergent creation stories, one often-cited account that has traditionally been interpreted in a way that man rules over nature and another largely overlooked account that suggests that nature rules over man? Alain Badiou (2003) discusses this problem when interpreting Corinthians 1:7:4, where the often-cited verse that a husband is to rule over the body of the wife appears, but is then immediately followed by the less frequently cited verse that wives are supposed to rule over the bodies of their husbands. Badiou (2003) notes the prevalence of this technique in the work of Saint Paul and "frequent symmetrization," a technique that he notes is reminder that what matters is a truth's universal becoming to be implicitly slipped into this inegalitarian maxim, it will, so to speak, be neutralized through the subsequent mention of its reversibility (104). The noted neutralization, does not mean that the marginalized term usurps the hegemonic term within the existing economy of power, but instead implies a suspension of the efficacy of this economy as a whole and of its capacity to function. In particular, it problematizes the whole notion of rule by one group over another.

We can read the contradiction between the first and the second creation stories in the same way. That is, we can read deconstructive reversal of the second creation story where nature appears to rule man instead of man ruling nature as neutralizing the entire economy of power that exists between man and nature and determines that one must rule the other. Such a reading gestures toward any entirely different, non-hierarchical economy of relations.

Incarnating Life

While the creation stories have played a crucial role in establishing humanity's place in nature, the other event that addresses these concerns and has to be a part of any discussion of this topic is, as we have already suggested, the incarnation. In fact, the incarnation is so crucial to Christian theology, Eric Daryl Meyer (2014) argues that "the whole Christion tradition labors under its weight" (147). Certainly, much rides on how we conceive of the incarnation. If the *logos* (or "the word") is placed in humanity alone and *not* in any other creature and is something separate from or opposed to creaturely life (*zoe*), then a radical dualism appears. Such a dualism, as Lynn White Jr. explained, allows humanity to claim an exceptional status that authorizes its dominion over the rest of creation. However, if the *logos* is in *all* creatures and not simply in humanity, then the ontological distinction that justifies the violent dominion of humanity over nature no longer exists, as a non-hierarchical picture emerges where the *logos* effectively becomes *all in all.*

A key text for working out these issues around the incarnation and the *logos* has traditionally been the beginning of the Gospel of John, a text that warrants closer examination.

> In the beginning was the Word, and the Word was with God, and the Word was God. The same was in the beginning with God. All things were made by him; and without him was not anything made that was made. In him was life; and the life was the light of men. And the light shineth in darkness; and the darkness comprehended it not. (John 1:1–15 [KJV])

While this passage states that the life that was in God was the light of *men*, once again, it certainly does not establish that this life is not in other creatures. Instead, it explicitly states that all things were made by him, there is *life* (*zoe*) in him, and this "life that is light" was imparted to all creatures. Moreover, by claiming that the Word (*logos*) imparts life (*zoe*) and not rationality, John's text represents a theology of life that breaks from the pagan equation of *logos* with rationality. As such, 1 John speaks of the "the Word of life." Meyer (2014) provides an interpretation on

John's text that recasts the human *logos* in light of this Word of Life. He explains,

> After asserting that everything in existence resonates with echoes of the Logos, having come into being through it, John narrows his view and writes that this Logos is life (*zoe*), and that this life is the light of human beings (*anthropon*). Human life (*zoe*) radiates as light from the Logos of God. But John's text is not all light and life. John quickly modulates into a minor key and writes of a darkness that refuses the light. The world of humanity, the *kosmos*, is the site of this darkness; humanity fails to recognize the Logos as its very life. (146)

But what precisely is it that humanity fails to recognize? Does humanity, like the prisoner in Plato's cave, cling to the darkness of its lower appetites and fail to recognize reason? For Meyer, this popular interpretation simply does not work. It is not that humans have failed to recognize *logos* understood as reason, speech, thought, argument, logic or discourse. Rationality is "hardly a marginal or undervalued aspect of human existence" (Meyer 2014, 146). In fact, it "does pretty well in the world of humanity" since for among other things, it has served as the basis for distinguishing us from all other creatures and been the very basis upon which we have asserted ourselves to be exceptional beings (146). So we can hardly be said to be in the dark about, to have rejected, or to have despised it. If anything, we have glorified it.

So what, then, is it that humanity has not recognized and even has scorned? The object of this misrecognition and contempt is the *Divine Logos*. But—and here is the crucial point—there is a difference between the *Divine Logos* and the human *logos*, and this difference is not one of quantity, but of quality. Indeed, the human *logos* is qualitatively different from the *Divine Logos*, and it is this difference that prevents human rationality from recognizing the light of the *Divine Logos* that *animates* all of existence by imparting life (*zoe*). This is precisely because the human *logos* defines itself over and against life, excluding the very life that the *Divine Logos* imparts. Therefore, the *Divine Logos* is a blind spot or darkness for the human *logos*, as it defies its grasp. Humans operating through the human *logos* simply cannot recognize the *Divine Logos*.

In his affirmation of *zoe*, understood as "animality," Meyer performs a deconstructive *reversal* of the hierarchy that the human logos establishes between reason and life. This deconstructive reversal is similar to the one performed by Genesis 2 and by Paul that was cited earlier. Meyer (2014) reverses the hierarchy inherent in the human logos when he claims that "the work of logos is more evident in *zoe* than in human thought and speech" (158). Animality is precisely what is discarded by the human *logos* as excrement, waste, or enemy. To this end, he notes Jesus is often recognized as an outsider to society and politics: "it is commonplace to recognize Jesus as outcast, the scapegoat, the refugee, whose life cannot be assimilated to the order of his society. Jesus is not far from the figure of *homo sacer*, exposed to death outside of the city (not murdered) for the sake of the political order" (2014, 158).

Moreover, according to Meyer, Jesus's identification with animality actually takes a very concrete form and is the subversive force of the kenotic unworking of the human *logos*. He explains,

> God's Logos appears as the fleshmeal around which humanity unites, the sacrificial lamb slaughtered for ritual purity, the scapegoat cast out by the fury of human sin, and the symbolic lion whose ferocity lends courage to the disheartened. One might ask whether the Logos of God appears *in the place of the animal* to endorse eating, slaughter, and experimentation, or to loose the knots holding these cultural structures together. (2014, 158, our italics)

By appearing in the place of animality, the *Divine Logos* subverts the human *logos* and the existing form of humanity. That is, it empties a form of humanity that opposes itself to animality in a kenotic un-working that takes as its object the division between man and life (*zoe*) on which the human *logos* and subsequent claims to dominion are based. While God may have been incarnated in humanity, he did so in order to hollow out a human subject based on the human *logos* and in order to impart life to this subject. Moreover, he did this by coming back *in the place of the animal* (*zoon*), by becoming the sacrificial lamb that, according to Rene Girard puts an end to all sacrifice (Girard 1998, 171).

By putting an end to sacrifice, Christ puts an end to a sacrificial structure of humanity, a structure where one becomes human only by identifying and violently opposing oneself to animality. Only a living being who says "no" to and breaks from animality becomes a rational, moral, and civilized human being. However, the incarnation goes after the very division between the human *logos* and animality, the division that serves as a basis of both the human *logos,* a *logos* premised on the human's alienation from nature and that, therefore, prevented from affirming the life (*zoe*) imparted by the *Divine Logo*s. The incarnation brings life to humanity precisely by undoing the division brought about by the Fall that divided humanity from nature and subjected to the command of the human *logos*.

But how does it do this? Because the incarnation takes place in a human who occupies the place of the animal within existing power relations, it creates a zone of indiscernibility between the human and the animal that renders the general economy of power relations inoperative. That is, like the reversibility of power relations in Paul's comments of husbands and wives, the point is not to keep the same economy of relations in place and to simply privilege the marginal term, but to suspend the entire economy of power relations in which both terms appear. As Jacques Derrida, explains, "Deconstruction does not consist in moving from one concept to another, but in reversing and displacing a conceptual order as well as the nonconceptual order with which it is articulated" (Derrida 1988, 21). These deconstructive reversals are not ends in themselves, but interventions that blur the divisions that need to be firmly in place for the existing economy to function, and in doing so, open toward the possibility of another economy "yet to come," another qualitatively different *logos* that ushers in another qualitatively different community. In terms of Christian theology, the effect of Christ occupying the place of the animal is to open the possibility of the messianic kingdom, or as animal theologians call it, the peaceable kingdom (Phelps 2002, 95).

The Messianic Kingdom

Christianity finds this *Divine Logos* set forth in its eschatological vision established by the prophetic tradition of Isaiah, the Book of Revelation, and above all others, Saint Paul. Drawing on this tradition, Tripp York

(2012) explains "a vital part of Christianity's lense, or grammatical framework revolves around time—specifically, that being a beginning, middle, and end" and that "faithful Christian existence" is called upon to live "a resurrected life in light of Christianity's eschatological nature (eschatological meaning that we live with the end in view)" (156). In the beginning, creation was at peace with itself; however, the fall from the peaceful state establishes the secular, not as a separate sphere, but as a mode of time (157). Secular time is the time between the beginning and the end, or the time that is "ripe with violence, sickness, war, death, war, famine, and destruction" (158). However, once this time passes away, the end resembles the beginning as "all will be subsumed into its original purpose" (158). In other words, the end of time does not simply denote the last moment in a line of successive instants, but also denotes the original purpose or "end" of creation (158).

The book of Isaiah not only states that "He shall feed his flock like a shepherd: he shall gather the lambs with his arm, and carry them in his bosom, and shall gently lead those that are with young" (Isa. 40:11 [KJV]), but also that the peaceable kingdom that will restore the broken relationship between humanity and animality. This kingdom is described as:

> The wolf also shall dwell with the lamb, and the leopard shall lie down with the kid; and the calf and the young lion and the fatling together; and a little child shall lead them. And the cow and the bear shall feed; their young ones shall lie down together: and the lion shall eat straw like the ox. And the sucking child shall play on the hole of the asp, and the weaned child shall put his hand on the cockatrice' den. They shall not hurt nor destroy in all my holy mountain: for the earth shall be full of the knowledge of the Lord, as the waters cover the sea. (Isa. 11.6–9 [KJV])

In this kingdom, humans and animals exist peacefully without violence, as "neither harm nor destroy" each other. Therefore, J.R. Hyland (2000) rightfully insists that "the animal kingdom is an integral part of the kingdom to come," adding that "animals and humans would enjoy the blessed existence of a restored creation" (3). Likewise, for Paul, the whole of creation cries out in groaning pains waiting for its restoration in this peaceable kingdom (Rom. 8:18–22 [KJV]).

This messianic dimension of Christianity, so critical in the work of Saint Paul, is not a call for utopia to be realized on earth, but the persistent call for the undoing of earthly powers, an ethical concept of Christianity that calls all Christians to welcome the coming of messiah and the peaceable kingdom. Tripp York challenges Christians to embrace an eschatological ethics. Hence, he asks,

> Is it not both our burden and privilege to embody Isaiah's vision so that others may catch a glimpse of that which we claim to be true? While it is true that the kingdom is not yet fully realized, it is up to Christians to be faithful to their understanding of creation, so that others will know that there is an alternative to this vision of a world at war with itself. Otherwise how can anyone know that there is a different way of interpreting creation that would seem to be a senseless, wasteful, and cruel practice of the cold and indifferent cycle called nature. (2012, 160)

From this eschatological perspective, this orientation toward the peaceable kingdom forms a normative compass, not hubristic proclamations of dominion, which function as a license to violently abuse and destroy life and as a means for humanity to re-inscribe its own alienation from nature.

This messianic tradition has been embraced by secular thinkers such Theodore Adorno, Jacques Derrida, Walter Benjamin and Giorgio Agamben as the means to think about ethics and justice. Adorno argues that viewing injustices from the perspective of the messianic kingdom was the only way to displace the existing powers of the world that we live in. According to Adorno,

> The only philosophy which can be responsibly practiced in face of despair is the attempt to contemplate all things as they would present themselves from the standpoint of redemption. Knowledge has no light but that shed on the world by redemption: all else is reconstruction, mere technique. Perspectives must be fashioned that displace and estrange the world, reveal it to be, with its rifts and crevices, as indigent and distorted as it will appear one day in the messianic light. (1978, 247).

After reading this quotation, we might ask ourselves what it means to contemplate the animal (and all living beings and all of nature for that matter) in light of its end, or as Adorno puts it "as it will appear one day in the messianic light," and ask ourselves if there is an ethical demand to do just this. Certainly, inside of the Christian eschatological ethics we have outlined there is. If Christian faith is precisely an eschatological faith, then Christian ethics involves orienting itself to this kingdom that bears more a superficial resemblance to Kant's kingdom of ends, a kingdom where all sentient beings will be ends in themselves and veganism will be, as it was in paradise, not an imperative, but a reality. But meanwhile, for Christians, living in light of this reality to come is an imperative, the necessary orientation for ethical action, a way to participate in the work of the peaceable kingdom which is the purpose or end of all creation. So how exactly does killing animals and eating them for pleasure fit into this picture?

Conclusion: Towards Creation's End

Efforts to treat animals as ends and not means have been forthcoming in the Christian sphere as many of its members observe the violence in the world around them, and, as a result, seek direction in the creation accounts of Genesis 1 and 2. Examples of this desire for restoration are illustrated in the faith statements found in *Every Living Thing: How Pope Francis, Evangelicals and other Christian Leaders are inspiring all of us to Care for Animals* (2015). Spanning 12 denominations, this collection offers insight into concepts of dominion, both contemporary and historical perspectives on humanity's relationship to creation, and the actions currently being taken to ensure that issues involving animals and the natural world receive faith-mandated attentiveness. In this work, which includes several of the ideas argued for in this chapter (namely our ideas on the *imago dei*, servanthood, and the *Divine Logos*), there are two major positions present: animal welfare and abstinence from animal products. While animal welfare is accepted by nearly every denomination, abstinence from animal products is less widely advocated for. We have included examples from both ends of the spectrum below.

Seventh-day Adventists, who "call for respect of creation, restraint in the use of the world's resources, reevaluation of one's needs, and reaffirmation of the dignity of created life" (Gutleben 2015, 137), and some members of the Evangelical Lutheran Church of America, which asserts that "the sphere of moral consideration must encompass all of nature, not simply the immediate circle of human beings" (Gutleben 2015, 72), translate their concern for creation into vegetarian diets and lowered meat consumption. On the other end of the spectrum, however, are denominations that overwhemingly call for the regulation of animal use. U.S. leaders in the Roman Catholic Church, for example, state that "certain farming practices, including the factory farming of animals… should be carefully regulated and monitored so that environmental risks are minimized and animals are treated as creatures of God" (Gutleben 2015, 130). Similarly, the Lutheran Church-Missouri Synod finds that "We can choose to eat those animals that were allowed to live as God intended prior to giving their lives for us" (Gutleben 2015, 94). These latter examples, and the prevalence of animal welfare reforms as opposed to reforms for veganism (or in this example, vegetarianism or reduced meat consumption), hint that humanity's destruction of animal life might partially hinge on perceived need rather than pure desire.

We have integrated this snapshot of Christian thought and activity because it illustrates the effects of Genesis 9, the final obstacle to cementing veganism as an imperative for Christians (as well as others who uphold Judeo-Christian derived perspectives). Is it possible that those who accept animals as also sharing the *imago dei*, who recognize the servant position as dominion's true form, and who hold that the incarnation balanced the power relations between humans and animals, might still be justified in resisting veganism? The answer to this question lies in assessing the Genesis 9 decree that animals have been lawfully permitted into the human diet and in determining the suitability of adopting principles of the messianic kingdom into the present.

The fateful allowance of Genesis 9 occurs after God orchestrates a flood to destroy all land-dwelling creatures. The cause of this event is His perception "that the wickedness of man was great in the earth, and that every imagination of the thoughts of his heart was only evil continually" (Gen. 6:5 [KJV]). The survivors of this catastrophe are limited to those inhabiting an ark: one extended human family, one male and female of every unclean animal species, and seven individuals of every bird and clean animal species (Lev. 11:2–47 [KJV]). Importantly, after the waters abate, God informs the family's patriarch and his sons that:

> the fear of you and the dread of you shall be upon every beast of the earth, and upon every fowl of the air, upon all that moveth upon the earth, and upon all the fishes of the sea; into your hand are they delivered.
>
> Every moving thing that liveth shall be meat for you; even as the green herb have I given you all things. (Gen. 9:1–2 [KJV])

A revised version of Genesis 1:28–29, this account becomes the mandate of the fallen world after the Flood. In the words of Matthew Halteman: "the killing of animals for food must be conceived as a disruption of shalom. God may allow it as a concession to sin, but the allowance is tinged with the tragedy of paradise lost" (Halteman 2019, 91). While before this passage, there is evidence of animal death in the forms of sacrifice (Gen. 4:4 [KJV]) and animal skin garments (Gen. 3:21 [KJV]), this declaration signifies the end of animals possessing a natural trust in humans and the dawn of a whole new level of bloodshed.

Like the one of Isaiah 11, the paradise depicted in Genesis 1 and 2 is currently beyond reach. However, the potential to foster a version of it is not. Unlike when animal consumption was permitted in Genesis 9 (after a flood that, no doubt, left substantial damage in its wake), humans, particularly those in the West, now have the resources to flourish without the use of animal products. Success in this area is confirmed through the validation of veganism by several major health organizations and medical communities, including: Harvard Health Publishing (2020), the Heart and Stroke Foundation (2018), and Mayo Clinic (2019). When it is correctly planned, the Academy of Nutrition and Dietetics (2016) deems veganism as "healthful, nutritionally adequate" … and "appropriate for individuals during all stages of the life cycle, including pregnancy, lactation, infancy, childhood, and adolescence," as well as "for athletes" (Melina et al. 2016, 1). As established by this information, consuming animal products has largely joined the acts of performing animal sacrifice and donning animal-derived clothing on the West's list of unnecessary practices.

Despite the facts that the conditions leading up to this event are intrinsically tied to human sin and that veganism in the West is a healthy and attainable lifestyle, mainstream Christianity holds that because humans are *allowed* to consume animals, this consumption is held as being perpetually acceptable until creation's renewal. If individual Christians refuse to alter their behavior, life prior to the peaceable kingdom will retain a peculiar resemblance to David VanDrunen's Genesis 8 and 9 based theology of natural law. In addition to providing directives for the human diet, these passages, which form the Noahic covenant, contain the promise that "while the earth remaineth, seedtime and harvest, and cold and heat, and summer and winter, and day and night shall not cease" (Gen. 8:22 [KJV]). VanDrunen argues that this covenant does not promise consummation for either humanity or earth, nor to eradicate the introduced evil. Instead it "envisions only the ongoing performance of the task (human dominion) in order to sustain human existence and to maintain an uneasy and partial peace in society" (VanDrunen 2010, 139). When Christians hold that animal consumption is acceptable in *all* conditions, they do not, in the fullness of their ability, usher in more of the peace that creation groans for or progress towards God's ultimate will.

For Christians, proceeding towards an attainable version of the messianic kingdom begins with considering the arguments that we have presented. The recognition that both humans and animals have been imparted with life (*zoe*) by the *Divine Logos* and that they both originate from the same soil means that there can be little argument for treating animal death as a trivial matter. The possibility of the *imago dei* being inherent in all creatures intensifies the demand for preserving animal lives. Furthermore, the incarnation's balancing of powers ensures that animal interests, including the interest to live as opposed to merely avoiding suffering, must be considered. Together these principles place animals firmly within the kingdom of ends. Lastly, because much of humanity's current killing of "food" animals is not necessitated, there is little reason for destroying these lives. To kill simply for the pleasure of doing so is to practice the egocentric brand of dominion that we called attention to earlier. As Andrew Linzey expresses, "*Where we do have the moral freedom* to live without killing, without recourse to violence, there is a *prima facie* case that we should do so. To kill without the strict conditions of necessity is to live a life with insufficient generosity" (Linzey 2003, 233).

Thus, according to the errors of human exceptionalism that we have identified, the alternative Christian traditions that we have examined, and the affirmations of the vegan diet that we have illustrated, there is little justification for Western Christians, as well as others who adopt concepts from Judeo-Christian tradition, to treat veganism as anything less than a moral imperative.

Bibliography

Adams, Carol J. 2012. What about Dominion in Genesis? In *A Faith Embracing all Creatures: Addressing Commonly Asked Questions About Christian Care for Animals*, ed. Tripp York and Andy Alexis-Baker, 1–12. Wipf and Stock Publishers: Eugene, OR.

Adorno, Theodor. 1978. *Minima Moralia: Reflections from Damaged Life*. London: New Left Books, Verso Editions.

Agamben, Giorgio. 2004. *The Open: Man and Animal*. Redwood City: Stanford University Press.

Badiou, Alain. 2003. *Saint Paul: The Foundation of Universalism*. Redwood City: Stanford University Press.

Barth, Karl. 1958. *Church Dogmatics 3.1: The Doctrine of Creation*. London: T & T Clark.

Brown, Harold O.J. 2009. Of Animal and Human Rights. *equip.org*, April 20. https://www.equip.org/article/of-animal-and-human-rights/. Accessed 3 Aug 2019.

Clough, David. 2009. All God's Creatures: Reading Genesis on Human and Non-Human Animals. In *Reading Genesis After Darwin*, ed. Stephen C. Barton and David Wilkinson, 145–162. New York: Oxford University Press.

Crane, Susan. 2013. *Animal Encounters: Contacts and Concepts in Medieval Britain*, 1–2. Philadelphia: University of Pennsylvania Press.

Cunningham, David S. 2009. The Way of All Flesh. In *Creaturely Theology*, ed. Celia Deane-Drummond and David Clough. London: SCM Press.

Deckers, Jan. 2016. Might a Vegan Diet Be Healthy, or Even Healthier? In *Animal (De)liberation: Should the Consumption of Animal Products Be Banned?* London: Ubiquity Press.

Derrida, Jacques. 1988. Signature, Event, Context. *Limited Inc*, 1–21. Evanston: Northwestern University Press.

Eagleton, Terry. 2010. *On Evil*. New Haven: Yale University Press.

Francione, Gary L., and Anna E. Charlton. 2015. *Eat Like You Care: An Examination of the Morality of Eating Animals*. Lexington: Exempla Press.

Girard, R. 1998. *The Girard Reader*. New York: Crossroad.

Gutleben, Christine. 2015. *Every Living Thing: How Pope Francis, Evangelicals and Other Christian Leaders are Inspiring All of Us to Care for Animals*. Canton: Front Edge Publishing.

Halteman, Matthew. 2019. Meat and Evil. In *Evil: A History*, 88–96. New York: Oxford University Press.

Heart and Stroke Foundation of Canada. 2018. Vegetarian Diets. https://www.heartandstroke.ca/get-healthy/healthy-eating/specific-diets/vegetarians. Accessed 8 Jan 2020.

Hibbert, Theodore. 1996. *Direction: A Mennonite Brethren Forum* 25 (2): 16–25. https://directionjournal.org/25/2/rethinking-dominion-theology.html

Hobgood-Oster, Laura. 2008. *Holy Dogs and Asses: Animals in the Christian Tradition*. Urbana: University of Illinois Press.

Hyland, J.R. 2000. *God's Covenant with Animals: A Biblical Basis for the Humane Treatment of All Creatures*. Brooklyn: Lantern Books.

Kant, Immanuel. 1997. *Groundwork of the Metaphysics of Morals*. Trans. Mary Gregor. Cambridge: Cambridge University Press.

Lazareth, William H. 2001. *Christians in Society: Luther, the Bible, and Social Ethics*. Minneapolis: Fortress Press.

Linzey, Andrew. 1994. *Animal Theology*. Champaign: University of Illinois Press.

———. 2003. The Bible and Killing for Food. In *The Animal Ethics Reader*, 227–234. London/New York: Routledge.

———. 2009. *Creatures of the Same God: Explorations in Animal Theology*. Brooklyn: Lantern Books.

Mayo Clinic. 2019. Vegetarian Diet: How to Get the Best Nutrition. https://www.mayoclinic.org/healthy-lifestyle/nutrition-and-healthy-eating/in-depth/vegetarian-diet/art-20046446. Accessed 7 Jan 2020.

McManus, Katherine. 2020. With A Little Planning, Vegan Diets Can Be A Healthful Choice - Harvard Health Blog. https://www.health.harvard.edu/blog/with-a-little-planning-vegandiets-can-be-a-healthful-choice-2020020618766. 22 August 2020.

Melina, Vesanto, Winston Craig, and Susan Levin. 2016. Position of the Academy of Nutrition and Dietetics: Vegetarian Diets. *Journal of the Academy of Nutrition and Dietetics* 116: 1970–1980.

Meyer, Eric Daryl. 2014. The Logos of God and the End of Humanity: Giorgio Agamben and the Gospel of John on Animality. In *Divinanimality: Animal Theory, Creaturely Theology*, ed. Stephen D. Moore, 146–160. New York: Fordham University Press.

Mill, John S. 2002. *Utilitarianism*. Ed. George Sher, 2nd ed. Indianapolis: Hackett Publishing.

Montano, Steffano. 2019. Addressing White Supremacy on Campus: Anti-Racist Pedagogy and Theological Education. *Religious Education* 114 (3): 274–286.

Moore, Stephen D. 2014. *Divinanimality: Animal Theory, Creaturely Theology*. New York: Fordham University Press.

Newport, Frank. 2017. *2017 Update on Americans and Religion*. December 22. https://news.gallup.com/poll/224642/2017-update-americans-religion.aspx. Accessed 3 Feb 2018.

Phelps, Norm. 2002. *The Dominion of Love: Animal Rights According to the Bible*. Brooklyn: Lantern Books.

Prager, Dennis. 2017. *Are Humans More Valuable Than Animals?* January 9.

Rosamund, Raha. 2016. Animal Liberation: An Interview with Professor Peter Singer. *The Vegan*, Autumn.

Saint, Augustine. 1963. *De Trinitate*. Washington, DC: Catholic University Press of America.

Scully, Matthew. 2003. *Dominion: The Power of Man, the Suffering of Animals, and the Call to Mercy*. London: Macmillan.

Shannon, Laurie. 2009. The Eight Animal in Shakespeare; or, Before the Human. *PMLA* 124 (2): 476.

Singer, Peter. 2015. *Animal Liberation*. London: The Bodley Head.

Van Drunen, David. 2010. Natural Law in Noahic Accent: A Covenantal Conception of Natural Law Drawn from Genesis 9. *Journal of the Society of Christian Ethics* 30: 131–149.

White, Lynn, Jr. 1967. The Historical Roots of the Ecological Crisis. *Science* 155 (3767): 1203–1207.

York, Tripp. "Can the Wolf Lie Down with Lamb without Killing it: Onfronting the Not-So-Practical Politics of the Peaceable Kingdom." 2012. A Faith Embracing All Creatures: Addressing Commonly Asked Questions About Christian Care for Animals, ed. Andy Alexis-Baker. Eugene: Wipf and Tripp York. 150–165. Eugene: Wipf and Stock Publishers.

10

Feeling Bad? Veganism, Climate Change, and the Rhetoric of *Cowspiracy*

Alexa Weik von Mossner

In one of the most memorable moments of Kip Andersen and Keegan Kuhn's documentary, *Cowspiracy: The Sustainability Secret* (2014), Andersen watches on as a backyard farmer kills a duck. The man first catches two animals and lifts them into the air by their necks, saying without any sign of compassion that "these two go first." Then he presses the first duck on a wooden table and slams a hatchet down on its neck, severing it from the twitching torso. "That's going to be a little gruesome," he says casually as he holds the duck's maimed body over a plastic bowl to drain the blood. The camera moves closer, confronting the viewer with a closeup of the part where the duck's head is missing, showing the gooey mess dripping on the severed head inside the bowl, the animal's body still twitching, the eyes still looking alive, its bill still opening and closing. "It's just nerves," says the farmer assuringly, but somehow that doesn't sound convincing. The killing of the second duck is just a sound accompanying a black screen, and then the camera cuts to a blurry shot

A. Weik von Mossner (✉)
University of Klagenfurt, Klagenfurt, Austria
e-mail: Alexa.WeikvonMossner@aau.at

C. Hanganu-Bresch, K. Kondrlik (eds.), *Veg(etari)an Arguments in Culture, History, and Practice*, The Palgrave Macmillan Animal Ethics Series,
https://doi.org/10.1007/978-3-030-53280-2_10

of Andersen walking up and down behind the farmhouse, visibly shaken. "I don't think I could do it," he says after having composed himself, "and if I can't do it, I don't want anyone to do it for me." In the very next scene, he is seen in his van as he drives a rescued chicken—which had also been scheduled for slaughter—to an animal sanctuary.

Veganism is a deeply emotional affair and so are the cultural texts that advocate it. *Cowspiracy* has been remarkably successful in making viewers uncomfortable about the relationship between the consumption of animal products and another issue they tend to feel bad about: climate change. Unlike Shaun Monson's *Earthlings* (2005), which makes a moral argument about our connection to nonhuman animals, and also unlike their more health-oriented follow-up *What the Health* (2017), Andersen and Kuhn's first film uncovers a truth that many people are still unaware of: the fact that our personal food choices have an impact on the climate and the environment more generally. While the duck-slaughtering scene is used strategically to confront viewers with the truth behind the idealized idea of small-scale backyard farming, the film's main beef, so to speak, is with the carbon emissions and other catastrophic environmental impacts of industrial livestock farming and the fact that this issue is often neglected when we talk about tackling climate change.

Much of my research has been dedicated to the exploration of the emotionalizing strategies of environmental narratives from a cognitive ecocritical perspective that draws on research in neuroscience and cognitive narratology.[1] While such an approach can help us better understand the narrative strategies of a given text, it faces limitations when speculating about its impact on actual readers or viewers. In this essay, I thus want to combine my analysis of *Cowspiracy*'s narrative strategies with the experiences I made in teaching the film to a group of undergraduate students at the University of Klagenfurt in Austria. While both the teaching context itself and the demographic makeup of the class pose their own limitations, I found the results highly informative and at times surprising. Class discussions as well as anonymized questionnaires showed that students were affected by the film's linking of vast and seemingly distant environmental problems to their own diet. Many of them felt for the first time that their personal choices mattered in fighting climate change, but they were also overwhelmed by the prospect that this moral choice would involve changing their eating habits.

I will discuss the potential of such individual emotional responses in the light of scholarship on the "rhetorical form" of documentary film (Bordwell and Thompson 2008). Although they do not allow me to make any claims about *Cowspiracy*'s impact on viewers in general, my students' responses suggest that highlighting the multiple connections between meat consumption and global warming can be a powerful rhetorical strategy in vegan advocacy films, and thus an alternative to more commonly used arguments around animal welfare and human health. At the same time, however, the responses of my students suggest that the disturbing display of animal suffering and death can create an emotional "spillover effect" (Plantinga 2009, 184) that impacts how viewers process information about veganism and climate change. *Cowspiracy*'s narrative strategies thus may be mutually reinforcing on the cognitive and affective level.

Rhetorical Documentaries and Cognitive Film Theory

Environmental documentary films come in all forms and stripes, ranging from contemplative nature explorations to argument-driven advocacy films. As Helen Hughes has pointed out, "the central aspect of the environmental documentary is the subject of the environment, however conceived," and there are "many subgenres within this subgenre" (2014, 4). Due to their subject matter, however, most vegan advocacy films use what David Bordwell and Kristin Thompson have called "rhetorical form" (2008, 348). According to the two film scholars, rhetorical form is marked by four attributes: (1) it addresses the viewer openly, trying to move him or her to a new intellectual conviction, to a new emotional attitude, or to action; (2) the subject of the film is a matter of opinion; (3) the filmmaker appeals to viewer emotions; (4) the film attempts to persuade the viewer to make a choice that will have an effect on his or her everyday life (348–49). In sum, such films are upfront about their agenda and they try to convince and *move* their viewers to a new attitude and, ideally, behavior, through a combination of intellectual arguments and affective appeals. Typical examples in the realm of vegan advocacy include documentaries focusing on animal welfare and rights such as Monson's *Earthlings* and Chris Delforce's *Dominion* (2017) as well as a range of films that explore the relationship between diet and health, among them Joe Cross and Kurt Engfehr's *Sick, Fat, and Nearly Dead* (2010), Lee Fulkerson's *Forks Over Knives* (2011), Marisa Miller Wolfson's *Vegucated* (2011), Kenny and Jasmine Leyva's *The Invisible Vegan* (2019), and Andersen and Kuhn's *What the Health.*

Cowspiracy is no exception. If anything, it is a prime example of rhetorical form. Not only does it "address the viewer openly, trying to move him or her to a new intellectual conviction, to a new emotional attitude, or to action" (Bordwell and Thompson 2008, 348), it also frequently appeals to emotion and attempts to persuade viewers "to make a choice that will have an effect their everyday lives" (349). Unlike the other vegan advocacy films mentioned above, however, the choice *Cowspiracy* asks viewers to make is to stop eating meat and other animal products because animal agriculture, as Andersen puts it in the film, is "the leading cause of resource consumption and environmental degradation destroying the planet today." While the advocated choice—to switch to a plant-based diet—thus is the same as in other vegan advocacy films, the reasoning provided is somewhat different. The film deliberately does not foreground a moral argument about the rights and wellbeing of nonhuman animals (as is the case in *Earthlings* and *Dominion*), nor does it primarily appeal to viewers' more egoistical desire for their own and their loved ones' health (as is the case in the other films mentioned above). Instead, *Cowspiracy* highlights altruistic arguments, asking us to go vegan for the health of the planet and the sake of future generations. If you really want to do something meaningful about climate change and other forms of environmental degradations, the message runs, the change you should make in your everyday life is to switch to a plant-based diet.

It's an intriguingly simple answer to a complex issue that many people feel bad and yet utterly helpless about (Lertzman 2015). Arguably, however, it is also a risky rhetorical strategy for a vegan advocacy film in that it layers one potentially overwhelming issue—climate change—on top of another one—eliminating meat and dairy from one's regular diet. Much of *Cowspiracy* is dedicated to finding out why major environmental organizations such as the Sierra Club refrain from mentioning the impact of meat and dairy consumption on our carbon footprints. Arguably, the reason that keeps these organization from advocating a plant-based diet is the same reason that has led the makers of climate change documentaries to share that silence: most people do not want to be told by anyone that they should stop doing what they like to do, in particular not when it concerns something as deeply personal and affectively charged as their food choices. That's why climate change documentaries from Davis Guggenheims' *An Inconvenient Truth* (2006) to Fisher Stevens's *Before the Flood* (2016) either omit the carbon footprint of the meat and dairy industry entirely or keep their dietary recommendations to a minimum. *Cowspiracy*, by contrast, is very straightforward about the fact that the "intellectual conviction ... new emotional attitude" and "action" they want their viewers to embrace is veganism. In order to make that entreaty more palatable, it is wrapped into a (to some degree staged) detective story in which Kip Andersen portrays himself as a naïve young American who goes on a quest for knowledge.

Before I move on to the analysis of the film, I want to say a few words about my theoretical framework and why I believe it to be helpful for the analysis of *Cowspiracy*'s rhetorical structure and affective appeal. Cognitive ecocriticism developed out of a combination of the methods of cognitive narratology and film studies on the one hand, and an interest in environmental issues on the other.[2] Within film studies, cognitive approaches go back at least to the mid-1990s, with several researchers (M. Smith 1995, 2017; G. Smith 2003, Plantinga 2009, 2018a) showing a pronounced interest in the ways in which films engage viewer emotions. As Carl Plantinga puts it, "emotion and affect are fundamental to what makes films artistically successful, rhetorically powerful, and culturally influential" (2009, 5). With their attention to narrative and visual forms and a pronounced interest in affective strategies, cognitive film scholars tease out how particular shots, scenes, sequences and plotlines cue a range of emotion in viewers in support of a narrative and, sometimes, a political goal. Their toolbox is perhaps not so different from that of other film scholars in that these tools are used to analyze filmic elements such as character, *mise-en-scène*, plot, editing, sound, and music. What is different, however, is that these tools are informed by insights in neuroscience and cognitive psychology. Murray Smith notes that this does not mean that "the traditional methods and principles of the humanities … should be supplanted by a battery of scientific techniques." Rather, cognitive film studies advocates for and participates in "an integrated research culture, in which the questions posed determine what methods are brought into play, without the traditional barriers between disciplines and the worlds of science, engineering, the arts, and the humanities intervening" (2017, 4).

Much of this interdisciplinary research has been focused on fiction films, but recently, it has been extended to the analysis of documentaries. In their introduction to *Cognitive Theory and Documentary Film*, Catalin Brylla and Mette Kramer explain that cognitive investigations of nonfiction film "highlight the permeable boundary between fiction and documentary," suggesting even that "the long-established divide between documentary and fiction is counterproductive when studying particular documentary forms" (2018, 3). Whether we look at the mediation of reality, character engagement, or emotions and embodied experience, viewers engage with documentaries in much the same way they engage with fiction, and filmmakers use many of the same techniques to encourage that engagement. Rhetorical documentaries make this continuum particularly obvious since they are so intent on telling a convincing *story* and on appealing to viewers' hearts in order to make them change their minds about the issue at hand.

This is precisely why I believe that an approach that combines the insights of cognitive film theory with ecocritical sensibility is particularly well suited for the analysis of vegan advocacy films. As I have argued in my contribution to Laura Wright's *Through a Vegan Studies Lens* (2019, 29), such an approach can complement the cultural studies side of vegan studies by turning our attention to the ways in which texts and films invite us to *feel* about animals, food, and the relationship between the two.[3] However, as *Cowspiracy* demonstrates, there are also other ways in which vegan advocacy films can harness what the cognitive narratologist Suzanne Keen has called *strategic empathy*. Evoking Gayatri Spivak's notion of "strategic essentialism," Keen explains that "strategic empathizing occurs when an author employs empathy in the crafting of fictional texts, in the service of 'a scrupulously visible political interest'" (83).[4] *Cowspiracy* is a nonfiction film, but it similarly relies on strategic empathizing in the service of a political interest, not least through the aforementioned "detective story," which offers viewers a protagonist who, along with the viewer, learns about the true environmental costs of meat and dairy and is by turns shocked and amazed.

"People Don't Wanna Hear It": The Affective Appeals of *Cowspiracy*

Cowspiracy opens with a well-known quote by Martin Luther King Jr.: "In the end, we will not remember the words of our enemies, but the silence of our friends." The first images we see are from an interview with Sierra Club Deputy Executive Director Bruce Hamilton who, with visible restraint, enumerates the horrific prospects ahead of us in case of unmitigated climate change. The world's climate scientists, Hamilton explains,

> tell us that the ... safest we could hope to do without having perilous implications as far as drought, famine, human conflict, major species extinction would be at about two degrees Celsius increase in temperature. We are rapidly approaching that and with all the built-in carbon dioxide that's already in the atmosphere, we're easily going to exceed that ... You know when whole countries go under water because of sea level rise and whole countries find that there's so much drought that they can't feed their population and as a result they need to desperately migrate to another country or invade another country, then we're going to have climate wars in the future.

Hamilton's aggravated face is shown in an extreme closeup at this point, his words accentuated by ominous, upswelling music. Then we hear Kip Andersen's voice, asking casually: "And what about livestock and animal agriculture?" Hamilton takes a moment before he responds, thrown off by the question. The music score stops abruptly, highlighting his silence.

Then he catches himself. “What about it?” he asks with an awkward smile, “I mean …” The film does not allow him to finish his sentence, leaving his thought incomplete as it cuts away to the opening credits.

Hamilton, it is now established, is one of the “friends” in the King quote whose silence needs to be remembered. The silence of those who are genuinely concerned about climate change and yet refuse to consider, or even acknowledge, the impact of animal agriculture on our changing climate and the catastrophes it might bring about. The main theme of the film is thus established. While it also exposes some of the lies thrown at the American population by those who deny climate change for economic or other egoistic reasons, its main focus is on those who do take the issue seriously and yet refuse to talk about the emissions of the meat and dairy industry. Why this silence? Andersen asks throughout the film as he tries to expose the global “cowspiracy” that seeks to hide the truth from citizens and consumers.

Following in the footsteps of documentary filmmaker Michael Moore, Andersen inserts himself as a character into his own film, an average guy named Kip with “a cliché U.S. American childhood.” Life, he tells the viewer, “was simple”—until he had his environmental awakening as a teenager, brought about by another climate change documentary: Guggenheim’s *An Inconvenient Truth*, starring Al Gore. The film, he explains “scared the emojis out of me,” leading him to make up his mind “right then and there to change how I lived and to do whatever I possibly could to find a way for all of us to live together in balance with the planet sustainably forever.” Hyperbole aside, this is the kind of immediate impact that most politically motivated makers of environmental documentaries dream of: immediate attitude change followed by immediate and sustained behavior change in the direction advocated by the film. Andersen reports that, in response to Guggenheim’s film, he became an “obsessive-compulsive environmentalist” who changed his light bulbs, recycled or composted his trash, saved water and electricity, and biked everywhere.

The problem, according to Andersen, was that Gore did not have the guts to talk about the most “inconvenient truth” of them all, namely the climate impact of animal agriculture, and so despite his best efforts the situation kept “growing worse.” Arguably, this deliberately sets up a false causation, since it is impossible for an individual to combat climate change on his own, with or without a dietary change. But it nevertheless addresses the debilitating sense of powerlessness reported by people all over the world who do care about climate change but feel that their individual actions do not matter on the larger scale of things. *Cowspiracy* aims to address precisely this feeling of powerlessness by explaining to viewers why the truth has been hidden from them and then offering concrete guidelines for what they can do to have a positive impact on the climate while saving animal lives and doing something for their own health.

Like Michael Moore, Andersen uses voice-over to tell the story of his *own* quest for knowledge, his journey from well-meaning, but naïve do-gooder to someone who knows the truth and acts accordingly.[5] He is thus positioned as the central identification figure in the film, the guy from next door who, at the beginning of his quest, was *just like* the viewer and who now wants to share the important truths he has discovered. Whether this strategy is successful might depend not least on the identity of the actual viewer. Documentaries with a strong focus on the filmmaker himself, especially when foregrounding a white male perspective, run the risk of alienating some viewers. But as the global success of Moore's award-winning documentaries—and ultimately also that of *Cowspiracy*—demonstrates, it is a strategy that nevertheless resonates with many viewers across the world. I will get more deeply into this aspect in the final section when discussing the responses of my Austrian students.

However, putting (one of) the filmmaker(s) front and center is not the only "Michael Moore-ism" (Estrada 2014) to be found in *Cowspiracy*. What Andersen and Kuhn also have learned from Moore is that "the first rule of documentaries is: Don't make a documentary—make a MOVIE" (Moore 2014). Moore has famously declared the term "documentarian" to be dead (Moore 2014), arguing that whoever makes any kind of film is a filmmaker, and if that filmmaker "makes "a *movie*, people might actually go see your documentary! … They don't care whether you make them cry, whether you make them laugh, whether you even challenge them to think — but damn it, they don't want to be lectured, they don't want to see our invisible wagging finger popping out of the screen. They want to be entertained" (Moore 2014).[6] Andersen and Kuhn go about that "entertainment" using many tricks and mannerisms that we also know from Moore's films: countless phone calls and unannounced visits with harmless sounding and yet deliberately provocative questions; filming as much as possible (if need be, with a hidden or supposedly switched-off camera) those that do not want to talk or are outright hostile; and, as is already apparent in the opening scene of the film, using sound and music as commentary and emotional trigger. None of this can actually hide the "invisible wagging finger" popping now and then out of the screen, Moore's own advice notwithstanding. But it does make for an engaging viewing experience.

Andersen's voice-over tells viewers that his quest for knowledge began with a 2006 UN report that alerted him to the fact that "rearing cattle produces more greenhouse gasses than driving cars."[7] Puzzled by the news, he didn't understand why this burning issue was not featured on any of the "nation's largest environmental organizations' websites," including those of +360.org, Greenpeace, the Rainforest Action Network, Climate Reality, Amazon Watch, and the World Wide Fund For Nature (WWF). The next interview presented in the film, however, is with two representatives of the California state government. After informing viewers that "one quarter pound hamburger is worth 600 gallons of water" and that "domestic water use is only 5% of what is consumed in the U.S. whereas it's 55% for animal agriculture," Andersen is seen sitting across Manucher Alemi and Kamyar Guivetchi from the California Department of Water Resources. The two men go through their routine of offering the usual water-saving recommendations for private households in California when Andersen drops the bomb. Mentioning animal agriculture, he asks whether they can "comment on that at all, on how much that plays a role in water consumption and pollution." As in the first interview, there is a distinct pause. When Alemi and Guivetchi speak again, they seem to weigh each of their words, responding slowly, hesitantly, and guardedly as Andersen keeps pressing them for an acknowledgment of the massive water footprint of animal agriculture.

Andersen and Kuhn use the material to its fullest effect. The silences are certainly telling, but even more fascinating for viewers are the mercilessly long takes of two state officials scrambling for words. What the filmmakers are taking advantage of here is our natural ability for what psychologists call mind-reading or Theory of Mind (ToM). In the words of Lisa Zunshine, theory of mind is occurs when "we ascribe to a person a certain mental state on the basis of her observable action," making it "the default way by which we construct and navigate our social environment" (2006, 6). Showing a human face in extreme closeup invites such natural processes of mind-reading during the viewing of a film scene (Plantinga 1999, 240), even more so in long takes because they allow us to study closely the mimic expressions of the displayed person and attribute certain states of mind. In the case of the Alemi and Guivetchi interview, viewers likely will deduce from their strained facial expressions that they are desperately trying to figure out how to get out of this trap. Neither of the men denies the massive impact of animal agriculture on water usage, but it is more than obvious that they do not want to talk about this dicey issue on camera. Instead, they give evasive responses insinuating that both animal agriculture and meat consumption are outside their area of expertise, nervous smiles around their lips suggesting they are fully aware that they are being played by the guys who keep the camera running. And the more they scramble, the more it seems they have something to hide.

Along with the Al Gore, the Sierra Club, and a whole range of other respectable environmental organizations, the California state government is thus cast not as the *villain*, but in the only slightly more flattering role of the *quiet bystander* who deliberately looks the other way. Plantinga notes that "the fostering of allegiances and antipathies toward characters" is a "way in which documentary filmmakers characterize the people they represent. In cases such as the films of Moore … people are slotted into the roles we might expect in a well-made fiction film—as sympathetic protagonists and morally questionable antagonists" (2018b, 128). *Cowspiracy*, too, has its array of morally questionable antagonists, but only some of them are in line with what Plantinga calls the "'tradition 'of the villainous capitalist' seen in the films of Moore" (126). These are the big players within the powerful animal agricultural industry, who, as the journalist Will Potter puts it in the film, "have massive legal resources" and are willing to use these resources in all kinds of ways to make sure that "to keep people in the dark about what they are actually doing" in order to protect corporate profits. These "villainous capitalists" receive support from Federal law enforcement agencies such as the FBI which, as

Potter reminds viewers, has classified "animal rights and environmentalist activists [as] the number one domestic terrorism threat." Along with the former cattle ranger Howard Lyman, Potter is among the interviewees who tell Anderson, on camera, that he and Kuhn themselves might become a target of those antagonists, a warning that Andersen sees confirmed when financial backers pull out of funding their film project due to its "growing controversial subject matter."

The nonprofit environmental organizations featured in *Cowspiracy* are not in the same category as these corporate and governmental players. They are shown to be basically good people who are aware of what is going on but choose to be silent, thereby betraying the animals, the filmmakers, the viewers and, ultimately, humanity as whole on a catastrophically warming planet. We cannot know who the people behind all those organizations truly are or what exactly drives them to their positions, since the film does not make any attempt to show them as complex human beings with complex motivations, responsibilities, and liabilities. Instead, they are flat characters that are easy to dislike even when they acknowledge, as Alemi and Guivetchi do, that animal agriculture causes enormous environmental problems. As Plantinga points out, "the generation of allegiances and antipathies toward various characters in any narrative film is a form of moral or sociopolitical judgment that is stamped on the film by the narration's implicit approval or disapproval of one or more of the characters" (128). Alemi and Guivetchi are among those that *Cowspiracy*'s narrative disapproves of, along with the spokespeople of the aforementioned environmental organizations. "So bizarre," comments Andersen in the film, "I had supported these organizations for so long, and now I was met with silence."

Not all "friends," however, are silent. *Cowspiracy* crucially relies on the testimony of "a handful of environmental authors and advocates"—most of them journalists like Will Potter or nutrition experts such as Richard Oppenlander, Demosthenes Maratos, Will Tuttle, and Michael Pollan—who are willing to address the climate impact of livestock farming. They are also the ones who spell out the reason why large environmental organizations will not address the issue. "I think they focus-grouped it and it's a political loser," says Pollan when Andersen interviews him. "They are looking to maximize the number of people making contributions. And if they get identified as being anti-meat or challenging people on their everyday habits, that's something that's so dear to people that it will hurt with their fundraising." Tuttle takes a similar position. "They are businesses," he says about NGOs such as Greenpeace, AmazonWatch and the Sierra Club, "and they wanna make sure they have reliable sources of funding." Maratos puts it even more bluntly. "If you listen to the majority of the major environmental organizations, they are not telling you to do much ... It's better for their fundraising and better for their profile, to create a victim and perpetrator sort of plotline." The designated "perpetrator" in this scenario is mostly the fossil fuel industry and its carbon emissions; the "victims" are individual citizens who, individually, cannot do anything against that industry and are thus let off the hook. Talking about the environmental impact of the production of meat, dairy and other animal-based foods would change that plotline and the roles of the protagonists populating it. Telling that alternative story is precisely what Andersen and Kuhn aim to do throughout much of the rest of the film, relying on the expert testimony of outspoken "friends" and all kinds of numbers and statistics.

It is one of those numbers that quickly became the main focus point of critics of the film: the claim that 51% of global greenhouse gas emissions are from livestock, which the filmmakers took from a Worldwatch Institute report authored by Jeff Goodland and Robert Anhang. "How did Goodland and Anhang come up with 51%, rather than the scientific consensus that livestock are currently responsible for about 15% of global greenhouse gas emissions?" asks Doug Boucher in his review of the film on the Union of Concerned Scientists website (2016), pointing out the fact that the report was not peer-reviewed and proceeding to debunk its central claims. Andersen and Kuhn defend the number on the "Facts and Sources" page of the film's website, writing that "Animal agriculture is responsible for 18 percent of greenhouse gas emissions, more than the combined exhaust from all transportation" and reasserting the claim that "Livestock and their byproducts account for at least 32,000 million tons of carbon dioxide (CO_2) per year, or 51% of all worldwide greenhouse gas emissions." That number might still be too high, but that does not change the fact of animal agriculture's staggering impact on the environment and the climate. A peer-reviewed meta-analysis by Poore and

Nemecek, published in *Science* in 2018, estimates that the food supply chain is responsible for 26% of anthropogenic GHG emissions (988). The authors also point out that "meat, aquaculture, eggs, and dairy use ~83% of the world's farmland and contribute 56 to 58% of food's different emissions, despite providing only 37% of our protein and 18% of our calories" (2018, 990). The bottom line, then, is that *Cowspiracy* is painting a mostly accurate picture of the relationship between animal agriculture and climate change, even if it does get some of the numbers wrong.[8]

Arguably, it was not a smart move on part of the filmmakers to rely on a non-peer-reviewed report for one of the central claims of their film. That notwithstanding, however, the real reason why the film has been so controversial might not be so much what it exposes, but what it concludes from that exposure. After all, *Cowspiracy* does not stop at criticizing environmental organizations for their silence. Nor does it only offer vague and inadequate advice to viewers as we find it in the notorious end credits of *An Inconvenient Truth*. Instead, starting with the scene in which Andersen witnesses the slaughtering of the duck, *Cowspiracy* turns into a vegan advocacy film, offering very clear guidelines for what can be done by each individual viewer to tackle climate change and make the world a better place: adopt a plant-based diet and stop supporting the raising and killing of animals for meat, eggs, and dairy. To this purpose, it introduces another round of experts, among them the vegan physician Michael Klaper and Ethan Brown from Beyond Meat, a company that has made a splash by producing plant-based meat substitutes. But it leaves it to Howard Lyman to give the film's official response to all those silent environmentalist friends who avoid the issue: "You can't be an environmentalist and eat animal products. Period. Kid yourself if you want. If you want to feed your addiction, so be it. But don't call yourself an environmentalist."

According to psychologists, such concrete guidelines for ameliorating action are urgently needed in climate change communication, especially when such communication evokes powerful negative emotions such as anger, guilt, worry, and fear (Smith and Leiserowitz 2014). However, as Pollan and other interviewees in the film point out, our eating patterns are deeply ingrained by cultural and personal habits, and we can get quite resistant or even hostile when someone tells us that we must change them. The question, then, is whether *Cowspiracy* manages to present that information in a way that resonates with viewers. As a partial and preliminary answer, I want to relate some of the experiences I made teaching the film.

"It's Real Hard": Watching *Cowspiracy* in the Classroom

I taught *Cowspiracy* in the spring of 2018 as part of a higher division undergraduate seminar on "American Climate Change Cinema" at the University of Klagenfurt in Austria. This context is important in that our focus throughout the semester was on climate change—not on veganism or any food-related issues. We watched and discussed several films addressing the issue, including *An Inconvenient Truth*, *The Day After Tomorrow*, *Beasts of the Southern Wild*, and *Chasing Ice*. After screening *Cowspiracy*, I distributed questionnaires to the 20 students in my class (8 male, 12 female; and all but 2 of them between 20 and 25 years old) and gave them 20 minutes to respond. The responses were anonymous. After the completion of the questionnaire and a short break, we discussed the film for about an hour.

What was most striking in both the questionnaires and the following class discussion was the difference between students' reactions to *Cowspiracy* and every other film we watched in class. Although they expressed strong concern about climate change and the potential impact on their future throughout the semester, students felt that their individual actions will not make a difference; some of them said frankly that they will not stop using a car or flying to vacation destinations. Sitting through a film like *Chasing Ice*, which shows dramatic time-lapse footage of the retreating glaciers in the Arctic, did not change that attitude. After watching *Cowspiracy*, however, students' responses were notably different. All 20 students said that they would have watched the film to the end also if it hadn't been shown in a classroom setting. They all said they would recommend the film to a friend and, remarkably, they all believed that *Cowspiracy* has the potential to change *other* people's attitudes toward meat consumption. And although none of them announced that they would go vegan as a result of watching the film, almost all of them wrote that it had changed their attitude toward meat consumption "a little" or "a lot."[9] Only one student (24/f) wrote that the film had not changed her attitude towards meat at all, adding that she was already vegan before watching the film. This, however, was the exception. Of the 20 students, 19 wrote that they had been unaware of the issue. Many expressed intentions to eat less red meat or less meat in general, even as some of them added that "this is not easy at all," that they will only be making "small changes," or that the film "hasn't convinced [them] to switch to a vegetarian/vegan diet." A 27-year-old female student wrote that she "would like to try to stop eating meat but I do not think it could be possible for me," showing both awareness of the tenacity of her habits and expressing concern over it. Some students considered eating more vegetarian or vegan food when available.

Students realized that *Cowspiracy* is a vegan advocacy film and some of them were critical about its rhetoric, finding it "preachy at times" (20/f) or even "annoying" (21/f). They also commented on Anderson's journey from ignorance to knowledge, which according to the student that found the film too preachy, "felt like he was trying too hard to be Michael Moore." Nevertheless, students felt that the film was showing them important facts they could not deny, and they mostly related those facts to the interviews in the film. In response to the question of which elements in the film she liked best, a 20-year-old female student wrote that she most enjoyed the interviews with "these 'green' organizations. It is funny to see these people's reaction even [though] they know cattle ranching is a main factor contributing to global warming." The use of "humor to get across points" (20/f) seems to have been an important factor in students' enjoyment of these interviews. Another student (23/f) liked what she considered "the realness of it. It didn't sugarcoat anything and stated facts just the way they are." While this trust in the film's stating of "facts" was somewhat relativized in the class discussion that followed after the questionnaires had been completed, such statements show that *Cowspiracy*'s presentation of data—and opinions—can be very convincing.

Even more interesting, of the students who had seen the film before, several reported behavior change. "After watching the movie for the first time, I have been eating less meat," wrote a 25-year-old female student. A 21-year-old male student wrote that he is "vegan now. Took a while, but the 'problem' is you just can't ignore the facts. If you did, you'd be ignorant, and that's not the person I want to be." A 24-year-old female student who identifies as vegetarian wrote that she "never liked the idea of eating a dead animal, but before watching the film I wasn't aware of the impact of livestock on the climate/soil/ocean etc." It seems that, like other vegan advocacy films such as *Earthlings*, *Cowspiracy* can provide a turning point for some viewers that leads them to make the dietary change the film is promoting.

Particularly interesting in terms of the rhetorical strategies of vegan advocacy films is that although—unlike *Earthlings*—*Cowspiracy* mostly eschews the prolonged depiction of animal suffering and death in support of its argument for veganism, the filmmakers' decision to include one such scene seems to have made a particularly deep impression on my students. Asked which moment in the film had touched them the most, the scene that was mentioned again and again in their responses was the slaughtering of the duck which, as one student (f/20) put it, "is hard to watch."[10] Other students called it "graphic" (21/f), "explicit" (34/m), and "traumatic" (24/f). The second-most-often mentioned scene was the much happier moment when Andersen drives a rescued chicken to a farm animal sanctuary. One might even wonder whether the positive emotions cued by that scene which, as one student (25/m) put it, "just brings faith to humanity and that there are still people who care," were in part a consequence of the duck slaughtering scene through a spillover-effect. In Plantinga's definition, the spillover-effect is "the *relief* from strong negative emotions, which are *replaced* by pleasurable emotions that depend for their strength on the arousal caused by physiological spillover remaining from the prior negative emotion" (2009, 184; emphasis in original). After witnessing the graphic slaughter of the duck, it seems that some students were just as relieved as Anderson reports to be that he (and by extension they) doesn't have to do it again but can instead save the animal. All arguments, experts, and data notwithstanding, it seems it was the animal stories—both sad and happy—that really got to my students on the emotional level.[11]

I cannot know whether any of them followed through on their intentions to eat less meat after watching the film, nor do I know whether they wrote the truth. Aside from the small number of subjects, the downside of using questionnaires in a classroom setting is that—even when responses are anonymized—there is the chance that students write what they believe to be the "right" answer in the context of the seminar. What is unusual about my students' responses to *Cowspiracy*, however, is how many of them admitted their previous ignorance, the degree of their aggravation, and the sheer number of those who said they wanted to try to at least somewhat change their diet because of what they had seen in the film. In our subsequent discussion, it became clear that what made a difference for many of them is that there was suddenly something *tangible* and *concrete* they could do in their everyday lives to make a difference. Some of them admitted it was a scary thought. And some said openly what I also saw in the questionnaires: that they would like to eat less meat or even to go vegan. But that they weren't sure they could really do it because they just enjoyed eating meat too much.

My students' responses to *Cowspiracy* raise an array of pedagogical questions as well as questions about the efficacy of vegan advocacy films and food-related documentaries more generally. Natalie Dorfeld has shared her own experiences of teaching *Cowspiracy* alongside films like Robert Kenner's *Food, Inc.* (2008) and Lee Fulkerson's *Forks over Knives* (2011) at Florida Tech. While affirming the films' efficacy in engaging students in a discussion on the negative effects of livestock farming, and even in securing "their [theoretical] support for meat-free Mondays," Dorfeld reports that engagement did not necessarily translate into personal action (2019, 241). She speculates that the reason for her students' reluctance to put their actions where their mouths were, so to speak, lies in their youth and their middle-class backgrounds. "Many young people," writes Dorfeld, "don't think thirty years into the future, but are concerned with the here and now," which is why they are unable to see how their habitual meat-based lifestyle will affect them and the planet in the future (254). Dorfeld's experiences in a classroom in Florida are somewhat different from my experiences with teaching the film in Austria. While most of my students were younger than 25, it is unlikely that all of them were middle-class (in Austria, university education is free and students at the University of Klagenfurt come from a wide range of economic backgrounds). Many of them did feel bad in response to the film's exposure of the relationship between meat consumption and climate change, and they supported its central message. But they still may have trouble translating that into actual behavior change for a variety of reasons. What they mentioned in both the questionnaires and in our discussion is not so much outright denial but force of habit and a perceived inability to simply switch to a different diet, let alone lifestyle. The increasing availability of plant-based alternatives to their accustomed foods might make a difference for some of them, as does their concern about climate change.[12]

Conclusion

Echoing the words of Michael Moore about the nature and purpose of documentary film, the cognitive film scholar Dirk Eitzen has suggested that "the only way to expose audiences to the potential social benefits of documentaries is to woo them and draw them in" (2018, 109). It seems that *Cowspiracy* succeeded in such wooing, at least in my classroom and in the short term. This may in part be related to its graphic depiction of animal suffering as we also find it in other vegan advocacy films from *Earthlings* to *Eating Animals* and beyond. And it seems in part related to a clear us-versus-them structure, in which those who remain silent are put in the pillories for everyone to see and, appalled, shake their heads. Whether that treatment of Greenpeace and other environmental organizations is *fair* is another question. As Eitzen (2018, 107) points out with reference to Moore's notorious exchange with Charlton Heston in *Bowling for Columbine*, Moore did abuse Heston's trust in order to be able to embarrass him on camera and "create contempt, not understanding" for the aging actor and former National Rifle Association (NRA) president. Eitzen acknowledges that this is effective filmmaking but wonders whether it does not contribute to the binary logic that increasingly divides US society. The same could be said about *Cowspiracy*, which creates contempt for other environmentalists in order to convince viewers to make the right choice and go vegan.

But it is not only the film's somewhat simplistic characterization of its "silent friends" that at times is problematic. The characterization of its protagonist, too, raises questions. Plantinga is right, I believe, in insisting that, like in any good fiction film, "the generation of allegiances and antipathies toward various characters," is of central importance for the narrative efficacy of documentaries (2018b, 128), and in first-person documentaries such as *Cowspiracy*, this includes sympathies toward the filmmaker.[13] Plantinga is also right when he writes that documentaries nevertheless risk being perceived as disingenuous when the discrepancy between the actual filmmaker and his persona in the film becomes too obvious. The fact that Andersen was a committed vegan before he started working on the film must not be problematic, since the film claims to tell the story of how he got to that point. However, Andersen and Kuhn's second film, *What the Health*, in which Andersen stars once again as the gullible everyman who is amazed to find out about the various health benefits of a plant-based diet, demonstrates that this rhetorical strategy can get problematic when it becomes too formulaic. Interested viewers might ask themselves whether Andersen miraculously unlearned everything he had learned during the making of the first film, or whether perhaps he lived through two parallel versions of his life in which he went vegan for different reasons. This is not to deny the rhetorical efficacy of the "Moore method," if we can call it that. It is rather that Moore tends to deal with issues in his films that do not necessitate a radical lifestyle change for the filmmaker as it is displayed in both of Andersen and Kuhn's films. It is certainly not unheard of that people make the decision to go vegan more than once in their lives, but it rarely happens that they forget in-between what veganism is all about.

The question remains, then, how the moral, environmental, and health-related dimensions of veganism can best be conveyed. Dorfeld speculates that the only thing that will really make a difference is the force of circumstance. "As the world population grows, diabetes/obesity rates rise yearly, and our natural resources shrink," she expects that eating less or no meat "will become more an inevitability than a crusade" (2019, 254). This can be read as a capitulation or, perhaps, an acknowledgment of the limits of both vegan advocacy and ecocritical pedagogy. A film like *Cowspiracy* may have the power to change a few minds, as does our teaching, but we cannot expect that they will magically change a classroom's or a country's eating habits. At best, they can do their part in a process of cultural transformation that may be too slow to be meaningful, at least when it comes to our changing climate.

Notes

1. I developed this approach in *Affective Ecologies* (2017).
2. The term was first used by Nancy Easterlin, who has argued for the central importance of cognitive science for ecocritical research (2010, 257).
3. Empirical evidence gained by Franklin et al. (2013) through functional magnetic resonance imaging (fMRI) suggests that our empathic response to the display of animal suffering and human suffering is very similar. Research by Filippi et al. (2010) shows that it makes a difference in people's brain functional networks whether they are omnivores, vegetarians or vegans.
4. Spivak's concept of strategic essentialism refers to the "strategic use of positivist essentialism in a scrupulously visible political interest" ([1985] 1996, 214).
5. Moore has used this narrative strategy in many of his best-known documentaries, from *Roger and Me* (1989) to *Bowling for Columbine* (2002), *Fahrenheit 9/11* (2004) and *Sicko* (2007).
6. Eitzen reminds us that with the theatrical release of *Roger & Me* in 1989, Moore proved "that documentaries could be profitable as entertainment. HI film, a semi-satirical treatment of his attempts to get an interview with the then-CEO of General Motors about corporate layoffs in Flint, Michigan, received widespread acclaim and made nearly seven million dollars at the box office ... Still, it was famously snubbed for Oscar consideration because some claimed that it was not truly a documentary" (2018, 101).
7. The fact that the UN report was published in 2006 sheds doubt on Andersen's claim that he did "all the things Al [Gore] told us to do" for years before finding out about the environmental impact of livestock farming, since *An Inconvenient Truth* was released in 2005 and thus just one year earlier.
8. On the relationship between animal agriculture and climate change, see also Springman (2016) and Springman et al. (2016).

9. The four available answers to this question were: no / not much / a little / a lot.
10. The students' strong emotional responses to the slaughtering of the duck can be explained with embodied simulation theory (Gallese 2016) and trans-species empathy that allows us to feel with other suffering animals across species lines (Franklin et al. 2013; Panksepp and Panksepp 2013).
11. It should be noted, however, that the display of animal suffering alone must not be the most effective strategy for vegan advocacy films. A study conducted with a range of viral videos produced by Mercy for Animals showed that videos "focusing entirely on cruelty footage with scenes of farmed animal suffering, confinement, and abuse" were less effective than "videos comparing suffering farmed animals to happy farmed animals or non-farmed animals, such as dogs" when it comes to promoting a vegetarian or vegan diet (Caldwell 2017).
12. A 2018 study by Leanne Cooper suggests that environmental concerns are increasingly a motive for veganism.
13. As Ros et al. have observed, "first-person documentary … comes with its unique set of viewer expectations and emotional affects" (2018, 235).

Bibliography

Andersen, Kip, and Keegan Kuhn, dir. 2014. *Cowspiracy: The Sustainability Secret*. Los Angeles: A.U.M. Films/First Spark Media. Netflix.

———, dir. 2017. *What the Health*. Los Angeles: A.U.M. Films/First Spark Media. Netflix.

Bordwell, David, and Kristin Thompson. 2008. *Film Art: An Introduction*. 8th ed. New York: McGraw-Hill.

Boucher, Doug. 2016. Movie Review: There's a Vast Cowspiracy About Climate Change. *The Equation Blog*. Union of Concerned Scientists, June 10. https://blog.ucsusa.org/doug-boucher/cowspiracy-movie-review

Brylla, Catalin, and Mette Kramer. 2018. Introduction: Intersecting Cognitive Theory and Documentary Film. In *Cognitive Theory and Documentary Film*, ed. Catalin Brylla and Mette Kramer, 1–17. Basingstoke: Palgrave Macmillan.

Caldwell, Kristin. 2017. Which Kinds of Pro-Vegetarian Videos are Best at Inspiring Changes in Diets and Attitudes? *Mercy for Animals* Website, March 14. http://www.mercyforanimals.org/kinds-of-viral-videos-are-best-at-inspiring?utm_

content=buffere2727&utm_medium=Social&utm_source=Twitter&utm_campaign=AnimalCharityEv. Accessed 7 May 2017.

Cooper, Leanne. 2018. A New Veganism: How Climate Change Has Created More Vegans. *Granite: Aberdeen University Postgraduate Interdisciplinary Journal*. https://www.abdn.ac.ukpgrs/documents/Grantie%20Vol%202%20-%20Leanne%20Cooper.pdf

Cross, Joe, and Kurt Engfehr, dir. 2010. *Sick, Fat, and Nearly Dead*. El Segundo: Gravitas Ventures. DVD.

Delforce, Chris, dir. 2017. *Dominion*. DVD.

Dorfeld, Natalie. 2019. Meatless Mondays? A Vegan Studies Approach to Resistance in the College Classroom. In *Through a Vegan Studies Lens: Textual Ethics and Lived Activism*, ed. Laura Wright. Reno: University of Nevada Press.

Easterlin, Nancy. 2010. Cognitive Ecocriticism: Human Wayfinding, Sociality, and Literary Interpretation. In *Introduction to Cognitive Cultural Studies*, ed. Lisa Zunshine, 257–275. Baltimore: Johns Hopkins University Press.

Eitzen, Dirk. 2018. The Duties of Documentary in Post-Truth Society. In *Cognitive Theory and Documentary Film*, ed. Catalin Brylla and Mette Kramer, 93–111. Basingstoke: Palgrave Macmillan. Film Review: Cowspiracy: The Sustainability Secret (2014).

Estrada, Orietta. 2014. Film Review: *Cowspiracy: The Sustainability Secret* (2014). *The Dodo*, July 19. https://www.thedodo.com/film-review-cowspiracy-the-sus-635002053.html

Filippi, Massimo, Gianna Riccitelli, Andrea Falini, Francesco Di Salle, Patrik Vuilleumier, Giancarlo Comi, and Maria A. Rocca. 2010. The Brain Functional Networks Associated to Human and Animal Suffering Differ Among Omnivores, Vegetarians and Vegans. *Plos One*. http://journals.plos.org/plosone/article?id=10.1371/journal.pone.0010847

Franklin, R.G., A.J. Nelson, M. Baker, J.E. Beeney, T.K. Vescio, A. Lenz-Watson, and R.B. Adams. 2013. Neural Responses to Perceiving Suffering in Humans and Animals. *Social Neuroscience* 8 (3): 217–227.

Fulkerson, Lee, dir. 2011. *Forks over Knives*. New York: Virgil Films and Entertainment. DVD.

Gallese, Vittorio. 2016. Finding the Body in the Brain: From Simulation Theory to Embodied Simulation. In *Goldman and His Critics*, ed. Brian McLaughlin and Hilary K. Kornblith, 297–314. New York: Wiley-Blackwell.

Goodland, Robert, and Jeff Anhang. 2009. Livestock and Climate Change. *WorldWatch*, November/December. https://awellfedworld.org/wp-content/uploads/Livestock-Climate-Change-Anhang-Goodland.pdf

Guggenheim, Davis, dir. 2006. *An Inconvenient Truth*. Los Angeles: Paramount Classics. DVD.

Hughes, Helen. 2014. *Green Documentary: Environmental Documentary in the 21st Century*. Bristol: Intellect.

Keen, Suzanne. 2010. Narrative Empathy. In *Toward a Cognitive Theory of Narrative Acts*, ed. Frederick Louis Aldama, 61–94. Austin: University of Texas Press.

Kenner, Robert, dir. 2008. *Food, Inc.* Los Angeles: Magnolia Pictures. DVD.

Lertzman, Renee. 2015. *Environmental Melancholia: Psychoanalytic Dimensions of Engagement*. New York/London: Routledge.

Leyva, Kenny, and Jasmine, dir. 2019. *The Invisible Vegan*. Los Angeles: The Leyva Company.

Miller Wolfson, Marisa, dir. 2011. *Vegucated*. New York: Kind Green Planet. YouTube.

Monson, Shaun, dir. 2005. *Earthlings*. Malibu: Nation Earth. DVD.

Moore, Michael. 2014. Michael Moore's 13 Rules for Making Documentary Films. *IndieWire*. https://www.indiewire.com/2014/09/michael-moores-13-rules-for-making-documentary-films-22384/

Panksepp, Jaak, and Jules B. Panksepp. 2013. Toward a Cross-Species Understanding of Empathy. *Trends in Neuroscience* 36 (8): 489–496.

Plantinga, Carl. 1999. The Scene of Empathy and the Human Face on Film. In *Passionate Views: Film, Cognition, and Emotion*, ed. Carl Plantinga and Greg M. Smith, 239–255. Baltimore: Johns Hopkins University Press.

———. 2009. *Moving Viewers: American Film and the Spectator's Experience*. Berkeley: University of California Press.

———. 2018a. *Screen Stories: Emotion and the Ethics of Engagement*. Oxford: Oxford University Press.

———. 2018b. Characterization and Character Engagement in Documentary Film. In *Cognitive Theory and Documentary Film*, ed. Catalin Brylla and Mette Kramer, 115–132. Basingstoke: Palgrave Macmillan.

Poore, J., and T. Nemecek. 2018. Reducing Food's Environmental Impacts Through Producers and Consumers. *Science* 360 (6392): 987–992.

Quinn, Christopher Dillon, dir. 2018. *Eating Animals*. New York: IFC Films. DVD.

Ros, Veerle, Jennifer M.J. O'Connell, Miklos Kiss, and Annelies van Noortwijk. 2018. Toward a Cognitive Definition of First-Person Documentary. In *Cognitive Theory and Documentary Film*, ed. Catalin Brylla and Mette Kramer, 223–240. Basingstoke: Palgrave Macmillan.

Smith, Nicholas, and Anthony Leiserowitz. 2014. The Role of Emotion in Global Warming Policy Support and Opposition. *Risk Analysis* 34 (5): 937–948.

Smith, Murray. 1995. *Engaging Characters: Fiction, Emotion, and the Cinema.* Oxford: Oxford University Press.

Smith, Murray. 2017. *Film, Art and the Third Culture: A Naturalized Aesthetics of Film*. Oxford: Oxford University Press.

Smith, Greg M. 2003. *Film Structure and the Emotion System.* Cambridge: Cambridge University Press.

Spivak, Gayatri Chakravorty. (1985) 1996. Subaltern Studies: Deconstructing Historiography. In *The Spivak Reader*, ed. Donna Landry and Gerald MacLean, 203–235. London: Routledge.

Springman, Marco. 2016. Going Veggie Would Cut Global Food Emissions by Two Thirds and Save Millions of Lives—New Study. *The Conversation*, March 22. https://theconversation.com/going-veggie-would-cut-global-food-emissions-by-two-thirds-and-save-millions-of-lives-new-study-56655

Springmann, Marco, H. Charles J. Godfray, Mike Rayner, and Peter Scarborough. 2016. Analysis and Valuation of the Health and Climate Change Co-benefits of Dietary Change. *PNAS* 113 (15): 4146–4151.

Stevens, Fisher, dir. 2016. *Before the Flood.* Los Angeles: RatPac Entertainment.

Weik von Mossner, Alexa. 2017. *Affective Ecologies: Empathy, Emotion, and Environmental Narrative*. Columbus: Ohio State University Press.

———. 2018. How We Feel About (Not) Eating Animals: Vegan Studies and Cognitive Ecocriticism. In *Through a Vegan Studies Lens: Textual Ethics and Lived Activism*, ed. Laura Wright. Reno: University of Nevada Press.

Wright, Laura. 2019. Doing Vegan Studies: An Introduction. In *Through a Vegan Studies Lens: Textual Ethics and Lived Activism*, ed. Laura Wright, vii–xxiv. Reno: University of Nevada Press.

Zunshine, Lisa. 2006. *Why We Read Fiction: Theory of Mind and the Novel.* Columbus: Ohio State University Press.

11

Constituting Vegetarian Audiences: Orchestrations of Egocentric, Anthropocentric, Ecocentric Exigencies in Jonathan Safran Foer's *Eating Animals*

Oren Abeles and Emma Lozon

Introduction

Although eating is a biological, instinctual activity for humans, it is inextricable from our use of symbols and language. Sonja Foss (2018, 368) writes that "as we choose foods our family members or friends like and set the table and arrange the food on our plates in aesthetically pleasing ways, the simple act of eating to sustain ourselves is transformed into

O. Abeles (✉) • E. Lozon
Michigan Technological University, Houghton, MI, USA
e-mail: omabeles@mtu.edu; emlozon@mtu.edu

C. Hanganu-Bresch, K. Kondrlik (eds.), *Veg(etari)an Arguments in Culture, History, and Practice*, The Palgrave Macmillan Animal Ethics Series,
https://doi.org/10.1007/978-3-030-53280-2_11

symbolically laden messages about ourselves, our friends, and food." Food and culture are deeply interrelated, together shaping human identity. Dan Jurafsky (2014, 189) provides that "how we talk about food also reflects human aspirations: our desire to live a healthy, natural, authentic life, to identify with our family and culture, and our deep strains of optimism and positivity."

This multiplicity of meanings could make arguing for vegetarianism more complicated, as every vegetarian argument would inevitably have broader implications that reach beyond diet into nature, culture, politics, and tradition. Food choice raises questions not only of taste and nutrition, but of identities and closely held values. Given that "the vegetarian/vegan movement has identity shift on its plate" (Hahn and Bruner 2012, 49), challenging the centrality of meat can constitute a polarizing rhetorical situation. This is an issue that vegetarian and vegan studies scholars have frequently considered: how to communicate the benefits of plant-based diets without alienating omnivores (Romo and Donovan-Kicken 2012; Jorgensen 2015; Priestley et al. 2016).

Food's multiple connotations can potentially be reconfigured as a resource rather than a barrier, particularly when arguments for vegetarianism are directed toward non-vegetarian audiences. Because food is about much more than just food, rhetors can constitute amenable audiences who—although they may initially disagree about the validity of vegetarianism—already concur on other non-dietary values that food engages. For example, that both meat-eaters and vegetarians may understand food as deeply political provides a point of identification, a merging of divergent interests (Burke 1962, 20). This point of identification affords pro-vegetarian rhetors a way to appeal to those who, at the outset, may disagree with them. Such arguments are "constitutive" when they simultaneously try to sway their audience *and* reframe their audience's sense of identity—whether it is their ethical, personal, social, or even ecological identity—in ways that realign the audience's interests with the pro-vegetarian cause.

In this chapter, we use Jonathan Safran Foer's bestselling book on vegetarianism, *Eating Animals* (2009), as a case study to demonstrate how rhetors can leverage vegetarianism's broader implications to constitute receptive audiences. In *Eating Animals*, Foer combines memoir, journalism, and activism to explore how his experience as a new father expanded his perspective on the relationship between meat consumption, nourishment, family, culture, and ethics. His research into the cultural values surrounding animal consumption brings him into dialogue with various people involved in resisting or reproducing factory farming and industrial animal slaughter, leading him to conclude that factory-farmed meat is incongruous with his values. We argue that Foer rhetorically structures his narrative to make it relevant not only to those who would tend to agree with him, but also to those who are disengaged from the ethical and cultural meaning of food. He accomplishes this through a practice of "orchestrating appeals," a technique of overlaying rhetorical exigencies to demonstrate their simultaneous relevance to individuals, humankind, and the broader ecosystem. Following Craig Waddell (1994, 229), we discuss how juxtaposing these egocentric, anthropocentric, and ecocentric motives can constitute audiences across various ideological commitments. As Waddell points out, anthropocentric exigencies frequently encompass individualistic, egocentric concerns and likewise, large-scale ecological problems can be of concern to both individuals and humanity. In Foer's case, he uses vegetarianism's wider relevance to demonstrate how its ideals are encompassed by broader commitments already shared by non-vegetarians.

Waddell theorizes rhetorical orchestration through a critical reading of demographer Paul Ehrlich's reductive arguments for population control. He argues that Ehrlich's rhetoric fails because it neglects to constitute a larger audience—including those who might not be directly affected by the problem—by not placing population growth in broader ecological terms (1994, 229). Specifically, Waddell sees Ehrlich as mistakenly treating population growth in adversarial and egocentric terms, often in ways that leverage first-world paranoia and racial and gender stereotypes. The result is a limiting framework confined to existing egocentric exigencies. In contrast, Waddell advocates for encompassing egocentric appeals within broader anthropocentric and ecocentric exigencies to constitute a cohesive and amenable audience despite differing levels of investment and resistance.

Waddell's concept of orchestration resonates with allied concepts in contemporary rhetorical theory, particularly Burkean identification and constitutive rhetoric. In pluralizing values and demonstrating their intersections, orchestration of appeals offers branching pathways toward identification. Arguments that enable identification appeal to existent shared commitments and can lead audiences to new values, just as inviting alternatives can dislodge harmful identification (Burke 1937, 263). Layering appeals expands a concept beyond its singular frame to interlace a variety of other tenets. Some appeals may provide a point of access to others that are more abstract, for instance, anthropocentric appeals may facilitate ecocentric appeals (Waddell 1996, 156). As Doug Cloud notes, such an overlapping of appeals allows rhetors not only to lead diverse and divergent audiences toward a recognition of shared interests but also to "engage difference as a resource—a chance to access bodies of knowledge from which may come other solutions" (2016, 63).

Orchestration is generative for the project of constituting additional audiences. Similar to Cloud's harvesting of difference, for Erin Branch, "recalcitrant elements" are viewed as resources in constituting an audience, "potentially productive elements that a rhetor can harness and foster" (2015, 167). Drawing on Maurice Charland's (1987) treatment of constitutive rhetoric, Branch demonstrates this concept through the example of Julia Child finding common ground between the sensual, intellectual French style of cooking and pragmatic American food culture. Similarly, for Waddell (1994, 232) reconstituting an audience can "mean challenging the audience to think in new ways about how values it already holds might be extended such that they apply in a new context." In the case of vegetarianism, such an expansion and connection of perspectives can resonate with new audiences through the consonance of vegetarianism with their existing values.

To demonstrate the effectiveness of such discursive exchange, we turn first to discerning the way orchestration grounds the introductory exordium of Foer's argument, embedding his audience's shared appreciation for familial bonds to frame the case for vegetarianism. We then show how that introductory orchestration opens the door for a wider range of identifications, leveraging the audience's general appreciation for ethical, health, and environmental concerns and demonstrating their consonance with the more specific case for vegetarianism.

Vegetarianism's Familial Exigencies

Among Foer's main rhetorical challenges is that making a pro-vegetarian argument to non-vegetarian audiences can often create a sense of ethical confrontation and conflict. Exigencies for vegetarianism can often seem to require a binary choice in which eating meat is either acceptable or it isn't. While slogans like "Meat is murder" may epitomize the case for vegetarianism stridently, they speak to the stark moral dilemmas that are often confronted in food choice. This puts Foer in the difficult position of having to appeal to an audience of meat-eaters who might interpret his very arguments against meat-eating as indictments of the reader's own immorality or indifference. There is a real risk that abstract or parochial arguments against meat-eating devolve into what Quintilian (2001, 371) labeled *vituperation*, an inverse encomium in which individuals are singled out as exemplars of vice. That Foer can make his case for vegetarianism without devolving into such vituperation is a testament to the effectiveness of his carefully orchestrated appeals.

Orchestration is a strategy deployed from the book's beginning, an introductory chapter in which Foer discusses food's connection to familial bonds by recounting his relationship to his Holocaust survivor grandmother and the meat she insisted he eat. Despite that fact that his grandmother made meat a staple of her cooking, the now vegetarian Foer remembers his grandmother as "The Greatest Chef Who Ever Lived" and recalls one of her specialties, chicken and rice, as the "most delicious thing I have ever eaten" (2009, 4). It is a surprising way to begin a book centered on the values of vegetarianism, yet the goal of Foer's introduction is not to argue the immorality of meat consumption but to leverage his audience's existing beliefs about the personal, cultural, and, above all, familial significance of food.

In this way, beginning his book with an account of his grandmother's cooking is rhetorically sensible, particularly because her experience of the Holocaust highlights a seamless intersection between food consumption, family survival, and the persistence of cultural identity. For Foer's grandmother, food's meaning was overdetermined by her experience of the Holocaust, so that post-war bounty and sustenance would always represent the endurance of culture and family against a countervailing violence that manifested as forced starvation. As Foer puts it, "The story of [my grandmother's] relationship to food holds all of the other stories that could be told about her. Food, for her, is not *food.* It is terror, dignity, gratitude, vengeance, joyfulness, humiliation, religion, history, and, of course, love. As if the fruits she always offered us were picked from the destroyed branches of our family tree" (2009, 5).

While most readers will not have an intimate connection to the consequences of the Holocaust, many can easily accept that food has an important relation to the strength of familial ties and cultural traditions, particularly against disintegrative social forces. Thus, regardless of the audience's diet or family history, Foer's (2009, 3–17) personal account of the relationship between food and family creates an initial point of identification between himself and his audience, one that can support broader arguments in both the introductory chapter and beyond. In Waddell's terms, Foer effectively recomposes a personal exigency within a wider, anthropocentric context. Such an introduction accomplishes the kind of orchestration that Cicero recommends in his discussion of *exordium*, the initial stage of an argument in which "the mind of the hearer [is brought] into a suitable state to receive the rest of the speech; and that will be affected if it has rendered him well disposed towards the speaker, attentive, and willing to receive information" (2009, 209). Because Foer's subsequent narrative is grounded in widely held beliefs about the important connection of food to family and culture, his audience is prepared to identify with him before vegetarianism has even been discussed.

Moreover, because this exordium highlights the centrality of food in a post-Holocaust Jewish family, the introduction's bounded familial frame inevitably gestures beyond to a wider sphere of historic and ethical human concerns. In this way, food's profound familial significance can overlap with the kinds of wider ethical and social dynamics that we consider below. This, not coincidentally, is precisely how a vegetarian like Foer approaches their personal food consumption and its broader pertinence. Foer's audience is asked to start thinking like a vegetarian while also being shown that such thinking—linking personal food consumption to broader anthropocentric exigencies—is neither exclusively vegetarian nor very radical.

Foer closes this introductory account of food and family in a way that exemplifies the efficacy of orchestration as a rhetorical strategy. He recounts a story his grandmother told him about the final days of the war, when food was at its scarcest and she was forced to make difficult choices that balanced adherence to religious values with sustenance and survival:

> "A lot of people died right at the end, and I didn't know if I could make it another day. A farmer, a Russian, God bless him, he saw my condition and he went into his house and came out with a piece of meat for me."
> "He saved your life."
> "I didn't eat it."
> "You didn't eat it?"
> "It was pork. I wouldn't eat pork."
> "Why?"
> "What do you mean why?"
> "What, because it wasn't kosher?"
> "Of course."
> "But not even to save your life?"
> "If nothing matters, there's nothing to save." (2009, 16–17)

This anecdote is still located within the series of appeals that lead up to it—the way food intersects with family, identity, and history—but Foer adds an additional inflection that orchestrates a final alignment of these more personal (egocentric) introductory frames with the broader (anthropocentric) vegetarian ethic for which he will go on to argue. In the story, Foer's grandmother refuses meat and does so for similar kinds of commitments that can undergird a wider humanistic argument for vegetarian ism. Obviously, his grandmother is not being portrayed as a vegetarian, but in the introduction's final scene, an abstention from meat epitomizes the same personal, familial, and cultural logic with which the audience has already identified. Foer's anecdote has a focused persuasive appeal, but such persuasive convergence is the result of orchestrating several appeals into a single rhetorical effect: that of using his own personal story to make a more broadly human argument about the critical relationship between food and family.

The Ethics of Vegetarianism

Foer's grandmother's story foregrounds a further orchestrated appeal that will recur throughout the text: the argument that food choice requires ethical engagement. The decision to eat or not eat a particular piece of food—a judgment people often make without much consideration—is given a forceful ethical charge in the story's retelling, as Foer's grandmother faces the dilemma of choosing between sustenance or religion. In this way, before the topic of vegetarianism has been broached, the normally mundane act of eating is conceived as deeply ethical, a sphere in which humans make important choices about the kind of life they wish to live (Allen 2018, 183–184).

In his grandmother's case, the choice to forgo pork, while not directly connected to the ethics of meat consumption, is an ethical decision that concerns her personally, as opposed to the animal she might have consumed. Yet we should be quick to note that Foer is not using this example to suggest food choice only be based upon personal, egocentric values. Though one could argue for vegetarianism based on individualistic exigencies (and, as we discuss below, Foer does), what his grandmother's anecdote indicates is that food choice is a venue for an individual to decide ethical values in the first place. For Foer's grandmother, the dilemma she faced required her to choose among two competing notions of self-interest: sustenance or religiosity. Readers are invited to consider how eating can be the subject of such serious ethical deliberation. Food choice, and our awareness of its strong ethical charge, becomes an index of whether people are purposefully engaged in the ethical consequences of their own everyday lives.

Indeed, it is this sense of an ethical engagement with life that Foer goes on to leverage as he develops his broader argument for vegetarianism. Witness, for example, Foer's discussion of the writer Franz Kafka, a vegetarian who reflected on how vegetarianism was concomitant with a complete awareness of human existence. Conversely, Kafka also felt that eating meat required that we not only ignore the value of animal life, but that we disconnect from fundamental aspects of our own ethical being. As Foer puts it:

> Beyond this literal forgetting of animals by eating them, animal bodies were, for Kafka, burdened with the forgetting of all those parts of ourselves we want to forget. If we wish to disavow a part of our nature, we call it our "animal nature." We then repress or conceal that nature, and yet, as Kafka knew better than most, we sometimes wake up and find ourselves, still, only animals....
>
> Today, at stake in the question of eating animals is not only our basic ability to respond to sentient life, but our ability to respond to parts of our own (animal) being. This is a war not only between us and them, but between us and us. (2009, 37)

Foer, like Kafka, sees meat consumption as part of a broader ethical disregard that humans visit upon both animals and themselves; it requires that they deny and alienate themselves from basic aspects of their nature. By contrast, the ethical commitments of vegetarianism can stem from an ethical imperative to engage with life fully. In this analysis, food choice is ethically consequential because it dovetails with an appreciation for both human and animal life.

This exemplifies considerable orchestration. One could mistake Foer's (2009) argument as just an egocentric or anthropocentric appeal, as his case certainly suggests existential benefits for those humans willing to approach food choice ethically. To a degree this is the case, but at the same time Foer deconstructs the normal distinctions by which we separate human and animal life, to the point where "anthropocentrism" becomes something of a contradiction in terms. Our humanity decreases to the extent we think of human life as ontologically superior to animal life, and this awareness requires we take food choice as an ethical opportunity to engage life fully. This is the basic orchestrated insight Foer foregrounds in the double meaning of *Eating Animals*' title; "We are not merely animals that eat," he clarifies towards the book's conclusion, "but *eating animals*" (our emphasis; Foer 2009, 194). Here Foer's rhetorical strategy resonates with Waddell's theory of orchestration: egocentric and anthropocentric appeals are not ignored but rather "encompassed" (1994, 232) within a broader ethical commitment to life in a wider, more ecocentric frame, including both us (the eating animals) and the animals we eat.

Foer's orchestration makes vegetarianism part of this broader ethical frame, one which extends beyond food choice to include the way humans treat both animals and themselves. Take, for example, his discussion of working conditions in industrial slaughterhouses and the existential effects they have on slaughterhouse employees. Particularly compelling are the descriptions offered by the workers themselves, which connect their own cruelty toward animals with a kind of ethical numbness in which workers withdraw from the experience of life. One worker describes the psychic toll of working in the "stick pit," the area of the slaughterhouse where animals are killed and drained of blood with cuts to the neck:

> The worst thing, worse than the physical danger, is the emotional toll. If you work in the stick pit for any period of time, you develop an attitude that lets you kill things but doesn't let you care. You may look a hog in the eye that's walking around down in the blood pit with you and think, God, that really isn't a bad-looking animal. You may want to pet it. Pigs down on the kill floor have come up and nuzzled me like a puppy. Two minutes later I had to kill them--beat them to death with a pipe … When I worked upstairs taking hogs' guts out, I could cop an attitude that I was working on a production line, helping to feed people. But down in the stick pit I wasn't feeding people. I was killing things. (Foer 2009, 252)

Notice how the worker describes his experiences not just as a poor ethical choice but as an inability to experience any ethical consequentiality whatsoever: "you develop an attitude that lets you kill things *but doesn't let you care*." Then there is the worker's compelling two-sentence concluding summary of their experience, which uses figurative rhetoric to epitomize industrial slaughter as a violation of both human and animal dignity. It does this through a combination of two rhetorical figures: (1) antithesis (a juxtaposition highlighted by parallel phrases) and (2) oxymoron (placing incompatible terms adjacent to one another). The antithesis "I wasn't feeding people. I was killing things" efficiently encapsulates how industrial slaughter takes us from an ethical engagement with life and sustenance ("feeding people") to an opposing state of desensitization and violent objectification ("killing things"). That opposition is underscored by the oxymoron "killing things," which links a violent relation to life ("killing") with the worker's own inability to experience life itself (treating lives as "things").

Thus, what could be an argument that is just about the ways humans treat animals cruelly is also made into an argument about the generalized suffering experienced by all those involved in industrial slaughter, both humans and animals. This orchestration effectively describes the existential consequences of meat production (as in the "emotional toll" experienced by the worker quoted above), but it can be equally effective in describing the similarly unethical physical cruelty inflicted on both animals *and* humans. Consider, for example, Foer's account of the reasons why chicken slaughterhouses tend to employ undocumented immigrants and non-English speakers, groups who are less likely or able to complain about both animal cruelty and workplace safety. In doing so, Foer's rhetorical arrangement emphasizes a parallel set of physical consequences affecting both humans and animals.

> If your chicken operation is running at the proper speed…the birds will be handled roughly and, as I was told, the workers will regularly feel the birds' bones snapping in their hands…. No laws protect the birds, but of course there are laws about how you can treat the workers, and this sort of labor tends to leave people in pain for days afterward, so, again, be sure you hire those who won't be in a position to complain--people like 'Maria,' an employee of one of the largest chicken processors in California, with whom I spent an afternoon. After more than forty years of work, and five surgeries due to work-related injuries, Maria no longer has enough use of her hands to do the dishes. She is in such constant pain that she spends her evenings soaking her arms in ice water, and often can't fall asleep without pills. (2009, 132)

Here Foer takes the reader from the way workers' hands break chickens' bones to the way a worker's hands are themselves broken by the same cruel work. The parallel reinforces Foer's goal of getting readers to see anthropocentric concerns as part and parcel of a broader ethical sensibility and to consider how the lives of humans and animals might be ecocentrically intertwined.

Toward the end of the book, Foer summarizes these parallels in a way that makes clear his intention to orchestrate an argument that includes the ethical interests of both humans and animals. He writes:

> Farmers have lost--have had taken from them--a direct, human relationship with their work. The factory model has estranged them not only from how they labor (hack, chop, saw, stick, lop, cut), but what they produce (disgusting, unhealthy food) and how the product is sold (anonymously and cheaply). Human beings cannot be human (much less humane) under the conditions of a factory farm or slaughterhouse. It's the most perfect workplace alienation in the world right now. Unless you consider what the animals experience. (2009, 254)

Significantly, Foer draws a distinction between being "human" and "humane" while simultaneously maintaining their connection. "Humane" behavior refers to a kind of normative ethics that would (or should) prescribe how we treat animals. The quality of being "human," on the other hand, describes the capacity of workers to experience their own lives in their existential and ethical fullness (Allen 2018, 183–184). It consists of an ethic of living fully in opposition to the kind of deadening "alienation" that industrial slaughterhouses impose on their employees. These distinctions are not intended to separate the two kinds of ethics, but to orchestrate an appeal in which a potentially unrelated issue of inhumane animal cruelty is enmeshed in the ability of humans to fully experience their own humanity.

Foer's final two sentences bring that orchestration full circle. When he writes that the "alienation" of slaughterhouse workers is second only to that of the animals they slaughter, he extends the traditionally human condition of alienation to the experience of livestock. A fastidious Marxist might take exception to including animals within the sphere of "alienation,"[1] but this assertion comes towards the end of Foer's book and by this point his audience has been so carefully constituted that they are well prepared to credence a more ecocentric comparison of the existential suffering experienced by humans and animals.

Vegetarianism and the Intersections of Individual, Public, and Environmental Health

Foer orchestrates familial and ethical concerns with health-related and environmental considerations such that egocentric and anthropocentric appeals provide a point of access to more abstract ecocentric arguments for vegetarianism. While an appeal to individual health can be characterized as an egocentric concern, health interfaces with both anthropocentric and ecocentric foci, in areas such as pandemic prevention and environmental protection. Health-focused appeals for vegetarianism afford persuasive potential because of their personal and accessible nature (Jorgensen 2015, 13). Such appeals can communicate both the health benefits of plant-rich diets and the multiple health risks associated with factory-farmed animal products. Foer orchestrates both intersections by way of the direct and indirect connections between food and human health including not only the nutritional profile of food, but the interrelations between health concerns and ethical, ecological exigencies.

Foer's orchestration of egocentric, anthropocentric, and ecocentric appeals reveals the overlap between unethical and unhealthy food. Foer finds that signifiers of health like "fresh" and "free-range" fail to qualify as either healthy or ethical, showing how the unethical conditions of factory-farmed practices not only affect the health of the animals themselves, but also that of the humans who eat them. This discrepancy between terms and reality is most evident when it comes to chicken, a food often considered a healthy choice of sustenance because of its protein-fat ratio. Foer conveys that "pathogen-infested, feces-splattered chicken can technically be fresh, cage-free, and free range" (2009, 61). A small block printed rectangle printed on two pages of the text, exactly the size of a standard chicken cage, encapsulates the claustrophobic, confining space, bringing the audience face to face with the material reality of the size of the cages in which hens are contained (2009, 78–79). Cage-free birds, Foer clarifies, have virtually the same amount of space. Revealing how unhealthful and unethical conditions can be mutually implicated in each other, Foer underscores that "When we eat factory-farmed meat we live, literally, on tortured flesh. Increasingly, that tortured flesh is becoming our own" (2009, 139). Given this orchestration of health and ethics, the notion that "you are what you eat" resounds physically and symbolically in the descriptions of factory-farmed animals that are highly confined, drugged with various antibiotics, and tortured.

In deploying health as both an egocentric and anthropocentric concern, Foer communicates that unethical, inhumane conditions for factory-farmed animals pose a threat not only to individual health, but also to public health more broadly. That human health is contingent on animal health is most alarmingly communicated in the context of growing antibiotic resistance and pandemic threat that stem from unchecked factory farming practices. Foer summarizes that "the same conditions that lead 76 million Americans to become ill from their food annually and that promote antimicrobial resistance also contribute to the risk of pandemic" (2009, 137). Zoonotic pathogens, which travel between animal and human bodies, physically evidence the porous boundary between humans and animals, and the risks of obscuring humans' relationality with the wider ecosystem. It is through discussing the relationship between factory-farmed animals, personal health, and public health that the notion of externalized social costs becomes most vivid. While the grocery store price of factory-farmed meat is comparatively cheap, the real cost must include factory farming's consequences for human health and well-being. As Foer's inclusion of one turkey farmer's perspective[2] makes clear, "It's possible you can't afford to care, but it's certain you can't afford not to care" (2009, 113). This instance of *enantiosis* (the juxtaposition of claims that seem paradoxical at the outset) underscores the bind confronting consumers of animal products: the true toll of meat accrues downstream in more serious ways for the health of individuals and the community.

Foer is also able to encompass this overlap of individual and public health within a broader ecocentric frame by emphasizing the connection between factory farming and pollution. To this end, he discusses how this relationship manifests on a macroscopic scale in that ecological crises like climate change, exacerbated by livestock agriculture, pose growing threats to human well-being. He also demonstrates this relationship on more local scales, as when a vast quantity of unmanaged manure "seeps into rivers, lakes, and oceans—killing wildlife and polluting air, water, and land in ways devastating to human health" (Foer 2009, 174). Foer employs *topografia*, a vivid and compelling description of place, to capture the enormity of waste produced by Smithfield Foods, America's largest hog producer:

> the liquified waste is pumped into massive lagoons adjacent to the hog sheds. These toxic lagoons can cover as much as 120,000 square feet—as much surface area as the largest casinos in Las Vegas—and be as deep as 30 feet. The creation of these lake-sized latrines is considered normal and is perfectly legal despite their consistent failure to actually contain the waste. A hundred or more of these immense cesspools might loom in the vicinity of a single slaughterhouse. If you were to fall into one, you would die.... In 1995, Smithfield spilled more than twenty million gallons of lagoon waste into the New River in North Carolina. The spill remains the largest environmental disaster of its kind and is twice as big as the iconic Exxon Valdez spill six years earlier. (2009, 177–178)

Foer's use of topografia points to the need to localize environmental exigencies, defining and capturing what might otherwise seem like an elusive subject. Since the environment is, by definition, both diffuse and interconnected, its omnipresence can make it hard for rhetors to point to specific environmental exigencies and focused interventions. For instance, unlike threats to family and health, environmental hazards can appear more distant and abstract. George Lakoff identifies a deficit of frames capturing that the environment is "intimately tied up with other issue areas: economics, energy, food, health, trade, and security" (2010, 76). Complex issues like climate change that exert both human and ecological impacts that are reduced to the domain of the environment may not register in terms of closely held reference points. Thus, they lack the "rhetorical presence" that grounds global ecological problems in the more local and concrete settings relevant to individuals and humanity. Foer's narrative accounts of health consequences, both personal and environmental, push back against these tendencies, presenting readers with vivid descriptions in which these seemingly different spheres are revealed to be ineluctably interconnected.

This emphasis on the overlap of personal, public, and environmental health allows Foer to reach out both to environmentally minded audiences and to those who might be dismissed as recalcitrant audiences for ecocentric arguments. For the former, Foer writes that

> If one cares about the environment, and if one accepts the scientific results of such sources as the UN (or the Intergovernmental Panel on Climate Change, or the Center for Science in the Public Interest, or the Pew Commission, or the Union of Concerned Scientists, or the Worldwatch Institute…), one must care about eating animals. Most simply put, someone who regularly eats factory-farmed animal products cannot call himself an environmentalist without divorcing that word from its meaning. (2009, 59)

By creating a point of identification between an environmental orientation and deliberate food consumption, Foer appeals to the ecological values already explicitly held by some audiences and points to dietary choices as an overlooked area of sustainable living. However, not only does Foer appeal to audiences who already identify as environmentalists, he identifies the shared regard for the environment that most readers can agree upon. He offers that

> Whether or not you are in favor of offshore oil drilling, whether or not you believe in global warming, whether you defend your Hummer or live off the grid, you recognize that the air you breathe and the water you drink are important. And that they will be important to your children and grandchildren. Even those who continue to deny that the environment is in peril would agree that it would be bad if it were (2009, 73).

This discussion aligns the environment with fundamental values of health and family, supplying accessible points of reference for conceptualizing the ecosystem. Like the use of topographia, these points of reference afford rhetorical presence, making environmental facts about factory farming more immediate and meaningful. Engaging these egocentric, anthropocentric, and ecocentric appeals, Foer cultivates a diverse audience that, for a variety of reasons, is invested in preserving human and environmental health. Given the cultivation of readers' existing values and the emphasis on food's inexorable connection to the health of the broader ecosystem, food consumption becomes forwarded as a site for advocacy and engagement.

Because of the careful orchestration of environmental appeals in *Eating Animals*, audiences may be more receptive to the potentially more polarizing ecocentric arguments that Foer (2019) makes in his follow-up book on food, *We are the Weather*. There he more forcefully animates ecocentric exigencies for sustainable food choice, arguing for collective consumption of fewer animal products in order to reverse global warming (Foer 2019, 64). He works to destabilize what Eubanks (2015, 127) refers to as an environmental frame for climate change, which focuses primarily on the planet while backgrounding humans: "the planet isn't what we want to save. We want to save life on the planet—plant life, animal life, and human life" (Foer 2019, 202). In this way, Foer expands on and intensifies the environmental themes of *Eating Animals*, more directly connecting the abstract, overwhelming exigency of climate change to the concrete, personal choices that individuals make at each meal. Given a first encounter with Foer's more invitational exploration of vegetarianism in *Eating Animals*, a broader audience may be constituted to bridge the distance between belief and action when it comes to climate change.

Conclusion

Rather than imposing a burden or barrier, the interconnectedness of vegetarianism with other issues and social spheres offers a rhetorical asset. Connecting food choice to family, ethics, health, and the environment yields multiple points of identification, allowing Foer's argument to resonate more persuasively with audiences despite varied ideological commitments. The technique of orchestrating egocentric, anthropocentric, and ecocentric appeals draws rhetorical power from the extensive network of significance and values surrounding a vegetarian diet.

To persuade audiences to consider the wider relevance and benefits of vegetarianism, future rhetors might deploy its intersections with other personal and social dynamics. For instance, arguments could leverage economic exigencies, such as the potential for a well-planned vegetarian diet to be more cost-effective for individuals. They might likewise consider the externalized economic consequences that meat-eating has for society. Similarly, rhetors might explore alignments between food choice and social change movements like environmental justice and intersectional feminism, connections forwarded by Carol Adams (1996). In this way, rhetors can connect more concrete exigencies to those with increasing levels of abstraction associated with complex, global issues.

Of course, the effectiveness of pro-vegetarian orchestrations need not lead us to confine the technique's utility to the case for vegetarianism. As Waddell showed, orchestration can serve as a rhetorical intervention for a wide variety of polarizing issues. To give a contemporary example, consider the orchestration being deployed to advocate for the "Green New Deal," an ambitious legislative agenda that responds to climate change through investments in a range of social spheres that are not always part of an environmentalist agenda. The legislation calls for supporting unions, protecting indigenous communities, providing universal health care, and increasing access to higher education, amongst a host of other initiatives. As Demond Drummer (one of the Green New Deal's architects) argues, orchestrating these exigencies together "actually makes it easier to attack the [climate change] crisis" ("The Case" 2019). He goes on to assert:

> If we're talking about making investments to create millions of new high-paying [environmentally sustainable] jobs, policies like universal health-care, policies like universal child care make it easier for people to join in on that mobilization. So this isn't a [case of combining unrelated issues]. It's absolutely necessary because it's about reshaping the entire economy. A fossil fuel economy that's designed to exploit and extract requires disposable people and disposable places and what the Green New Deal says is 'No more disposable people. No more disposable places.'

Drummer's figurative use of isocolon ("No more disposable people. No more disposable places") and the multiple senses of "exploit" and "extract" mirror Foer's techniques, making a more expansive case that nonetheless increases the agenda's coherence through a rhetorical overlaying of ego-centric, anthropocentric, and ecocentric imperatives. Policy advocates would do well to consider how orchestrations like these allow them to raise the stakes of their arguments whilst simultaneously building larger, more effective coalitions.

By engaging these wider ecologies, orchestration of appeals attunes to how the spheres of life flow into each other. By mapping the intersections of food consumption with areas such as family, health, ethics, and the environment, vegetarian orchestration exemplifies engagement with interwoven exigencies for social change. Food consumption serves then as a generative intersection for critical reflection and dialogue about whether practices follow from holistic values. Given its deep interrelation with nature and culture, talking about food can ultimately serve as a conduit for a more compassionate, happier, more ethical life.

Notes

1. Marx specifically distinguishes between human and animal suffering when he defines alienation as an impoverishment of a human worker's *gattungswesen* or "*species* being" (1994, 62–63).
2. Referred to as "the last poultry farmer" because he avoids the industrial model of selective breeding and antibiotic use.

Bibliography

Adams, Carol J. 1996. *The Sexual Politics of Meat: A Feminist-Vegetarian Critical Theory*. New York: Continuum.

Allen, Ira J. 2018. Composition Is the Ethical Negotiation of Fantastical Selves. *College Composition and Communication* 70 (2): 169–194.

Branch, Erin. 2015. 'Taste Analytically': Julia Child's Rhetoric of Cultivation. *Rhetoric Society Quarterly* 45 (2): 164–184.

Burke, Kenneth. 1937. *Attitudes Toward History*. Berkeley: University of California Press.

———. 1962. *A Rhetoric of Motives*. Los Angeles: University of California Press.

Charland, Maurice. 1987. Constitutive Rhetoric: The Case of the Peuple Quebecois. *Quarterly Journal of Speech* 73 (2): 133–150.

Cicero, Marcus Tullius. 2018. *The Orations of Marcus Tullius Cicero*. Vol. 4. Frankfurt am Main: Outlook.

Cloud, Doug. 2016. Communicating Climate Change to Religious and Conservative Audiences: The Case of Katherine Hayhoe and Andrew Farley. *Reflections* 16 (1): 57–74.

Eubanks, Philip. 2015. *The Troubled Rhetoric and Communication of Climate Change*. New York: Routledge.

Foer, Jonathan Safran. 2009. *Eating Animals*. New York: Back Bay Books.

———. 2019. *We Are the Weather: Saving the Planet Begins at Breakfast*. New York: Farrar, Straus and Giroux.

Foss, Sonia K. 2018. *Rhetorical Criticism*. Long Grove: Waveland.

Hahn, Laura K., and Michael S. Bruner. 2012. Politics on Your Plate. In *The Rhetoric of Food: Discourse, Materiality, and Power*, ed. Joshua J. Frye and Michael S. Bruner, 42–57. New York: Routledge.

Jorgensen, Beth. 2015. To Meat or Not to Meat? An Analysis of On-line Vegetarian Persuasive Rhetoric. *Poroi* 11 (1): 1–19.

Jurafsky, Dan. 2014. *The Language of Food: A Linguist Reads the Menu*. New York: Norton.

Lakoff, George. 2010. Why It Matters How We Frame the Environment. *Environmental Communication* 4 (1): 70–81.

Marx, Karl. 1994. In *Karl Marx: Selected Writings*, ed. Lawrence Simon. Indianapolis: Hackett.

Priestley, Alexis, Sarah K. Lingo, and Peter Royal. 2016. 'The Worst Offense Here Is the Misrepresentation': *Thug Kitchen* and Contemporary Vegan Discourse. In *Critical Perspectives on Veganism*, ed. Jodey Castricano and Rasmus R. Simonsen, 349–371. London: Palgrave Macmillan.

Quintillian, Marcus Fabius. 2001. Institutes of Oratory. In *The Rhetorical Tradition*, ed. Patricia Bizzell and Bruce Herzberg, 2nd ed., 364–428. Boston: Bedford/St. Martin's.

Romo, Lynsey Kluever, and Erin Donovan-Kicken. 2012. 'Actually, I Don't Eat Meat': A Multiple-Goals Perspective of Communication About Vegetarianism. *Communication Studies* 63 (4): 405–420.

The Case for the Green New Deal. MSNBC. https://www.msnbc.com/all-in/watch/the-case-for-the-green-new-deal-1468170819829. Last Modified 29 Mar 2019.

Waddell, Craig. 1994. Perils of a Modern Cassandra: Rhetorical Aspects. *Social Epistemology* 8 (3): 221–237.

———. 1996. Saving the Great Lakes: Public Participation in Environmental Policy. In *Green Culture: Environmental Rhetoric in Contemporary America*, ed. Carl G. Herndl and Stuart C. Brown, 141–165. Madison: University of Wisconsin Press.

12

Beyond Diet: Veganism as Liberatory Praxis

Tara Roeder

In a 2018 report, the UK's Agriculture and Horticulture Development Board claimed that "[a]mong young people, veganism has become a badge of identity or a tribal marker, much as identifying as other youth tribes such as 'gym bro' or 'craft beer nerd'" (Bird 2018). The CEO of Ireland's National Dairy Council similarly described veganism as a "trend" that "will only last another two or three years" (Chiorando 2019). In 2019, the U.S. National Cattlemen's Beef Association authored a document detailing a multi-pronged approach to "aggressively" address the rise of plant-based beef alternatives, despite also assuring its audience that "[m]eat alternatives represent just a tiny fraction of pounds sold" (Frazier 2019). Reports such as these reflect the meat and dairy industry's growing anxiety about the rise of plant-based food markets, and often rely on tropes of veganism as a fleeting dietary "fad."

Reducing veganism to a trendy "lifestyle choice" is common in the animal agriculture industry for good reason. Such rhetoric conveniently

T. Roeder (✉)
St. John's University, New York, NY, USA
e-mail: roedert@stjohns.edu

C. Hanganu-Bresch, K. Kondrlik (eds.), *Veg(etari)an Arguments in Culture, History, and Practice*, The Palgrave Macmillan Animal Ethics Series,
https://doi.org/10.1007/978-3-030-53280-2_12

masks the potential of veganism as an anti-oppressive praxis, instead offering it as one dietary option among many (e.g. "Paleo," "Atkins," "Raw Food"). While consumers may indeed engage in plant-based eating habits for reasons that do not include animal rights, veganism by definition extends beyond the consumption of a plant-based diet. Defined by the Vegan Society as "a way of living which seeks to exclude, as far as is possible and practicable, all forms of exploitation of, and cruelty to, animals for food, clothing or any other purpose" ("Definition of Veganism"), veganism, for those who identify as such for ethical reasons, is a consistent commitment to transforming structural inequality. Because one of the major global sites of such inequality is the animal agriculture complex, it's not surprising that animal based food industries would acknowledge a rise in veganism as a threat. While dismantling the animal-based food industry is not the sole aim of veganism, it is certainly central to the movement. For humans who are invested in combatting speciesism—which includes not only an assumption of human supremacy but the belief that members of some non-human species are morally more important than others—veganism functions as an embodied practice rooted in resistance to the ideologies that justify harming, incarcerating, breeding, commodifying, experimenting on, displacing, and consuming non-human animals.

Despite ethical veganism's roots in anti-oppression, however, the perception of veganism as an individual identity marked by "elitist social privilege" (Wright 2015) lingers. This stereotype, which erases the voices of many in the vegan community as well as the voices of non-human animals, is one that I hope to challenge with this piece. The work of scholars and activists such as Carol Adams, A. Breeze Harper, pattrice jones, Aph Ko and Syl Ko, Christopher-Sebastian McJetters, Lauren Ornelas, and Sunaura Taylor, among many others, speaks back to the limited view of veganism promoted by the meat and dairy industries and instead proposes an ideological shift, advocating a radical veganism that plays a transformative role in creating a more sustainable, just future for non-human animals and humans alike.

My goal here is to explore the generative effects of moving away from the notion of veganism as an issue of personal consumption and to examine some of the relationships among speciesism and other harmful systems and practices such as misogyny, colonialism, ableism, food injustice, and environmental degradation. I argue that when we conceptualize and practice veganism as an act of resistance in the face of the dominant and ecologically destructive ideologies espoused within capitalist, white supremacist heteropatriarchy, we disrupt the rhetorical and institutional norms that oppress many humans alongside the non-human animals who currently function as the "ghosts in our machine" (Marshall 2013).

The "Who" in Veganism

At the start of his book *Beyond Words: What Animals Think and Feel*, researcher Carl Safina describes the experience of witnessing a group of dolphins surfacing alongside his boat. As the dolphins leap and call to each other, Safina allows "[him]self to ask them the question that is forbidden fruit: *Who* are you? Science usually steers firmly away from questions about the inner lives of animals [...] Permissible questions are 'it' questions: about where it lives, what it eats, what it does when danger threatens, how it breeds. But always forbidden is the one question that might open the door: Who?"(Safina 2015, 1). For many vegans I've encountered, a common question is similarly not *what* you eat, but *whom* you eat—or choose not to.

Colin Dayan reminds us in his entry on "personhood" in *Critical Terms for Animal Studies* that the "Cartesian division between mind and body put nonhuman animals squarely on the side of the body, emptied of consciousness, feeling, and awareness" (Dayan, 2018, 267). The dualistic binaries in western rhetoric and philosophy that have been used to justify the wide scale oppression of people of color, women, and humans with disabilities throughout recent centuries have also been used to relegate non-human animals to the status of "thing." As disability and animal activist Sunaura Taylor writes: "Animals are a category of beings that in the Western tradition we have decided that we rarely, if ever, have duties toward—we can buy them, sell them, and discard them like objects" (Taylor 2017, 108). The positionality of non-human animals as "things" is often used to legitimize their wide-scale oppression at the hands of human persons.

The systematic objectification of non-human animals is far from a universal cultural norm, however. Jainism and Buddhism, for example, have long histories of vegetarianism. Contemporary theorists such as Layli Philips offer ecowomanist perspectives "based on a holistic perception of creation encompassing humans and all living organisms plus the nonliving environment and the spirit world" (Philips 2010, 8). And the "human/animal" binary that haunts much of western thought itself is not present in a variety of First Nations metaphysics and ontologies (Laws 1994; Tallbear 2011; Sepie 2017).

The status of non-human animals as objects has also been challenged by western philosophers such as Jeremy Bentham, who, in 1789, questioned prevailing "rationality" based paradigms when he attempted to shift the basis for legal protection for non-human animals not in their ability to "reason," but in their ability to feel: "The question is not, Can they reason?, nor Can they talk? but, Can they suffer?" (Varner 2018, 366). Subsequent rhetoric surrounding the legal status of non-human animals has sometimes called for the extension of specific rights to some non-human animals by deeming them "persons" under the law. A small number of non-human animals have recently benefitted from such an approach: "Spain and New Zealand have extended personhood rights to great apes […] Trigueros del Valle, Spain, voted unanimously to define dogs and cats as 'non-human residents,' giving them rights similar to those of humans living there […] The Indian government ruled that cetaceans, such as whales and dolphins, are 'non-human' persons with their own specific rights" (Dayan 2018, 272). Such moves evidence a shifting perception of personhood and rights with positive implications for a select group of non-human animals.

Yet such piecemeal tactics have so far offered little to the majority of non-human animals, including those incarcerated and tortured on factory farms as they await slaughter for human consumption. While the fight to obtain personhood status for non-human animals within existing legal structures has concrete benefits for non-human animals deemed to have "sophisticated cognitive abilities—chimps, elephants, and dolphins, for instance" (Dayan 2018, 272), Taylor points out that the "characteristics that humans have used to measure cognitive capacity are no doubt signs of a certain kind of complex cognition, but they are not necessarily the only ways to measure intelligence, let alone value or worth" (Taylor 2017, 72). Granting rights to certain non-human animals on the basis of anthropocentric perceptions of intelligence leaves the paradigm that privileges "rationality" above other modes of thinking and knowing intact. In her essay "Addressing Racism Requires Addressing the Situation of Animals," Syl Ko meaningfully challenges the very choice to "honor the hyper-obsession with the 'person' or the 'individual' in the West and try to extend personhood or individuality to animals in order to rethink/reimagine animality" (Ko and Ko 2017, 40). The legal conference of "personhood"—a shifting category that has at times excluded certain classes of human beings as well as almost all non-human animals—to some classes of non-human animals is not enough to undermine the systems that continue to designate non-human animals as "vermin" to be eradicated, vehicles for human entertainment, or products for human consumption.

Central to any analysis of the objectification of non-human animals is the animal agriculture industry. Coming to understand individual non-human animals as having unique personalities, familial relationships, and emotions remains a crucial realization for many vegans. Yet the cognitive dissonance that enables some contemporary "animal lovers" to rescue cats while eating cows (a fellow vegan recently despaired at a humane society benefit with a completely non-vegan menu) is reinforced both through speciesism and the relative invisibility of animal-based food production in heavily industrialized areas. The animal agriculture industry, which routinely enmeshes non-human animals in oppressive structures alongside marginalized human beings—the Texas prison system, for example, has its own "agribusiness department," in which thousands of prisoners work without pay on factory farms (Abdulraf 2017)—is an illustrative example of how the liberation of non-human animals is connected to the liberation of the humans oppressed by the same system.

Overlapping Oppressions in Animal Agriculture

As I mentioned earlier, veganism is not solely a response to the role of non-human animals in food production/consumption. Vegans are also concerned with the fates of non-human wild animals killed and displaced by human "development" and climate change, the lives of non-human animals held captive and experimented on in universities and government labs, the non-human animals wounded and killed in the production of clothing, bedding, and sports equipment, those bred as "pets" and those designated as "pests," those imprisoned in zoos, circuses, and aquariums, and those used in "entertainment" events such as rodeos, horse racing, dog fights, and bullfights, among others. I choose to focus on food production in this section because of the daily implications of the violence entailed in breeding, raising, and killing non-human animals specifically for human consumption, and the ways in which speciesist ideologies and institutions routinely harm the human beings enmeshed in our industrial food systems, as well as our environment.

It is estimated that 99% of farmed non-human animals in the U.S. spend their lives on factory farms before being slaughtered (Gilliver 2019). Non-human animals on factory farms spend their short lives in intensive confinement and regularly face unspeakable abuse. Mercy For Animals has extensively documented routine factory farm practices such as ripping out the testicles and cutting off the tails of piglets without anesthesia; cramming mother pigs in gestation crates in which they cannot turn around or lie comfortably; removing calves from cows shortly after birth to be slaughtered for veal and packaging the breast milk produced by their grieving mothers to sell to humans; cramming egg laying hens into tiny cages in windowless sheds and searing off their beaks; electrifying turkeys who have been bred to grow so fast they suffer deformities; and suffocating, skinning, and dismembering living fish ("The Problem" 2019). Alongside this inherent mistreatment, numerous instances of sadistic abuse towards individual animals have been documented, including sexual abuse, beatings with instruments including pipes, and spraying paint up the nostrils of distressed pigs and cows (Adams 2017, 197–198; "Animal Equality Films…" 2018).

The objectification of the non-human animals involved in human food production is connected to the objectification of the factory farm and slaughterhouse workers who are often "funneled into such undesirable jobs due to class, disability, and immigration status" (Taylor 2017, 201). Carol Adams writes that "[o]ne of the basic things that must happen on the disassembly line of a slaughterhouse is that the animal must be treated as an inert object, not as a living, breathing being. Similarly the worker on the assembly lines becomes treated as an inert, unthinking object, whose creative, bodily, emotional needs are ignored" (Adams 2017, 33). A 2016 Oxfam study found that poultry workers are routinely "denied breaks to use the bathroom" and "urinate and defecate while standing on the line; they wear diapers to work; they restrict intake of liquids and fluids to dangerous degrees; they endure pain and discomfort while they worry about their health and job security" ("No Relief" 2016). Workers on factory farms additionally face a variety of long-term health risks from consistent exposure to "hazardous levels of particulate matter as well as ammonia and hydrogen sulfide gases" ("Factory Farm Workers" 2019), as well as antibiotic resistant pathogens ("Working Conditions…" 2019). Many employees are migrant workers who are "largely unaware of the inherent health hazards and social struggles they will encounter" in the industry, and employers find undocumented workers to be especially "ideal recruits because they are less likely to complain about low wages and hazardous working conditions" ("Factory Farm Workers" 2019). The industry's rush for profit not only leads to high rates of serious injury for these workers (Lowe 2016), but to further torture for the non-human animals who are sometimes boiled, skinned, and/or dismembered alive in haste (Warrick 2001).

Participating in slaughterhouse violence takes a significant toll on the human workers involved. When Leo Tolstoy famously wrote "As long as there are slaughterhouses, there will be battlefields," he connected two forms of socially sanctioned violence. The act of killing is largely viewed as justified in both war and in the slaughterhouse, and slaughterhouse workers, like soldiers, are encouraged to become desensitized "killing machines." Sunaura Taylor cites a former kill floor manager who says

"The worst thing is the emotional toll [...] Pigs down on the kill floor have come up and nuzzled me like a puppy. Two minutes later I had to kill them—beat them to death with a pipe. I can't care" (Taylor 2017, 185). Another worker at a poultry plant testifies "The more I [killed chickens], the less it bothered me. I became desensitized. The killing room really does something to your mind—all that blood, killing so many times, over and over again" (Taylor 2017, 187). It is no coincidence that, like soldiers, slaughterhouse workers are at an increased risk of developing PTSD (Von Alt 2017).

It should also come as no surprise that the violence that takes place on the killing floor does not remain contained there, but spills over into the communities in which such facilities are located. A 2009 study by Fitzgerald, Kalof, and Dietz found that "slaughterhouse employment increases total arrest rates, arrests for violent crimes, arrests for rape, and arrests for other sex offenses in comparison with other industries. This suggests the existence of a 'Sinclair effect' unique to the violent workplace of the slaughterhouse." A more recent 2014 study by Racine Jacques similarly found communities with slaughterhouses had a 166% increase in rape arrests (Wrenn 2014), further underscoring the well-documented connections between violence against non-human animals and violence against other humans, particularly women and children (Ascione et al. 2007; Flynn 2011).

Factory farms also pose serious environmental and health risks to the communities in which they are located. Multiple reports have emerged over the past several years on the effects of industrial hog farming on low-income communities of color in North Carolina, who are forced to contend with polluted drinking water as a result of overflowing manure lagoons, a spike in illnesses including asthma, and a pervasive odor that impedes quality of life (Hellersten and Fine 2017). The placement of factory farms in and near low-income communities of color is no coincidence; a 2014 study by Wing and Johnston found that "the state's industrial hog operations disproportionately affect African Americans, Hispanics and Native Americans. That pattern, they concluded, 'is generally recognized as environmental racism'" (Hellersten and Fine 2017). As A. Breeze Harper writes: "Think of all the toxic waste coming out of the agribusiness industry. Where does it end up? It doesn't end up in the backyard of Beverly Hills, but where there are working class people of color [...] A lot of waste is going in my backyard and causing my community lots of health disparities and suffering" ("The *Satya* interview..." 2007).

The lack of access to affordable whole foods is also a problem that disproportionately affects low-income Americans of color, who are more likely to live in food landscapes dominated by the non-human animal based "fast food" industry. Angela Davis connects speciesism to food apartheid in black communities when she describes how "sentient beings endure pain and torture as they are transformed into food for profit, food that generates disease in humans whose poverty compels them to rely on McDonald's and KFC for nourishment" (Loria 2016). Fellow vegan A. Breeze Harper's collection *Sistah Vegan: Black Female Vegans Speak on Food, Identity, Health, and Society* locates the food and health inequalities experienced by many black U.S. communities at the intersections of speciesism and institutionalized racism. The volume chronicles the food and activist practices of a diverse array of black vegan women, many of whom experience veganism as a form of resistance to oppressive dominant ideologies and make connections between colonialism and contemporary U.S. dietary patterns that rely heavily on the bodies and secretions of captive non-human animals.

On top of the harm that animal agriculture causes non-human animals and human workers and communities, the industry is also "the second largest contributor to human-made greenhouse gas (GHG) emissions after fossil fuels and is a leading cause of deforestation, water and air pollution and biodiversity loss" ("Animal Agriculture's Impact on Climate Change"). Because the plight of non-human animals in our food industries is intrinsically connected with that of human beings and the environments in which we live, approaches that link food justice and human and environmental health and safety alongside the health and safety of non-human animals may be especially effective in promoting wide-scale veganism. Contemporary initiatives such as the Food Empowerment Project, founded by Lauren Ornelas, provide integrated models that simultaneously benefit humans and non-human animals—the organization calls attention to the abuse of farmed non-human animals as well as produce workers' rights, and promotes access to healthy food in low-income areas. Their Vision includes "a food system free from the exploitation of humans and the environment and with equitable access to healthy, sustaining food for all communities; where non-human animals are not seen as food but as individuals with lives, personalities, friendships, and family and are free from harm and exploitation; and where workers, communities, and the environment are always protected and treated with dignity, respect, and appreciation" ("Mission and Values" 2019).

The activist and culinary work of Latina vegans such as Jocelyn Ramirez (founder of Todo Verde), Gabriela Alvarez (Founder of Liberation Cuisine), and Ysanet Batista and Merelis Catalina Ortiz (founders of Woke Foods) conceptualizes plant-based food practices as a response to the "colonization that brought ingredients such as beef, dairy, or bread into the Latino diet. The Africans and the Indigenous were consuming mainly plant-based meals and using foods they were growing and harvesting themselves on the land" (Ferreira 2018). Alvarez points out that European colonist settlements in the Americas "were sustained by free labor of African slaves, many of whom worked in agriculture. African and indigenous descendants are now statistically living in food apartheids" (Ferreira 2018). For some contemporary vegans, the return to a plant-based diet signifies resistance to a legacy of European colonialism that harms both human health and the natural environment. Choctaw and Cherokee scholar Rita Laws cites the disruption of the plant-rich food practices of tribes such as the Choctaw by European colonialism and asks: "What would this country be like today if the ancient ways were still observed? I believe it is fair to say that the Indian respect for non-human life forms would have had a greater impact on American society. Corn, not turkey meat, might be the celebrated Thanksgiving Day dish. Fewer species would have become extinct, the environment would be healthier, and Indian and non-Indian Americans alike would be living longer and healthier lives. There might also be less sexism and racism, for many people believe that, as you treat your animals (the most defenseless), so you will treat your children, your women, and your minorities" (Laws 1994).

Rhetoric and praxis that connects the liberation of humans with that of non-human animals challenges the misconception that veganism is a fleeting dietary trend embraced solely by the privileged, as well as the assumption that a single-issue veganism that does not address the underlying systems that enable factory farming, environmental destruction, and food inequity to thrive will be enough to secure the liberation of non-human animals.

Liberatory Connections

In her groundbreaking book *The Sexual Politics of Meat*, feminist and animal rights activist Carol Adams writes that "we have to stop fragmenting activism; we cannot polarize human and animal suffering since they are inter-related" (2017, xxvi). A number of scholars and activists have made critical connections among veganism and feminism (Carol Adams, Cathryn Bailey, Josephine Donovan), black liberation (A. Breeze Harper, Aph Ko and Syl Ko, Christopher-Sebastian McJetters), disability liberation (Geertrui Cazaux, Sunaura Taylor), queer liberation (Christopher-Sebastian McJetters, pattrice jones, Nathan Runkle, Rasmus R. Simonsen), and decolonization (Aph Ko and Syl Ko, Rita Laws, Margaret Robinson). Such perspectives challenge the biases in single-issue veganism while simultaneously resisting speciesism, locating the marginalization of non-human animals within the same frameworks that facilitate the marginalization of certain classes of human beings, and connecting, without conflating, their oppressions—and hence their liberation.

We often encounter rhetoric in which moral baselines are drawn around the boundaries of what constitutes "human." In the context of human on human violence, statements such as "treated like an animal" signify an especially callous disregard for human life and dignity. The term "treated like an animal" is often an apt one to describe the brutality visited upon oppressed human beings in a white supremacist patriarchy in which non-human animals themselves are routinely exposed to unjust incarceration, forced family separation, rape, torture, and murder. Embedded in the phrase "treated like an animal" is the knowledge that not only does the human state perpetrate acts of violence on the bodies of non-human animals on a regular basis, but also that many of these acts of violence are sanctioned as acceptable because of the hierarchy inherent in the "human/animal" binary discussed before.

Within a dualistic western taxonomy, comparing human beings to (general or specific) non-human animals functions as an act of othering, and has historically been used to oppress indigenous peoples, people of color, people with disabilities, women, and queer people. Within the context of European colonialism, for example, the "doctrine of *terra nullius*, the idea that Settlers found 'empty land that could be used as their own' (McMillan & Yellowhorn, 2009), reframes Indigenous people as part of the flora and fauna […] instead of as a people with a claim to the land on which [they] had lived" (Robinson 2017, 77). pattrice jones further notes that "[s]een by both Catholic conquistadors and Protestant pilgrims as a sign of godless animality, same-sex pleasure was ruthlessly suppressed throughout the subjugation of the Americas" (jones 2010, 198). The association of indigenous peoples with animality was routinely used by settler colonialists to justify their brutal oppression.

Because the marginalized status of non-human animals is widely accepted, designating certain groups of humans as "animal-like" sets the groundwork for a related process of marginalization. In her book *Beasts of Burden: Animal and Disability Liberation*, Sunaura Taylor explores complex connections among disability and animality. "In my life I have been compared to many animals," she writes, "These comparisons have been said both out of mean-spiritedness and playfulness. I remember knowing that my kindergarten classmates meant to hurt my feelings when they told me I walked like a monkey […] I understood that being told I was like an animal separated me from other people" (Taylor 2017, 103). This separateness is made concrete in institutions like human "sideshows" in which "animality was front and center, with the most demeaning of animal comparisons being reserved for people of color and for intellectually disabled people" (Taylor 2017, 104). Sideshows, like the human zoos that preceded (and sometimes co-existed) with them, both legitimized and capitalized on "scientific racism, imperial expansion, colonization, and fear of disability" (Taylor 2017, 104).

Because of their doubly marginalized positions, disabled non-human animals themselves may provoke complex reactions among humans—sometimes evoking public pity and/or calls for "mercy killing," while also being particularly prone to abuse in the food industry. Taylor points out that "ableism helps construct the systems that render the lives and experiences of both nonhuman animals and disabled humans as less valuable and as discardable, which leads to a variety of oppressions that manifest differently" (Taylor 2017, 59). She argues that the idea that non-human animals are "voiceless," a concept and rhetoric shared both by some humans who commodify them and by some humans who fight alongside them, is an ableist assumption, and compellingly reminds us that non-human animals in fact "speak to us every day when they cry out in pain or try to move away from our prods, electrodes, knives, and stun guns. Animals tell us constantly that they want to be out of their cages, that they want to be reunited with their families, or that they don't want to walk down the kill chute. Animals express themselves all the time, and many of us know it" (Taylor 2017, 63). Listening to and amplifying the expressions of non-human animals can thus challenge the paradigms of both able-ism and speciesism.

Like disability theory, feminism is also a powerful tool for analyzing dominant conceptions of animality, specifically the ways in which they are gendered. Carol Adams famously traces the "sexual politics of meat" through the rhetoric of a wide variety of western texts, including contemporary advertising campaigns, to explore the way non-human animals are feminized, female humans are animalized, and meat-eating is associated with masculine virility. The politics of meat, as theorized by Adams, makes evident both "[n]aked assumptions of power over non-dominant beings as well as an almost desperate attempt to reinforce a gender binary" (Adams 2017, 183). It is also classed and raced: "The humanized human in Western culture has often been white male, the one who had the right to vote and own property. Casting individuals as animalized humans is usually influenced by race, sex, and class. Animalizing discourse is a powerful tool in oppression" (Adams 2017, 203). Its use in distancing humans who vary from the European "ideal" (i.e. white, male, able-bodied, heterosexual, wealthy) has been well-documented throughout history. Such distancing tactics can also, significantly, be used to mask routine male violence: "Animalizing discourse [...] substitutes for an analysis of why violence against women happens; that is rapists and batterers or others who commit acts of violence are often animalized (called 'brutes,' 'animals,' etc.), when in fact they are acting like humans, in that their violence is deliberate and often planned" (Adams 2017, 203)[1].

Institutions that are generally seen as oppressing human beings, such as misogyny, intertwine with speciesism in complex ways. A photo courtesy of Nathan Runkle included in the most recent afterword to *The Sexual Politics of Meat* shows a sow trapped in a small cage on the largest pig farm in Iowa with the words "Fat selfish bitch" written by a human worker above her (Adams 2017, 196). Feminists have long critiqued the sexual objectification that occurs when human women are regarded as "meat" and challenged the use of animal-derived pejoratives such as "pig" or "bitch" to describe human women. Here, a term originally used to label certain female non-human animals that has since been deployed as an animalizing term for human women is now applied to another non-human animal to evoke her perceived (human) female "selfishness" in the course of her torture. And pattrice jones writes that "[m]ost of the stereotypes by which we excuse the exploitation of animals began as justifications for animal husbandry, the success of which depends entirely on the ability to control reproduction" (jones 2010, 198)—an obsession that echoes the ways in which men have historically attempted—and relentlessly continue to attempt—to control the reproductive systems of human women, especially women of color.

The rhetoric of "animality" has been used as a justification for some of the greatest human on human atrocities, and it has also been used as a justification for vivisection, factory farming, and the forced captivity of non-human animals. The instability of the human/animal divide itself, however, has increasingly come under scrutiny by animal researchers and critical animal and posthuman theorists who interrogate its very foundations from multiple perspectives. In Syl Ko's compelling critique, "the notions of 'human' and 'animal' are racially constituted. The racial hierarchy tracks not just a color descent but also a species descent. At the top of the hierarchy sits the white male human and at the bottom sits the shady and necessarily opposite figure of 'the animal.' These two poles signify two contrary moral statuses—the closer your category is to the white male human, the more you 'matter.' The closer your category is to the shady, vague 'animal,' the less you 'matter'" (Koand Ko 2017, 55). In order to dismantle oppression, she argues that "we need to go beyond the racial categories and subvert their anchor: the human/animal divide" (Koand Ko 2017, 56).

Campaigns for human social justice sometimes invoke claims of shared humanity—a rhetorical move that distances humans from non-human animals for reasons that are easy to understand given the spaces that most non-human animals occupy in human culture. A 2016 Odyssey editorial about women's right to safety while walking in public, for example, is entitled "Stop The Catcalling —Women Are Not Animals: Just because YOU are a dog doesn't mean that I am one" (Ricciuti), while signs reading "We are not animals" have appeared at Black Lives Matter rallies. While undoubtedly effective in pointing out the abusive treatment of minoritized human communities, such rhetoric, Ko's argument implies, cannot lead to ultimate liberation as long as the white supremacist (and able-ist) "human/animal" divide itself remains intact.[2]

Because of the ways in which it has historically been deployed by those in power against both non-human animals and marginalized humans, challenging the hierarchical human/animal divide is a transgressive act. In "A Queer Vegan Manifesto," Rasmus R. Simenson claims that veganism's emotional involvement "with species other than the human directly expresses a desire to transverse not to say disrupt the boundaries that uphold and police the categories that separate the human from the non-human" (2012, 54). Even the mildest of such transgressions may literally be punishable by law—activist Anita Krajnc was arrested and charged with "interference with the use, enjoyment and operation of property" for giving water to dehydrated pigs on their way to slaughter in 2015 (Pallota 2016). The driver who pressed charges was recorded telling her "You know what, these are not humans, you dumb frickin' broad" (Waring 2015).[3] In 2013 Amy Meyer was the first to be arrested under one of many controversial "ag-gag" laws designed, under pressure from the agricultural lobby, to prevent the documentation of factory farm operations. Meyer had recorded a video from a public street of what she described as a "live cow who appeared to be sick or injured being carried away from the building in a tractor [...] as though she were nothing more than rubble" (Potter 2013). Though these charges were also ultimately dropped, cases such as these bear witness to the threat that mere documentation of its daily practices poses for the animal agriculture industry.

The dissemination of factory farm and slaughterhouse footage may indeed highlight uncomfortable similarities between the non-human animal "others" who suffer in the production of human "food" and the non-human animal classes with which many humans have chosen to empathize—for example, "pets." It may call into question the capricious nature of such distinctions, as well as disrupt the alienation of a sizeable percentage of humans in industrialized societies from the sentient, suffering sources of their food. Lori Gruen configures "entangled empathy" as empathy in which "we are attentive to both similarities and differences between ourselves and our own situation and that of the fellow creature with whom we are empathizing" (Gruen 2018, 148). Mercy for Animals founder Nathan Runkle significantly talks about veganism as a practice that includes "extending our circle of compassion," connecting his own experiences as the victim of a random hate crime with the violence experienced on factory farms ("The Gay Animal: Nathan Runkle" 2011). And movements such as "Mothers Against Dairy" are predicated on crossing the boundaries of species in their advocacy for bodily sovereignty and non-interference in mothering relationships (Mothers Against Dairy 2019). Within such frameworks, empathy acts as a critical tool in dismantling false distinctions between whose lives and autonomy matters, and whose can be discarded.

Sunaura Taylor configures veganism as "an embodied act of resistance to objectification and exploitation across difference—a corporeal way of enacting one's political and ethical beliefs daily" (Taylor 2017, 202). The approaches to veganism outlined by many of the theorists cited in this section are invested in liberating both humans and non-human animals from the "human/animal" divide and thus dismantling *all* hierarchy-based oppression. Christopher- Sebastian McJetters describes the politics of what he calls radical veganism—a political movement that centers non-human animals and recognizes them as "persons of a marginalized community," and simultaneously "elevates the perspectives of marginalized persons who can speak authentically about our experiences of oppression as they relate to undoing speciesism," "supports intersectional feminism," "necessitates that we deconstruct white supremacy," and is definitively "anti-capitalist" (2017). Contrary to rhetoric of veganism as diet that dominates some mainstream conversations, McJetters and others invoke veganism's potential as a transformative and inclusive liberation movement. There are a pluralism of approaches to deconstructing the systems that lead to the oppression of non-human animals. Numerous individuals and grass roots organizations participate in actions designed to disrupt widespread silences around the commodification of non-human animals such as holding protests, documenting violence against non-human animals, and organizing slaughterhouse vigils, as well as running sanctuaries, rescues, and community gardens. Inclusive activism that centers non-human animals while taking into account institutionalized human inequity (e.g., food injustice) underscores the potential of veganism as a politicized social justice movement. Like the aforementioned Food Empowerment Project, whose mission encompasses the humans as well as the non-human animals harmed by our food production systems, A Well Fed World connects hunger relief with environmental and non-human animal protection in its local empowerment initiatives and institutional alliances. Its research focuses "on the ways in which the livestock sector is a major driver of climate destruction resulting in increased global food insecurity and disparities" ("Mission" 2019). Vine Sanctuary, founded by pattrice jones and Miriam Jones (both of whom have a history of activism around issues such as class, race, gender, disability, and sexuality), was founded on the realization that "commodities that begin with animal anguish end up impoverishing people and wrecking ecosystems" ("Connections" 2019). Jones and jones urge "animal advocates to become more cognizant of social and environmental issues while urging environmentalists and social justice activists to include animals within their spheres of concern" ("Connections" 2019). And the Vegan Advocacy Initiative, founded by Linda Alvarez and Liz Ross, is a "people of color, animal advocacy, and food justice organization" that seeks to incorporate animal rights into food and environmental justice movements ("Home" 2017). Organizations such as these, as well as digital spaces including *Black Vegans Rock*, *Crip HumAnimal*, *Striving With Systems*, and *The Sistah Vegan Project* build coalitions across movements and offer inclusive perspectives on our relationships with non-human animals that challenge the systems that compartmentalize and hierarchize humans, non-human animals, and the rest of our environment.

Resisting Veganism

Politicized veganism is a resistance movement, but also a largely resisted movement. Because veganism is a threat to dominant social structures including human supremacy, heterosexism, and capitalism, proponents of veganism may often encounter sometimes vociferous opposition from those invested in the systems that perpetuate and rely on the commodification of non-human animals. A 2015 study by MacInnis and Hodson found that "only drug addicts" were viewed more negatively than vegans. The study concluded: "Unlike other forms of bias (e.g., racism, sexism), negativity toward vegetarians and vegans is not widely considered a societal problem; rather, [it] is commonplace and largely accepted" (Reynolds 2019). In 2007, Anthony Bourdain famously labeled vegetarians "the worst kind of terrorists" (Potter 2007), and in 2018 food critic William Sitwell, editor of *Waitrose Food*, responded to a pitch for a series of plant-based recipes by a vegan freelancer with an email that read: "How about a series on killing vegans one by one. Ways to trap them? How to interrogate them properly? Expose their hypocrisy? Force-feed them meat? Make them eat steak and drink red wine?" (Higgins 2018).

One only has to spend a bit of time on social media to encounter a variety of anti-vegan rhetoric. Some of this rhetoric relies on the construction of non-human animal consumption as "natural"—for example, "circle of life" arguments that falsely equate lions hunting to survive with humans purchasing factory-farmed non-human animal parts at a supermarket, or faulty analyses of how human teeth are "designed" to eat "meat." Others may be offended by the fact that veganism questions assumptions of human supremacy/exceptionalism and use scriptural justifications for human dominance. Some, as Carol Adams has long documented, equate meat-eating with masculine virility and veganism with "feminizing" threat[4].

On a 2019 segment of Fox News's "The Five" punctuated by the panelists eating cow burgers and ribs, co-host Jesse Waters mocked vegan Democratic candidate Cory Booker: "Trump is blessed with the best enemies of all time [...] He's the McDonald's President and he's running against a vegan. [...] This guy runs on a vegan campaign, Trump is going to say he's going to take away your hamburgers, he wants to take away your steak, Booker wants to take away your hot dog on the Fourth of July. This guy is going to be obliterated and it's going to be amazing watching the takedown." The hostility with which veganism is met in this context is linked to its perceived threat to conservative views of masculinity. As Simenson points out, "as a minority, (especially male) vegans will be rendered deviant by normative society [...] [R]efusing meat [...] does not only involve taking a stance against patriarchal culture, as Adams suggests; it is also, specifically, a way of resisting heteronormativity, since meat-eating for men and, perhaps to a lesser degree, women is tied to the rhetorical as well as the actual reproduction of heterosexual norms and practices" (2012, 55). We can see evidence of this in the rise of right-wing epithets such as "soy boy," which equates veganism with a lack of traditional masculinity, as well as in recent rhetoric from right-wing politicians (including Ted Cruz, Rob Bishop, and former White House advisor Sebastian Gorka) that asserts that "the left" is attempting to "outlaw" meat consumption (Reynolds 2019). It's not surprising that the MacInnis/Hodson study found that "respondents from a rightwing background, who seek to uphold traditional gender values, see something alarmingly subversive and worthy of derision" in veganism, especially when practiced by men (Reynolds 2019).

As I mentioned at the start of this essay, veganism is also an obvious threat to the animal agriculture industry. As the consumption of plant-based alternatives rises, the animal agriculture industry's anxiety manifests itself in a variety of ways, from ag-gag laws to ramped up advertising to attempts to control the language employed by vegan and vegetarian companies. Under pressure from animal agriculture lobbyists, lawmakers in 24 U.S. states have pushed for legislation that would make it illegal for companies to market plant-based products as "meat" or "milk." Republican Arkansas state representative David Hillman was quoted as saying "I want my rib-eye steak to have been walking around on four feet at one time or another" (Bromwich and Yar 2019). And *VegNews* reports that the European Union has proposed similar legislation that would "ban vegetarian companies from using terms like 'burger,' 'sausage,' and 'steak' to describe their products," instead suggesting terms such as "veggie discs," "meatless tubes," and "seitan slabs" (2019, 22).

Resistance to veganism does not solely come from places of gender and/or capitalist anxiety, however; vegans may also encounter resistance from those within otherwise liberal or leftist social justice movements who fight other forms of oppression. Carol Adams, reflecting on her activism in the women's movement, writes that "everywhere I went I encountered disconnections—battered women's advocates eating hamburgers while talking about peace in the home; biographies of feminists that failed to consider the vegetarianism of their subject; peace-activist potlucks with dead animals" (Adams 2017, xxiv). While many of the scholars and activists cited in this chapter have made connections among speciesism and other forms of oppression, many others do not yet include non-human animals in their analyses. In response to vegan actor Joaquin Phoenix's 2020 Oscars acceptance speech, in which he said that "whether we're talking about gender inequality or racism or queer rights or indigenous rights or animal rights, we're talking about the fight against injustice," for example, writer Harriet Hall criticized the "galling juxtaposition" and took umbrage that Phoenix "compare[d] marginalised groups to bovine animals" (Hall 2020).[5]

In addition to speciesist rationales and bad faith arguments about plants having feelings—a bizarre justification for consuming non-human animals that at the very least fails to take into account the amount of plants harvested to feed "livestock"—there are meaningful reasons why a variety of humans may feel excluded by some mainstream, single issue vegan movements. When celebrities promote expensive plant-based "lifestyles," for example, they may alienate people with low incomes who assume that veganism is unaffordable. When prominent white vegans sidestep issues such as food injustice and environmental racism in their analyses, or PETA relies on the objectification of women and people of color or the employment of fatphobia in its campaigns against animal cruelty, they re-enforce stereotypes that veganism is exclusionary, or that vegans "care more about animals than humans."

To dismiss veganism as the purview of a privileged elite or anti-speciesism as politically disconnected from human liberation movements, however, de-centers the voices of vegans of color, vegans with disabilities, queer vegans, vegan women's rights activists, and poor, incarcerated, and homeless vegans, among others, as well as the voices of non-human animals themselves. Statements such as "it's impossible to ever be fully vegan anyway" may be justified in a late-stage capitalist landscape in which wealth disparity is on the rise, medications are routinely tested on non-human animals, and non-human animals are harmed in crop harvesting. Rather than being a commitment to an unattainable "purity," however, the very definition of veganism calls us to avoid the exploitation of species other than our own "as far as practicable and possible." Veganism does not require the consumption of expensive products, nor do radical vegans ignore the exploitation of crop harvesters. Radical veganism in fact calls us to work for a more just and sustainable world for *all* earthlings. In the words of Christopher-Sebastian McJetters, "until conservative vegans and neoliberal nonvegans stop trying to distance human interests from the civil rights of our animal cousins and understand that oppression does not occur in isolation, we are collectively doomed. And until all of us are free, NONE OF US are free" ("Sorry Conservative Vegans" 2018).

Conclusion

There's a joke most vegans have probably heard that goes "How do you know someone is vegan? Don't worry; they'll tell you." The stereotype of the "preachy vegan," like that of the "feminist killjoy" is prevalent in mainstream U.S. culture, in which empathy and an engagement with social justice can become a source of mockery. Yet a commitment to liberatory veganism compels many of us to add our voices to those of the non-human animals who, in the words of Sunaura Taylor, "speak to us every day" (2017, 63).

By positioning and enacting veganism as a social justice movement in dialogue with other social justice movements, I hope to avoid the trap of framing veganism in terms of neoliberal rhetoric about individual choice/personal consumption. When we posit veganism as an individualized, dietary, consumerist lifestyle choice, we leave intact the underlying, binary logic that enables supremacist models to thrive. It may also lead us to ignore the multiple ways in which non-human animals are abused outside the food industry in sectors such as medical research, "entertainment," and industrial development. And a focus on consumption may lead some westerners to demonize cultures in which certain non-human species that are privileged in the U.S. or Europe, for example, are captured or consumed, rather than seeing all non-human animal abuse as part of the same oppressive system.

I've recently become interested in work such as that of Resistenza Animale, an Italian collective that "documents the acts of rebellion of animals on farms, in slaughterhouses, in zoos and in circuses" (Cazaux 2019). Moving away from assumptions of non-human animals as voiceless beings who need humans to speak *for* them, they instead amplify the voices of non-human animals as an act of solidarity. By framing our role as that of co-conspirators with the non-human animals who every day rebel against their own oppression, we resist speciesism and re-enforce a commitment to veganism as an act of resistance, rather than a prescriptive diet.

Though it was literally the voices of non-human animals who called me to become vegan, my research into inclusive vegan praxis over the years has enabled me to understand the complex ways in which the oppression of non-human animals is connected to the other oppressions I was and am committed to resisting as a feminist and inclusive educator. I became aware that people of color were disproportionately affected by the environmental and health effects of government-subsidized big agriculture, and were more likely to hold the dangerous jobs involved in slaughtering and dismembering the non-human animals that are ironically often served at social justice events at the institution at which I teach. The more I learned about the colonialist history of animal agriculture, the ways in which able-ism and speciesism intertwine, and the multiple correlations between the abuse and subjugation of women and children and that of non-human animals, the more convinced I became that a truly inclusive social justice praxis would encompass both non-human animals and human beings, alongside a concern for the environments in which we live.

Supremacist hierarchies and late-stage capitalism benefit by fragmenting both oppression and activism. When we resist systems in which some lives are valued more than others and look at veganism as part of the struggle against the commodification of the planet and the beings who inhabit it, however, we can see a larger picture: one in which the fight for liberation from the structures that exploit the most vulnerable among us connects, rather than divides, humans and the non-human animals with whom we share our world.

Notes

1. Thus, for example, the context in which the term "pig" might be inappropriately used to describe a human man who brags about "moving on" a woman "like a bitch."
2. Many vegans have also pointed out the limitations of "animal welfare" approaches in which the "animal" category is also left intact, but through which humans seek "better conditions" for the nonhuman animals designed to be slaughtered for food and held captive in zoos and laboratories.
3. A judge later dismissed the charges.
4. One of the numerous advertisements she analyzes in *The Sexual Politics of Meat* reads: "Putting Together A BBQ: +374 Man Points. Cooking Tofu Sausages On It: -417 Man Points" (2014, 186).
5. Hall's response, in turn, was criticized by vegan activists such as Mc Jetters, who wrote in a blog post called "In Response to Harriet Hall: The White Savior is You, Not Joaquin Phoenix" that "Phoenix merely recognized bovine animals as marginalized persons themselves. It is only insulting to the bigoted imagination that someone should even consider bovine animals to be marginalized persons at all […] I lose nothing by expanding the scope of my justice to include other animals."

Bibliography

Abdulraf, Zahrah. 2017. The Reality of Prison Farms: Modern Plantations. *Caged Bird*, April 29. https://www.cagedbirdmagazine.com/single-post/2017/04/29/The-Reality-of-Prison-Farms-Modern-Plantations

Adams, Carol. 2017. *The Sexual Politics of Meat: A Feminist-Vegetarian Critical Theory*. New York: Bloomsbury.

Animal Agriculture's Impact on Climate Change. *ClimateNexus*. https://climatenexus.org/climate-issues/food/animal-agricultures-impact-on-climate-change/

Animal Equality Films Workers Beating Pigs on a Lincolnshire Farm. 2018. *Animal Equality*, May 23. https://animalequality.org.uk/news/breaking-animal-equality-filmsworkers-beating-pigs-on-a-lincolnshire-farm/

Ascione, Frank R., et al. 2007. Battered Pets and Domestic Violence: Animal Abuse Reported by Women Experiencing Intimate Violence and by Nonabused Women. *Violence Against Women* 13 (4): 354–373. https://journals.sagepub.com/doi/10.1177/1077801207299201.

Bird, Susan. 2018. Is Veganism an Instagram Fad'? The Meat Industry Seems to Think So. *Care2 Causes*, August 21. https://www.care2.com/causes/is-veganism-an-instagram-fad-the-meat-industry-seems-to-think-so.html

Black VegansRock. 2019. http://www.blackvegansrock.com/

Bromwich, Jonah Engel and SanamYar. 2019. What's Meat Got to Do With It? *The New York Times*, July 25. https://www.nytimes.com/2019/07/25/style/plant-based-meat-law.html

Bruek, Julia Feliz. Introduction. In *Veganism In An Oppressive World: A Vegans of Color Community Project*, 1–32. Sanctuary Publishers.

Cazaux, Geertrui. 2019. Crazy Horses. Intersections Between the Animal Liberation and the Antipsychiatric Movement. *CripHumAmimal*, July 3. https://criphumanimal.org/2019/07/03/crazy-horses/?fbclid=IwAR3I9M5TGP1J3hVdbDqRCGncmgSQS_ZAy97hUpFgATuwG-01s7D1NCbWYAM

Chiorando, Maria. 2019. 'Vegan Fad Will Only Last 2 More Years' Insists Dairy Boss. Plant Based News, March 5. https://www.plantbasednews.org/culture/vegan-fad-only-2-more-years-dairy-boss

Connections. 2019. *Vine Sanctuary*. http://vine.bravebirds.org/connections/

Dayan, Colin. 2018. Personhood. In *Critical Terms for Animal Studies*, ed. Lori Gruen, 267–279. Chicago: University of Chicago Press.

Definition of Veganism. 2019. *The Vegan Society*. https://www.vegansociety.com/go-vegan/definition-veganism

Factory Fam Workers. 2019. *Food Empowerment Project*. https://foodispower.org/human-labor-slavery/factory-farm-workers/

Ferreira, Johanna. 2018. Decolonizing Foods: This Is How Latinas Are Reclaiming Their Ancestral Diets. *HIPLATINA*, July 1. https://hiplatina.com/decolonizing-foods/

Fitzgerald, Amy J., Linda Kalof, and Thomas Dietz. 2009. Slaughterhouses and Increased Crime Rates. *Organization & Environment* 22 (2): 158–185. *Sage Journals*. https://journals.sagepub.com/doi/abs/10.1177/1086026609338164.

Flynn, C.P. 2011. Examining the Links Between Animal Abuse and Human Violence. *Crime, Law and Social Change* 55: 453–468. https://doi.org/10.1007/s10611-011-9297-2.

Frazier, Kendal. 2019. There Is No Alternative to Beef. https://www.ncba.org/CMDocs/BeefUSA/Media/National%20Cattlemen/There%20is%20no%20alternative%20to%20BEEF.pdf

Gilliver, Liam. 2019. 99% of US Farmed Animals Live On Factory Farms, Study Says. *Plant Based News*, April 22. https://www.plantbasednews.org/culture/factory-farms-study

Gruen, Lori. 2018. Empathy. In *Critical Terms for Animal Studies*, 267–279. Chicago: University of Chicago Press.

Hall, Harriet. 2020. Joaquin Phoenix's White Male Savior Speech at the Oscars Was Painful to Watch. *The Independent*, February 10. https://www.independent.co.uk/voices/oscars-winner-joaquin-phoenix-speech-white-male-saviour-animal-rights-a9328071.html

Harper, A. Breeze, ed. 2010. *Sistah Vegan: Black Female Vegans Speak on Food, Identity, Health, and Society*. New York: Lantern Books.

Hellersten, Erica, and Ken Fine. 2017. A Million Tons of Feces and an Unbearable Stench: Life Near Industrial Pig Farms. *The Guardian*, September 20. https://www.theguardian.com/us-news/2017/sep/20/north-carolina-hog-industry-pig-farms

Higgins, Abigail. 2018. Why Do People Hate Vegans So Much? *Vox*, November 10. https://www.vox.com/future-perfect/2018/11/2/18055532/vegans-vegetarian-research-uk

Home. 2017. *Vegan Advocacy Initiative*. http://veganai.org/

jones, pattrice. 2010. Afterword: Liberation as Connection and the Decolonization of Desire. In *Sistah Vegan: Black Female Vegans Speak on Food, Identity, Health, and Society*, ed. A. Breeze Harper, 187–201. New York: Lantern Books.

Ko, Aph, and Syl Ko, eds. 2017. *Aphro-ism: Essays on Pop Culture, Feminism, and Black Veganism from Two Sisters*. New York: Lantern Books. E-book.

Laws, Rita. 1994. Native Americans and Vegetarianism. *Vegetarian Resource Group Journal*, September. https://www.vrg.org/journal/94sep.htm#native

Loria, Joe. 2016. Angela Davis: Feminist, Civil Rights Activist, and Vegan. *Mercy for Animals*, September 22. https://mercyforanimals.org/angela-davis-feminist-civil-rights-activist

Lowe, Peggy. 2016. Working 'The Chain,' Slaughterhouse Workers Face Lifelong Injuries. *The Salt*, August 11. https://www.npr.org/sections/the-salt/2016/08/11/489468205/working-the-chain-slaughterhouse-workers-face-lifelong-injuries

Marshall, Liz. Director. 2013. *The Ghosts in Our Machines*. Indiecan Entertainment.

McJetters, Christopher-Sebastian. 2018a. Radical Veganism. *Christopher Sebastian*, July 8. https://www.christophersebastian.info/post/2018/07/08/radical-veganism

———. 2018. Sorry Conservative Vegans. Animal Rights Is Political…And It Leans Left. *Christopher Sebastian*, July 29. https://www.christophersebastian.info/post/2018/07/29/sorry-conservative-vegans-animal-rights-is-political-e2-80-a6and-it-leans-left 2018

———. 2020. In Response to Harriet Hall: The White Savior Is You, Not Joaquin Phoenix. *Christopher Sebastian*, February 11. https://www.christophersebastian.info/post/in-response-to-harriet-hall-the-white-savior-is-you-not-joaquin-phoenix

Mission. 2019. *A Well Fed World*. https://awfw.org/mission/

Mission and Values. 2019. *Food Empowerment Project*. https://foodispower.org/mission-and-values/

Mothers Against Dairy. 2019. https://mothersagainstdairy.org/

No Relief. 2016. Oxfam Report. https://www.oxfamamerica.org/static/media/files/No_Relief_Embargo.pdf

Pallota, Nicole. 2016. Activist Faces Criminal Charges for Giving Water to Thirsty Pigs. *Animal Legal Defense Fund*, August 27. https://aldf.org/article/activist-faces-criminal-charges-for-giving-water-to-thirsty-pigs/

Philips, Layli. 2010. Veganism and EcoWomanism. In *Sistah Vegan: Black Female Vegans Speak on Food, Identity, Health, and Society*, ed. A. Breeze Harper, 8–19. New York: Lantern Books.

Potter, Will. 2007. Chef Calls Vegetarians 'Terrorist Scum.' *Green Is the New Red*, November 29. http://www.greenisthenewred.com/blog/chef-calls-vegetarians-terrorists/314/

———. 2013. First Ag-Gag Prosecution: Utah Woman Filmed a Slaughterhouse from the Public Street. *Green Is The New Red*, April 29. http://www.greenisthenewred.com/blog/first-ag-gag-arrest-utah-amy-meyer/6948/

Reynolds, George. 2019. Why Do People Hate Vegans? *The Guardian*, October 25. https://www.theguardian.com/lifeandstyle/2019/oct/25/why-do-people-hate-vegans

Ricciuti, Emilia. 2016. Stop The Catcalling—Women Are Not Animals. *Odyssey*, May 23. https://www.theodysseyonline.com/stop-catcalling-women-animals

Robinson, M. 2017. Intersectionality in Mi'knaw and Settler Vegan Values. In *Veganism In An Oppressive World: A Vegans of Color Community Project*, ed. Julia Feliz Brueck, 71–88. Sanctuary Publishers.

Safina, Carl. 2015. *Beyond Words: What Animals Think and Feel*. New York: Picador.

Sepie, Amba. 2017. More Than Stories, More Than Myths: Animal/Human/Nature(s) in Traditional Ecological Worldviews. *Humanities* 6 (4): 78. https://www.mdpi.com/2076-0787/6/4/78/htm

Simenson, Rasmus R. 2012. A Queer Vegan Manifesto. *Journal for Critical Animal Studies* 10 (3): 51–81. https://pdfs.semanticscholar.org/03bf/0597d636227fcfab322350e1b5ecd4aeb2f4.pdf

Slaughterhouse Workers. 2019. *Food Empowerment Project*. http://foodispower.org/human labor-slavery/slaughterhouse-workers/

Striving With Systems. 2019. https://strivingwithsystems.com/

TallBear, Kim. 2011. Why Interspecies Thinking Needs Indigenous Standpoints. *Fieldsights*, November 18. https://culanth.org/fieldsights/why-interspecies-thinking-needs-indigenous-standpoints

Taylor, Sunaura. 2017. *Beasts of Burden: Animal and Disability Liberation*. New York: New Press.

The Gay Animal: Nathan Runkle. 2011. *Our Henhouse*. https://vimeo.com/21624081

The Problem. 2019. *Mercy for Animals*. https://mercyforanimals.org/the-problem

The *Satya* Interview with Amie Breeze Harper. 2007. *Satya*, March. http://www.satyamag.com/mar07/harper.html

The Sistah Vegan Project. 2019. http://sistahvegan.com/

Varner, Gary. 2018. Sentience. In *Critical Terms for Animal Studies*, ed. Lori Gruen, 267–279. Chicago: University of Chicago Press.

VegNews. 2019. Vol. 119, Summer 2019.

Von Alt, Sarah. 2017. This Is How Factor Farms Discriminate Against Communities of Color. *Mercy for Animals*, February 27. https://mercyforanimals.org/this-is-how-factory-farms-discriminate-against

Waring, Olivia. 2015. Woman Charged with Criminal Mischief for Giving Water to Pigs Bound for Slaughter. *Metro*, October 21. https://metro.co.uk/2015/10/21/woman-charged-with-criminal-mischief-for-giving-water-to-pigs-bound-for-slaughter-5454066/

Warrick, Jo. 2001. 'They Die Piece by Piece'. *The Washington Post*, April 10. https://www.washingtonpost.com/archive/politics/2001/04/10/they-die-piece-by-piece/f172dd3c-0383-49f8-b6d8-347e04b68da1/?utm_term=.5e118621e41c

Working Conditions in Factory Farms. 2019. *Beyond Factory Farming*. http://www.beyondfactoryfarming.org/get-informed/labour/working-conditions-factory-farms

Wrenn, Corey. 2014. Rape & Slaughter: Empirical Correlations. *Vegan Feminist Network*, September 9. http://veganfeministnetwork.com/rape-slaughter-empirical-correlations/

Wright, Laura. 2015. *The Vegan Studies Project: Food, Animals, and Gender in the Age of Terror*. Athens: University of Georgia Press.

Index[1]

A

Abolitionist approach, 215, 217, 218
Adams, Carol J., xxv, 197, 198, 208, 210, 210n1, 219, 225, 288, 292, 297, 301, 303, 304, 308–310
Affect studies, 33
Agricultural industry laws
  ag-gag laws, 37, 48n20, 305, 309
Agriculture, xviii, xxvi, 27, 39, 48n20, 248, 251, 252, 254–258, 265n8, 284, 291, 292, 295–301, 305, 309, 312
Albertoni, Pietro, 10
Alcoholism
Algeria, 28
Alpha-gal syndrome (AGS), 143–165
Andersen, Kip, 245, 246, 248, 249, 251–254, 256–258, 261, 264, 265n7
Animal, xvii–xix, xxi, xxiii–xxviii, 3, 26, 63, 96, 120, 144, 178, 195–210, 215, 245, 273, 291
Animal Liberation, xxi, 216
Anthropatriarchy, 33
Anthropocene, 152, 154, 155
Anti-veg(etari)an discourse, xvii–xxviii, 26, 41, 187
Appeals
  affective, 248, 249, 251–259
  orchestrated, 275, 278

[1] Note: Page numbers followed by 'n' refer to notes.

C. Hanganu-Bresch, K. Kondrlik (eds.), *Veg(etari)an Arguments in Culture, History, and Practice*, The Palgrave Macmillan Animal Ethics Series,
https://doi.org/10.1007/978-3-030-53280-2

Appetite, 144, 196, 199, 203, 204, 209, 226, 232
Argument, xix–xxvii, 35, 37, 38, 64, 65, 70–73, 85, 96, 102, 109, 158, 210, 210n1, 215, 216, 218–222, 224, 232, 240, 246–248, 261, 272–279, 281–283, 286, 287, 289, 305, 308, 310
Associazione Naturista Italiana, 16
Associazione Vegetariana d'Italia, 12

B
Barnes, Djuna, 195–210
Baron Felix Volkbein, 196, 200
Beef, xxvii, 29, 130, 148–150, 162, 173, 175, 176, 178–182, 187, 191n11, 197, 198, 246, 291, 300
Benazeer, Suraiyya, 26, 33, 44
Beyond Meat, 258
Bhabha, Homi K., 198, 199, 202
Binary constructions, 202, 206, 209, 210

C
Carnivore, 96, 162–165, 181, 185
Cartesianism, 227, 228
  Cartesian rationalism, 227
Cattle, 7, 71, 130, 145, 173, 175–178, 181, 185, 191n8, 191n11, 192n13, 254, 256, 260
Cetuximab, 148, 149
Christian ethics, 225, 226, 237
Clerici, Alessandro, 10, 11
Climate change, xix, xxiii, xxiv, xxviii, 41, 43, 45, 151, 245–265, 284–288, 296, 299
Cognitive film theory, 247–251
Cohen, Stanley, 80, 81
Colonialism, 293, 299, 300, 302
Constitutive rhetoric, 274
Cowspiracy, 120
Cultural identity, 41, 198, 202, 275
Culture, xi, xv, xviii, xix, xxii, xxiii, xxviii, 27–29, 35–39, 41, 64, 67, 75–84, 94, 99, 106–108, 112, 125, 127, 136, 144, 151, 160, 162, 171, 173, 174, 178, 179, 181, 186–189, 190n3, 190n4, 197–199, 207, 209, 219, 220, 227, 228, 250, 272–274, 276, 289, 303, 305, 309, 311, 312

D
Deleuze, Gilles, 198
Derrida, Jacques, 196–198, 200, 205, 209, 234, 236
Desire, xxv, 9, 85, 120, 131, 195, 199–201, 203–207, 219, 226, 237, 238, 248, 272, 305
Dhabīḥah/*zabiha* (slaughtered animal through *halāl* practice), 30–32, 38, 46n11, 47n12
Di Salaparuta, Duca Enrico Alliata, 19
Disability, 293, 297, 301–303, 307, 310
Documentary film, 247, 263
Dominion, 248

E

Earthlings, xxviii, 246, 248, 261, 263
Ecocriticism
  cognitive ecocriticism, 249
EcoPeace Middle East, 43, 45
Eid (high religious holiday), 30
Eid al-Adha (Feast of the Sacrifice), 26, 33–38, 40, 48n18
Empathy, 71, 81, 125, 127, 145, 151, 165, 251, 266n10, 306, 311
Environment/environmentalism
  environmental advocacy, 49n27
  environmental safety, 43
  environmental sustainability, 26, 28, 42–44
Equality Act 2010 (UK), 27

F

Factory farms, 37, 66, 120, 127, 128, 130, 165, 282, 294–298, 305, 306
Faith-based communities and cultures, 45
Fake meat, 13
Fascism, 6, 12, 15–22
Fatwā/fatāwa (Islamic religious ruling and scholarly opinion on a matter of Islamic law), 40, 44, 48n23
First-Amendment, 176, 177, 189
Food
  influencer, 173, 174, 188
  justice, 189, 190n3, 299, 307
  libel laws, 189
Foodie, 171–174, 187, 188, 189n1, 190n3
Francione, Gary, 217, 218
Free speech, 173–176, 178, 181, 182, 187, 189

G

Gav Hauma, 77
Genesis, 207, 219–224, 229, 230, 233, 237–240
Gilboa, Tal, 63, 72, 73, 75–77, 79
Gore, Al, 252, 255, 265n7
Grillo, Niccolò, 20
  See also Licò Nigro
Guattari, Félix, 198

H

Hadith (tradition), 30, 41, 44
Halāl (that which is permissible under Islam), 29–35, 37–39, 41, 44, 48n20
Harām (that which is forbidden under Islam), 29, 31, 32
Harper, A. Breeze, 292, 298, 299, 301
Hindu nationalism, 33
Holocaust, 67–71, 74, 75, 81, 83, 85, 275, 276
Human exceptionalism, xviii, xxvii, 215–241
Hunger, xxiii, 21, 195–210, 307
Hybridity, 198, 199, 202, 206

I

Identification, 64, 68, 71, 233, 253, 272–274, 276, 286, 287

Identity, xxii, xxiii, xxvii, xxviii, 26, 28, 39, 44, 94, 99, 146, 151, 185, 196, 197, 201–206, 210, 253, 272, 277, 291, 292
Ideograph, 67, 68, 70, 75, 83, 85
IDF, *see* Israeli Defense Forces
An Inconvenient Truth, 249, 252, 258, 259, 265n7
India, 32, 46n2
Islam, xxvii, 26–45, 47n13, 86n5, 122, 219
Islamophobia, 28, 34, 37, 45, 46n2
Israeli Defense Forces (IDF), 64, 82
Italian cookbooks, 12
Italian cuisine, 5–7

J

Jones, pattrice, 301, 302, 304, 307

K

Ko, Aph, 292, 295, 301
Ko, Syl, 292, 295, 301, 304
Kuhn, Keegan, 245, 246, 248, 253, 254, 256, 257, 264

L

Legitimation strategies, xvii–xxviii
Lent, 107–110, 114n9
Liberation, 295, 300, 301, 305, 306, 310, 313
Licò Nigro, 20
L'Organisation Mondiale de la Santé Animale (OIE), 36

M

Mad cow disease, xxiii, xxiv, 173, 174, 191n6
Malaysia, 32
McJetters, Christopher Sebastian, 292, 301, 306, 311
Meat
  in Islam, xxii, 26–45
  meatlessness, 3–22
  meat substitutions, 13–15, 19
Middle East and North Africa (MENA), 26–28, 39, 42–43, 45, 48n27
Monson, Shaun, 246, 248
Moore, Michael, 252–255, 260, 263, 264, 265n5, 265n6
Morals, xvii, xviii, xxi, xxii, xxiv, xxvii, xxviii, 9, 69, 80, 82, 84, 109, 120, 122–125, 127, 130, 133–136, 145, 182, 216–218, 222, 229, 234, 238, 246–248, 256, 264, 275, 301, 304
Mussolini, Benito, 5, 15, 20

N

National identity, 44, 107
Natural diet, xxi, 8
Nazi Germany, 15, 17, 20
Netanyahu, Benjamin, 63, 76–78
Niyamas, 122, 123, 129–136
Non-human animals, xviii, xix, xxvii, xxviii, 29, 216, 292–308, 310–313
Nora Flood, 195, 200, 239

O
Object-Oriented Ontology (OOO), 143–165
O'Connor, Matthew, 196, 202–208, 210
Oprah, xxiii, 171–189
Orchestrated appeals, 275, 278
Other, The, 199, 200, 202, 203, 207, 210
 othering, 29, 36–38, 41, 44, 45, 197, 199, 210, 302

P
Patanjali, 120–136
Patriarchy, 185, 301
Periodic vegetarianism, 108
PETA, xviii, xxvii, 310
 PETA-India, 26, 32, 33, 44
Phenomenology, 144, 146, 153, 154
Piccoli, Ettore, 10, 11
Politics of sight, 33, 36, 37
Posno, 107–111, 113, 114n9
Public health, 8, 15, 42, 44, 284

Q
Queer, xxvii, 195–210, 301, 302, 310
Qur'an, 29–31, 35, 38, 42, 46n9

R
Regional Animal Welfare Strategy (RAWS) for the Middle East, 36
Religion
 religious identity, 196
Robin Vote, 195, 199
Rossi, Felice, 10

S
Sacrifice, 21, 26, 34, 35, 37, 38, 40, 47n13, 47n15, 233, 234, 239
Schleien, Lior, 77, 82
Serbia, xxii, xxvii
Sexual Politics of Meat, xxv, 197, 301, 303, 304
Sharia (Islamic law), 30
Sierra Club, 249, 255, 257
Singer, Peter, xxi, 216–218
Slaughter, 26, 30–38, 40, 47n12, 68, 124, 125, 134, 217, 218, 233, 246, 261, 273, 280–282, 294, 305
Slaughterhouses, 9, 32, 36, 48n19, 125, 165, 280–282, 285, 297, 298, 305, 307, 312
Società Vegetariana d'Italia, 8, 14
Speciesism, xxv, 155, 216–218, 292, 295, 299, 301, 303, 304, 310, 312
Sutras, 122, 123, 125, 126, 128, 131–136

T
Tactic, xxvii, 160, 181–184, 294, 303
Taylor, Sunaura, 292–295, 297, 298, 301–303, 306, 311
Texas, 146, 173–177, 180, 181, 183–187, 191n5, 295

Texas Beef Group
  Texas Beef Group v. Oprah Winfrey, 173
Tunisia, 28, 49n27

U

UN Environment Programme, 43
Unione Naturista Italiana, 16

V

van Leeuwen, Theo, xx, xxi, xxiii, xxiv
Vegan, xviii, 26, 63, 93, 120–136, 146, 174, 197, 215, 247, 272, 292
Veganism
  as a moral imperative, xxvii, 215–217, 219, 241
Vegan Muslim Initiative, The, 36, 42
Vegetarian, xvii, 3–5, 7–11, 16, 27, 40, 64, 86n5, 93, 120–136, 143–165, 172, 181, 188, 189, 238, 260, 261, 266n11, 271–289, 308, 309
Vegetarianism, xvii, xx, xxii, xxiv–xxvi, 3–22, 26, 28, 29, 33, 39, 40, 44, 45, 76, 120–129, 132, 133, 135, 136, 144, 145, 151, 153, 158, 162, 165, 210, 238, 272–288, 294, 310
Veggie libel laws, 173–177, 187, 190n5, 191n5
Voit, Carl, 9

W

What the Health, 120, 130, 246, 248, 264
White, Lynn Jr., 226–228, 231

Y

Yamas, 122–129, 136
Yoga, xxii, 120–123, 128, 136
Yourofsky, Gary, 66, 68, 70, 74, 75

Z

Zabiha, see Dhabīḥah

The manufacturer's authorised representative in the EU is Springer Nature Customer Service Centre GmbH, Europaplatz 3, 69115 Heidelberg, Germany. If you have any concerns regarding our products, please contact ProductSafety@springernature.com

Printed and bound by CPI Group (UK) Ltd, Croydon, CR0 4YY

30/09/2026

02225460-0010